Introduction to ANALYTICAL GAS CHROMATOGRAPHY

CHROMATOGRAPHIC SCIENCE

A Series of Monographs

Editor: **JACK CAZES**

*Waters Associates, Inc.
Milford, Massachusetts*

Volume 1: Dynamics of Chromatography
J. Calvin Giddings

Volume 2: Gas Chromatographic Analysis of Drugs and Pesticides
Benjamin J. Gudzinowicz

Volume 3: Principles of Adsorption Chromatography: The Separation of Nonionic Organic Compounds
Lloyd R. Snyder

Volume 4: Multicomponent Chromatography: Theory of Interference
Friedrich Helfferich and Gerhard Klein

Volume 5: Quantitative Analysis by Gas Chromatography
Josef Novák

Volume 6: High-Speed Liquid Chromatography
Peter M. Rajcsanyi and Elisabeth Rajcsanyi

Volume 7: Fundamentals of Integrated GC-MS (in three parts)
Benjamin J. Gudzinowicz, Michael J. Gudzinowicz, and Horace F. Martin

Volume 8: Liquid Chromatography of Polymers and Related Materials
Jack Cazes

Volume 9: GLC and HPLC Determination of Therapeutic Agents (in three parts)
Part 1 edited by *Kiyoshi Tsuji and Walter Morozowich*
Parts 2 and 3 edited by *Kiyoshi Tsuji*

Volume 10: Biological/Biomedical Applications of Liquid Chromatography
Edited by Gerald L. Hawk

Volume 11: Chromatography in Petroleum Analysis
Edited by Klaus H. Altgelt and T. H. Gouw

Volume 12: Biological/Biomedical Applications of Liquid Chromatography II
Edited by Gerald L. Hawk

Volume 13: Liquid Chromatography of Polymers and Related Materials II
Edited by Jack Cazes and Xavier Delamare

Volume 14: Introduction to Analytical Gas Chromatography: History, Principles, and Practice
John A. Perry

Other Volumes in Preparation

Introduction to ANALYTICAL GAS CHROMATOGRAPHY

History, Principles, and Practice

JOHN A. PERRY
*Regis Chemical Company
Morton Grove, Illinois*

MARCEL DEKKER, INC. New York and Basel

Library of Congress Cataloging in Publication Data

Perry, John A. [date]
   Introduction to analytical gas chromatography.

   (Chromatographic science ; v. 14)
   Includes bibliographical references and indexes.
   1. Gas chromatography. I. Title.
QD79.C45P47   543'.0896   80-27790
ISBN 0-8247-1537-3

57298

Lanchester Polytechnic Library

COPYRIGHT © 1981 by MARCEL DEKKER, INC.   ALL RIGHTS RESERVED

Neither this book nor any part may be reproduced or transmitted in any form or by any means, electronic or mechanical, including photocopying, microfilming, and recording, or by any information storage and retrieval system, without permission in writing from the publisher.

MARCEL DEKKER, INC.
270 Madison Avenue, New York, New York 10016

Current printing (last digit):
10 9 8 7 6 5 4 3 2 1

PRINTED IN THE UNITED STATES OF AMERICA

To Jay, who started it;
to Inger, who saw it through;
and to my students

PREFACE

On the invitation of Dr. Jay Curtice, now Chairman of the Department of Chemistry of Roosevelt University, in the spring of 1964 I began teaching a 16-week graduate-level evening credit course in gas chromatography. Since then it has been presented at least once each year.

Responding to my questionnaires, students in the course have described themselves as full-time working chemists. Ranging in age from 21 to over 70 years, they have an average age of perhaps 28. Usually, they have earned at least a bachelor's degree, have not had experience in gas chromatography, and have registered only for this course.

This book was written for people such as these, to provide them with orientation and insight in language they find clear and comprehensible. For perspective, a topic is often presented against its history. Throughout, reasons and principles are stressed.

The book is selective in choice of topic and depth of treatment. For each topic, once it has been chosen, the relevant technical background is furnished. The topic is then developed in a way that is understandable to readers usually thoroughly ill at ease with mathematics and electrical circuitry. In discussions, for instance, names always accompany symbols. Derivations can be read.

Because the treatment is cumulative, the book should be read from front to back. However, the chapters on qualitative and quantitative analysis (Chaps. 13 and 14) can be understood without first reading the chapter on derivatization.

I acknowledge the friendly and effective assistance of Miss Anne O'Donnell and of Dr. Peter Klein in the treatment of gas chromatography-mass spectrometry. Mr. E. M. Bens reviewed the chapter on supports, helped clarify and strengthen the section on the preferred diatomaceous earth, and furnished additional valuable (and beautiful) photographs. Dr. Mathieu H. J. van Rijswick reviewed the chapter on quantitative analysis and commented generously and constructively on my treatment of his high-precision algorithm. I thank both of these experts for their help.

I thank Miss Bernadine Palka, who has read the complete text several times, suggested many improvements, caught innumerable errors. The book is the better for her.

I am deeply grateful for the manifold cooperation of the Regis Chemical Company of Morton Grove, Illinois. And my gratitude to my wife, Inger, is boundless.

John A. Perry

CONTENTS

Preface   v

Notation   xi

Chapter 1.   Gas Chromatography: A Simple Description   1

              1.1   The Gas Chromatogram   2
              1.2   Sample Introduction   3
              1.3   Theoretical Plates   4
              1.4   The van Deemter Equation   6
              1.5   Maximum Allowable Sample Size   9
              1.6   Column Performance   9
              1.7   The Selective Stationary Phase   10
              1.8   Peak Movement   11
              1.9   Relative Retention Times   13
              1.10 A Recapitulation and an Overview   13
              1.11 A Preliminary Guide to the Literature of Gas Chromatography   13
                    References   15

Chapter 2.   Some Language: A Few Terms and Symbols   17

              2.1   Gas Holdup Time   17
              2.2   Handling Retention Data: Symbols and Language   19
              2.3   The Partition Coefficient, Partition Ratio, and Phase Ratio   20
              2.4   Relative Retention   24
              2.5   Calculating Some Numerical Values Relevant to These Concepts   24
                    References   26

Chapter 3.   Sample Introduction   27

              3.1   Sample Size   27
              3.2   Plug Introduction   28
              3.3   Experimental Methods   29
              3.4   The Sample Introduction Port   35
                    References   36

viii   Contents

Chapter 4.   Resolution: Separability *versus* Plate Number   37

             4.1   Introduction   37
             4.2   Two Auxiliary Methods   37
             4.3   Resolution   38
             4.4   Studies of Three Independent Areas   43
             4.5   The Direction of Gas Chromatography   52
                   References   53

Chapter 5.   The Stationary Phase   55

             5.1   Introduction   55
             5.2   Solubility   55
             5.3   Adsorption on the Gas-Liquid Surface   58
             5.4   Stationary Phase Characterization   59
             5.5   Preferred Stationary Phases   63
             5.6   Solvent Selection   70
                   References   75

Chapter 6.   The Jones-van Deemter Equation: Derivations   77

             6.1   Diffusion   77
             6.2   Gaussian Peak Shape   79
             6.3   The Jones-van Deemter Equation   82
             6.4   A Review, without Comment, of the Derivations   93
                   References   96

Chapter 7.   The Jones-van Deemter Equation: Comments   97

             7.1   The Terms of the Jones-van Deemter Equation   98
             7.2   The Reduced Plate-Height Equation   111
                   References   112

Chapter 8.   Supports   114

             8.1   Diatomaceous Earth   114
             8.2   Other Supports   131
             8.3   Molecular Sieves   134
             8.4   Gas-Liquid *versus* Gas-Solid Chromatography   135
             8.5   The Bonded Stationary Phase   136
                   Recent Developments   139
                   References   139

Chapter 9.   Detectors   142

             9.1   The General Requirements for a Widely Useful Detector 142
             9.2   The Hot-Wire Thermal Conductivity Detector (TCD)   146
             9.3   The Flame Ionization Detector (FID)   156
             9.4   The Electron Capture Detector (ECD)   164
                   References   175

Chapter 10. The Column   178

             10.1 The Conventional Packed Column   179
             10.2 The Open Tubular Column   188
             10.3 The Glass Open Tubular Column   204

Contents ix

10.4	The Evaluation of Column Performance	211
	Recent Developments	216
	References	218

Chapter 11. The Column Temperature   223

11.1	Introduction	223
11.2	Column Behavior as a Function of Temperature	224
11.3	Stationary Phase Stability as a Function of Temperature	249
	References	250

Chapter 12. Derivatization   252

12.1	Introduction	252
12.2	Abbreviations	252
12.3	Comparison of Common Reagents and Derivatives	253
12.4	Hydroxy Compounds	256
12.5	Carbonyls	261
12.6	Acids	262
12.7	Amines	265
12.8	Amino Acids	272
12.9	Amides and Imides	274
12.10	Carbohydrates	277
12.11	Steroids	280
12.12	Miscellaneous	285
	Acknowledgment	287
	References	287

Chapter 13. Qualitative Analysis   295

13.1	Identification of Eluted Components	295
13.2	Identification by Inference from Sample Pretreatment	311
13.3	More on Gas Chromatography-Mass Spectrometry	317
	References	336

Chapter 14. Quantitative Analysis   341

14.1	Introduction	341
14.2	Quantitation without an Integrator	341
14.3	Area Percent	342
14.4	Response Factors	342
14.5	Sample Size *versus* Detector Range	346
14.6	Trace Detection	347
14.7	Potentiometric Recorders	352
14.8	Integrators	358
14.9	Digital Technology Applied to Gas Chromatography	361
14.10	Precision and Accuracy	388
	References	391

Author Index   395

Subject Index   413

NOTATION

For each symbol, the location of its definition or first appearance follows its explanation.

a   1.   Constant expressing fraction of corrected retention volume that may be used as the maximum acceptable sample volume (pages 9, 28):

$$V_{max} = a \; \frac{V_M^o + KV_L}{\sqrt{n}} = a \; \frac{V_R^o}{\sqrt{n}}$$

       2.   Constant in equation for vapor pressure p^o (Chap. 11 only). P. 224, Eq. 11.3.

A   The eddy diffusion term of the Jones-van Deemter equation. P. 6.

ADC   Analog-to-digital converter. P. 362.

A_{min} The minimum detectable peak area as a function of digital noise. P. 377, Eq. 14.10.

b   Constant in equation for vapor pressure p^o. P. 224, Eq. 11.3.

B   The axial molecular diffusion term of the van Deemter equation. P. 7.

c   Constant in equation for vapor pressure p^o (Chap. 11 only). P. 224, Eq. 11.3.

c_L Constant in the C_L term of the Jones-van Deemter equation. P. 86, Eq. 6.30.

c_2 Constant in the C_2 term of the Jones-van Deemter equation. P. 92, Eq. 6.63.

$c_1 \ldots c_{12}$ Constants.

C   The overall term that expresses resistance to mass transfer in the Jones-van Deemter equation. P. 8.

C_{gas} This symbol, cited on p. 232 in a quotation, has the meaning of our C_G.

C_{liq} This symbol, cited on p. 232, in a quotation has the meaning of our C_{LP}.

C_{max} The weight-per-volume concentration in the mobile phase of solute at the peak apex (page 346, Eq. 14.1):

$$C_{max} = \frac{\sqrt{n}}{V_R^o} \; w_I \; \sqrt{2\pi}$$

xi

c_t	The solute concentration at time t following injection at concentration C_O into an insufficiently hot injection port that produces exponential rather than plug sample introduction. P. 199. Eq. (10.16).
C_D	The maximum solute weight-per-volume concentration in the mobile phase that does not exceed the linear dynamic range of the detector. P. 346.
C_G	The term of the Jones-van Deemter equation that expresses resistance to mass transfer in the gas phase. P. 91. Eq. (6.54).
C_L	The weight-per-volume concentration of solute in the liquid stationary phase. P. 20. P. 21, Eq. (2.8).
C_{LP}	The term of the Jones-van Deemter equation that expresses resistance to mass transfer in the liquid stationary phase. P. 88, Eq. (6.40).
C_M	The weight-per-volume concentration of solute in the gas mobile phase. P. 21, Eq. (2.8).
C_0	Initial sample concentration on sample injection. P. 199, Eq. (10.16).
C_2	The term of the Jones-van Deemter equation that expresses resistance to mass transfer due to velocity distribution in the mobile phase. P. 92, Eq. (6.64).
C_3	The term of the Jones-van Deemter equation that expresses interaction of the nominally independent C_G and C_2 terms. Not a very important term.
d	Internal diameter of an open tubular column. P. 214.
d_f	The thickness of the annular coating of the stationary phase in a wall-coated open tubular column. P. 190, Eq. (10.15).
d_p	Particle diameter. P. 6
d_p^*	The particle diameter as used in the reduced plate-height equation. In the case of gas-liquid chromatography, d_p^* refers to the pore size of the inert support rather than to the inert support particle diameter. P. 111, Eq. (7.13).
d_c	The internal diameter of a column. P. 9.
d_G	The effective thickness of the gas phase; the shortest distance through the carrier gas from one surface of the stationary phase to another. P. 89, Eq. (6.45).
d_L	The effective thickness of the liquid stationary phase. P. 8.
db	Decibels. A decibel is equal to 10 times the base-10 logarithm of a power or rejection ratio. For example, 60 db implies a ratio of one million.
D	Diffusivity; dimensions: $(distance)^2/time$. Defined p. 199, Eq. (10.18). The diffusivity is identically the diffusivity coefficient; see p.78, Eq. (6.1).
D	Detectability: the smallest quantity that will cause a detector response equal to twice the noise level. P. 143.
D_m	The diffusivity of the mobile phase, as used in the equation for reduced plate height. P. 111, Eq. (7.15).
D_G	The diffusivity of the carrier gas. P. 84, Eq. (6.21).
D_L	The diffusivity of the liquid stationary phase. P. 8.
e	The base of the natural logarithms. First used, p. 171.
e_b	Bucking voltage. This is the voltage that is developed by a potentiometric recorder within itself to equal the signal being measured, so that the difference ($e_b - e_i$) approaches zero as the recorder approaches balance. P. 352.
e_i	The voltage of a source of potential. P. 151.

Notation

$e_{i\text{-span}}$	An electric potential just sufficient to cause a full-scale pen deflection of a potentiometric recorder. P. 354.
e_o	The voltage delivered by a voltage divider.
e_r	A reference voltage within a potentiometric recorder in terms of which an incoming voltage is measured. P. 352.
E	Electron affinity, pertaining to electron capture. P. 171.
E_d	Dispersion interaction energy. P. 57.
E_i	Dipole-induced dipole interaction energy. P. 57.
E^*	The concentration of an electron-capturing species in an electron capture detector. P. 171.
F	Flow rate of column effluent; p. 145. Gas flow rate entering mixing chamber; p. 199, Eq. (10.17). And gas flow rate, p. 240.
F_a	Flow rate of added gas in flame ionization detector, p. 145 (see also V_e, effective detector volume, p. 145).
F_c	The temperature-corrected carrier gas flow rate, measured at the column exit pressure and corrected to column temperature. P. 18.
h	The height equaivalent to a theoretical plate: $h = L/n$. Longer symbol: HETP. P. 5, Fig. 1.5.
h	The reduced plate height. P. 111, Eq. (7.13).
H	The height equivalent to an effective theoretical plate: $H = L/N$. Longer symbol: HEETP. P. 213.
ΔH_s	Differential heat of vaporization of solute from infinitely dilute solution. P. 225. Eq. (11.7).
ΔH_v	Heat of vaporization of the pure organic liquid. P. 224, Eq. (11.4).
Hz	Hertz: one cycle per second.
i	Current of electricity.
I	Kovats Retention Index. P. 60. Eq. (5.1).
I_b	The standing current of an electron capture detector. P. 165.
I_e	The current observed in an electron capture detector while an electron-capturing species is in the detector. P. 171.
$I_{PT(i)}$	Kovats Retention Index of substance i from programmed temperature data. P. 246, Eq. (11.30).
I_X	Ionization energy for species X. P. 57.
ΔI	Kovats Retention Index difference. P. 60.
j	Pressure gradient correction factor. P. 19.
k	Capacity factor: $k = K/\beta = p/q = t'_R/t_M$. Also called partition ratio or, less frequently, the mass distribution ratio. P. 22. Eq. (2.10).
k_B	Boltzman constant. P. 56.
k_{TC}	Coefficient of thermal conductivity. P. 149, Table 9.1.
Δk_{TC}	Thermal conductivity coefficient difference. P. 149, Table 9.1.
K	Partition coefficient: $K = C_L/C_M = (w_L/V_L)/(w_M/V_M) = (w_L/w_M)(V_M/V_L) = k\beta$. Also called distribution coefficient. Pages 21, 22.
K_a	Adsorption partition coefficient. P. 59.

K_H		Henry's Law constant. P. 225, Eq. (11.7).
K^*		Coefficient of electron capture. P. 171.
ℓ		A distance, the travel through which causes a given type of peak spreading during molecular "random walking." P. 83. Eq. (6.14).
L		Column length. See Fig. 1.5, p. 5.
m	1.	Constant.
	2.	In the filtration method of packing preparation, the ratio m of solution volume retained on the support to the weight of the support: $V_S = mW_S$. P. 182; see also p. 183, Eq. (10.3).
M		Subscript; refers to mobile phase. P. 20.
n	1.	Number of theoretical plates: $n = 16\,(t_R/w_b)^2$. P. 4.
	2.	Number of carbons in an homologous series (Chap. 11 only). P. 224, Eq. (11.1).
	3.	In the filtration method of packing preparation, the ratio n of solution volume V_S retained on the support to the total volume V_T of the solution (Chap. 10 only). P. 182, Eq. (10.4).
n		Number of opportunities for causing broadening. Chap. 6 only. See, for instance, p. 83, Eq. (6.15).
n_1, n_2		Numbers of interdiffusing molecules in derivation of Einstein's Law of Diffusion. (Chap. 6). P. 79. Eq. (6.2).
N	1.	Number of theoretical effective plates: $N = 16\,(t'_R/w_b)^2 = n/[k/(1+k)]^2$. P. 41, Eqs. (4.9), (4.12).
	2.	Net number of molecules crossing area O in derivation of Einstein's Law of Diffusion (Chap. 6 only). P. 79. Eq. (6.2).
N_C		Effective carbon number: the number of carbons a n-paraffin would have to have to yield the same response from a flame ionization detector as a given molecular species. For n-heptane, $N_C = 7.00$. P. 161.
O		The designation and magnitude of an area, in the derivation of Einstein's Law of Diffusion. P. 78. Eq. (6.1).
p	1.	The weight fraction of solute in the stationary phase. P. 22. Eq. (2.11).
	2.	Partial pressure (Chap. 11 only). Pages 226, 227.
$p°$		Standard vapor pressure for a given molecular species. P. 224. Eq. (11.3).
Δp_{ne}		Pressure drop necessary for a separation. P. 214. Eq. (10.28).
P_1, P_2		Equivalent grams of stationary phase on supports 1 and 2, in equivalent loading. P. 129. Eq. (8.1).
P_i		Carrier gas pressure at the inlet of a column. P. 19.
P_o		Carrier gas pressure at the outlet of a column. P. 19.
PP		Performance parameter: $PP = \Delta p_{ne} t_{ne}$. P. 214. Eq. (10.28).
q		The weight fraction of solute in the mobile phase. P. 22. Eq. (2.11).
r	1.	Electrical resistance. P. 151.
	2.	Internal radius of column (Chap. 10 only), P. 190. Eq. (10.14).
	3.	Heating rate in programmed temperature gas chromatography (Chap. 11 only). P. 240.
	4.	Effective distance of molecular separation (Chap. 5 only). P. 57.

Notation

R		Gas constant. P. 224. Eq. (11.4).
R_S		Resolution. P. 41. Eq. (4.13).

$$R_S = \frac{\sqrt{N}}{4} \frac{\alpha - 1}{\alpha}$$

R_I		Intrinsic resolution. P. 245. Eq. (11.29).
R_{IT}		Isothermal intrinsic resolution. P. 244. Eq. (11.27).
R_L		Linear dynamic range of a detector. P. 347.
$[R]$		The time-dependent resolution of Struppe. P. 215. Eq. (10.33).
s		The sensitivity of a detector in grams per second. P. 143.
SF		The Glueckauf separation factor. P. 42; see Fig. 4.2.
t	1.	The time for a molecule of diffusivity D to diffuse distance Δ: $D = \Delta^2/2t$. Eq. (6.4), p. 79.
	2.	The time interval after sample injection (Chap. 10 only). P. 199, Eq. (10.16).
t_e		The time at which a peak is found to end. P. 377, Eq. (14.11).
t_s		The time at which a peak is found to start. P. 377, Eq. (14.11).
t_{ne}		The minimal time necessary for a separation. P. 214, Eq. (10.28).
t_G		The time spent by a solute molecule in the gas phase while passing through a column. P. 84. Eq. (6.23).
t_L		The time spent by a solute molecule in the liquid stationary phase while passing through a column. P. 86. Eq. (6.31).
t_M		The gas holdup time: The elapsed time from sample injection to peak apex of an unretained solute. P. 17.
t_M°		The time from sample injection to peak apex of an unretained solute, corrected for pressure gradient: $t_M^\circ = j\, t_M$.
t_N		The net retention time: $t_N = t_R^\circ - t_M^\circ$.
t_R		The time from sample injection to peak apex; the unadjusted, uncorrected, observed retention time. P. 4.
t_R'		The adjusted retention time: $t_R' = t_R - t_M$. P. 24.
t_R°		The corrected retention time: $t_R^\circ = j\, t_R$.
T		Absolute temperature. P. 171.
T_a		In Eq. (14.6), p. 348, the isothermal temperature for elution, following sample injection at a lower temperature T_0.
T_b		Boiling point. See Trouton's rule. P. 228, Eq. (11.18).
T_0	1.	In programmed temperature gas chromatography, the *initial temperature*. This is the column temperature at the beginning of a temperature program and may be the column temperature on sample injection. P. 240. Eq. (11.23)
	2.	In Eq. (14.6), p. 348, the injection temperature of a cold column or pre-column, as used in enhancement of trace detection.
T_r		In programmed temperature gas chromatography, the *retention temperature*. This is the column temperature at which a solute is eluted during a temperature program. P. 240. Eq. (11.23).
T'		In programmed temperature gas chromatography, the *significant temperature*, at which a programmed temperature process is equivalent to an isothermal. P. 241.

u	1.	Carrier gas velocity, measured. P. 87. Eq. (6.34).
	2.	Dipole moment (Chap. 5 only). P. 57.
$\bar{u}$		Average carrier gas velocity. P. 4.
$\bar{u}_o$		Optimum carrier gas velocity. P. 5, Fig. 1.5.
u_{opt}		Optimum carrier gas velocity. P. 98. Eq. (7.3).
u_L		The velocity of a peak apex along a column. P. 86. Eq. (6.33).
u_S		The velocity of a peak apex with respect to the carrier gas. P. 90. Eq. (6.50).
v		In the equation for reduced plate height, the reduced velocity: $v = d_p^* \, u/D_m$. P. 111. Eq. (7.14).
V	1.	Retention volume. P. 240.
	2.	Mixing chamber volume (Chap. 10 only). P. 199. Eq. (10.17).
	3.	Effective volume of a flame ionization detector *without* added gas (Chap. 9 only). P. 145.
V_a		The isothermal retention volume for elution temperature T_a. P. 348. Eq. (14.6).
V_e		The effective internal volume of a detector. P. 145.
V_{eff}		The effective internal volume of a detector (same meaning as V_e). P. 203. Eq. (10.21).
V_g		The specific retention volume. P. 224. Eq. (11.2).
V_{max}		That volume of a sample that will not cause any more than 10% peak broadening with respect to any smaller volume of it. $V_{max} = a\, V_R^o/\sqrt{n}$, but the "constant" a is determined by experiment, namely, by injecting successively smaller samples until resolution no longer improves with further decrease in sample size. P. 9.
V_r		The isothermal retention volume for elution temperature T_r (see Eq. (14.6)). The elution temperature is usually an arbitrarily chosen temperature for isothermal elution from a column used as a cold trap; the elution temperature is thus to be distinguished from the retention temperature, which is associated with a temperature program. P. 243.
V_F		The filtrate volume $(V_T - V_S)$ of solution from filtration impregnation. P. 184. Eq. (10.12).
V_L		Volume of stationary phase in the column. P. 9. See also p. 24, Eq. (2.21)
V_M		The retention volume for air, uncorrected for gas compressibility. It is used in determining V_R'. P. 19. Eq. (2.2).
V_M^o		The column dead volume: $V_M^o = j\, V_M$. P. 20. Eq. (2.4).
V_N		The net retention volume, adjusted for column gas volume and corrected for gas compressibility: $V_N = V_R^o - V_M^o$. It is the net retention volume that is directly and linearly related to the stationary phase volume V_L by the partition coefficient K: $V_N = K\, V_L$. P. 20. Eq. (2.6).
V_R		The unadjusted, uncorrected, retention volume. $V_R = F_o\, t_R$. P. 19. Eq. (2.1)
V_R'		The adjusted retention volume, uncorrected for gas compressibility: $V_R' = V_R - V_M$. P. 20. Eq. (2.5).
V_R^o		The corrected retention volume. P. 19. Eq. (2.3).
V_S		In filtration impregnation, the volume of solution that is retained on the support. P. 182. On p. 184, Eq. (10.12).
V_T		In filtration impregnation, the total volume of solution used. P. 184, Eq. (10.12).

Notation

V_{T_r}	Retention volume measured isothermally at retention temperature. P. 244.
V_0	The isothermal retention volume corresponding to a column temperature T_0 on sample injection. P. 243. See also Eq. (14.6), p. 348.
w	Weight.
w_b	Peak width at base, where the base is defined as the distance between the intercepts of the peak tangents with the interpolated baseline. P. 4.
w_f	The width of a digital matched filter. P. 375.
w_p	The peak width detected as part of the van Rijswick algorithm. P. 377.
W_L	The weight of stationary phase in a column. P. 24, Eq. (2.21).
W_P	In filtration impregnation, the weight of the packing. P. 182. Eq. (10.1).
W_S	In filtration impregnation, the weight of the support. P. 182. Eq. (10.1).
$W_{T,L}$	In filtration impregnation, the total weight of stationary phase to be used. P. 183. Eq. (10.11).
w_D	The maximum weight of a given solute that may be injected without exceeding the linear dynamic range of the detector. P. 346, Eq. (14.2). P. 347, Eq. (14.3).
w_L	Weight of solute dissolved in the stationary phase: $k = w_L/w_M$. P. 22, Eq. (2.9).
w_M	Weight of solute in the mobile phase: $k = w_L/w_M$. P. 22, Eq. 2.9.
w_S	The minimum detectable weight of a trace component. P. 347. Eq. (14.4).
x	In filtration impregnation, the weight fraction of stationary phase in a packing. P. 183, Eq. (10.2).
z	Carbon number of a n-paraffin, in connection with the Kovats Retention Index. P. 60. Eq. (5.1).
Z^*	The temperature-independent factor in the equation describing the temperature dependence of the electron capture coefficient. P. 171.
α	Relative retention. $\alpha = t'_{R2}/t'_{R1} = K_2/K_1 = k_2/k_1$. P. 24. Eq. (2.19).
α_p	Polarizability. P. 57.
β	Phase ratio, also called column characteristic. $\beta = V_M/V_L$. P. 22. Eq. (2.12).
γ	Tortuosity factor. P. 84. Eq. (6.23).
δ	Solubility parameter. P. 70.
λ	Packing irregularity factor. P. 83. Eq. (6.16).
ρ_L	Density of stationary phase (in bulk). P. 24. Eq. (2.21).
σ	Standard deviation, the square root of the variance. P. 80.
σ_T^2	1. The total variance, which is the sum of the variances arising from independent causes. P. 80. Eqs. (6.5) and (6.6). 2. The plate height contribution from the effective internal volume of a detector (Eq. (10.21) only). P. 203.
τ	Response time, p. 199, Eq. (10.16). This is identically the time constant used in Eq. (10.20), p. 199; and is also the effective time interval in Eq. (10.21), p. 203. The concept is the same in each case.
τ_D	The time constant for concentration decrease at the entrance to a mixing chamber. P. 199, Eq. (10.18).
τ_G	The typical time required for a solute molecule to diffuse through the mobile phase to a gas-liquid interface. P. 88. Eq. (6.42).

τ_L		The typical time required for a solute molecule to diffuse through the stationary phase to a gas-liquid interface. P. 86. Eq. (6.30).
τ_M		The time constant for concentration decrease at the exit of a mixing chamber P. 199. Eq. (10.17).
τ_2		The time during which unit peak spreading due to velocity distribution takes place. P. 91. Eq. (6.56).
Δ	1.	The distance diffused in time t according to Einstein's Law of Diffusion: $D = \Delta^2/2t$. P. 79. Eq. 6.4. See also p. 78.
	2.	Sampling interval (Chap. 14 only). P. 377, Eq. (14.7); p. 377, Eq. (14.8); and p. 377, Eq. (14.10).
Δp		Pressure required to produce a desired carrier gas flow. P. 180.

Introduction to
ANALYTICAL GAS
CHROMATOGRAPHY

Chapter 1

GAS CHROMATOGRAPHY: A SIMPLE DESCRIPTION

The word "chromatography," formed from the Greek "chroma" meaning "color," and "graphein" meaning "to write," was coined by Tswett [1, 2] around 1900 to describe his process of separating mixtures of plant pigments. He washed the pigments down a column of adsorbent powder, thus separating them into colored (chroma) bands (graphein) on the powder. Such a separation of the components of a mixture for qualitative or quantitative analysis, or for isolation and recovery of the components, is the desired end of any type of chromatography.

In gas chromatography (Fig. 1.1), the mixture (solute) to be separated is vaporized and swept over a relatively large absorbent or adsorbent surface inside a long narrow tube, or *column*. The sweeping is done by a steady stream of inert "carrier" gas, which serves only to move the solute vapors along the column. The different components are moved along the column at different rates and, under proper circumstances, become separated. The analyst arranges matters so that the separation is complete within a minimum time and can then be detected and recorded.

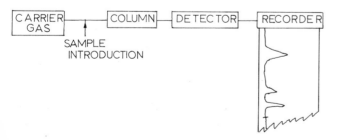

FIGURE 1.1 The gas chromatograph usually consists of a regulated supply of carrier gas, a means for sample introduction, a column for separating the components of the sample, a detector for detecting and signaling the components as they emerge in sequence from the column, and a recorder for measuring and recording the signal from the detector.

1.1 THE GAS CHROMATOGRAM

A typical gas chromatogram has a series of peaks (Fig. 1.2). The times from sample injection to the tops of the peaks are known as *retention times*; each component has a characteristic retention time for a given set of experimental conditions. These retention times serve as a means of qualitative analysis. The sensitivity of detection for a given detector varies according to peak height; that is, it depends on how well the peak can be differentiated from the baseline.

The utility of the gas chromatographic process depends on separating one peak from another. Thus, gas chromatographic technology is aimed at producing *one peak per component, maximum height per peak, minimum peak width, minimum retention times*. The phenomenally rapid adoption and implementation of gas chromatography stem from the ease with which almost anyone can quickly obtain usable separations and the great utility of the attendant information for many branches of science.

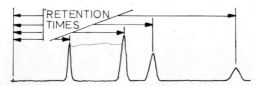

FIGURE 1.2 A typical gas chromatogram has a series of peaks. The times from sample injection to the tops of the peaks are known as *retention times*. These can identify the components producing the peaks. Ideally, there should be one peak per component, minimum peak width, maximum peak height, and minimum retention times.

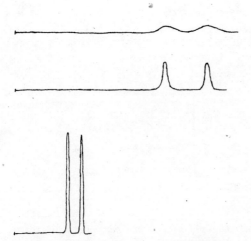

FIGURE 1.3 Separating peaks is easier when they are narrow. This is also beneficial for sensitivity and speed of analysis.

Sec. 1.2 Sample Introduction

To reiterate: The gas chromatogram is the result of passing the components of a mixture through a *column*, each component at its own velocity. It is a series of peaks (the narrower the better), each peak if possible distinct in position and meaning. Each component is qualitatively identified by its retention time and quantitatively estimated by its peak area.

Separating peaks is easier when the peaks are narrow, and sensitivity and speed of analysis also benefit (Fig. 1.3). Therefore, in the following examination of sample introduction and of the column, attention will continually revert to keeping peaks narrow.

1.2 SAMPLE INTRODUCTION

The technique of and equipment for effecting sample introduction are crucial. The sample should be introduced to the column as a plug, that is, practically instantaneously. And because column efficiency decreases sharply with increasing sample size, samples must be small or even minute.

Liquid samples usually are 1 µℓ or less in volume and are introduced by microsyringe, with satisfactory reproducibility. The microsyringe needle is pushed through a rubber or silicone rubber septum, the plunger is tapped or pushed home quickly, and the syringe is withdrawn. Gas samples may range up to several milliliters in volume and are introduced in several ways—sometimes by gas syringe but usually from a special valve. In a chamber of such a valve, the gas sample can be temporarily stored; the valve is then turned and the sample is swept out of the long, narrow chamber as a plug.

In either case, the sample must be introduced to the column as a plug, that is, deposited practically instantaneously on the very top of the column. For one thing, this means that the sampling chamber must be so designed that the sample does not mix with the carrier gas but is swept out of the chamber as a plug. Obviously, the sampling chamber must have minimum volume and no dead space or blind alleys. Because gas chromatography does not concentrate but rather only separates components, if sampling somehow spreads out each component, then effecting the separation is just that much more difficult and sensitivity is lower as well.

Another reason for the sample introduction-caused axial spreading of components, that is, peak broadening, is that a liquid sample cannot evaporate instantaneously (this is a fundamental limitation to the attainable speed of separation in gas chromatography). The sampling chamber should therefore be kept at a temperature perhaps 50°C higher than the highest boiling point of any component introduced. Here again, a minimum sample size may aid in permitting plug introduction.

The points stressed so far about sample introduction are these: Samples must be introduced not over any significant duration of time but all at once. Within the instrument, the design should cause the suddenly injected sample to be carried to the column without significant mixing with the carrier gas. This requires (1) that the inlet be hot enough to guarantee immediate flashing of a liquid sample; (2) that any gas introduction chamber be long and narrow enough to allow the gas sample to be swept out as a plug; (3) that the sample introduction system be free of dead spaces--spaces not continually in and dynamically swept out by the carrier gas stream; and (4) that the tubing connecting the sample introduction point or points to the column be considerably narrower than the column itself, although diameter changes should be made smoothly rather than discontinuously.

The sample arrives at the column, if possible, as a gaseous plug. All the sample components are contained within and spread throughout the plug. In the ideal case, no mixing of the plug components with the carrier gas has occurred at the head of the column. As the sample proceeds along the column, the components, selectively retarded and therefore moving at different rates, spread. Each component vapor packet progressively loses its plug shape and, as time goes by, and in the absence of nonideal effects, acquires the shape of a broader and broader Gaussian distribution curve. This provides a means of describing a column, assuming the rest of the gas chromatograph allows no further spreading of the component vapor packet. We now consider this means of description.

1.3 THEORETICAL PLATES

The term *theoretical plate* must now be introduced. The word *plate* refers to a plate in a distillation column, where ascending vapors are condensed by descending liquids. At each higher plate, the equilibrium liquid becomes richer in more volatile components, leaner in less volatile components. The change in composition between a volatile liquid and the vapors in equilibrium with it corresponds to a theoretical plate.

A column is said to have $n = 16(t_R/w_b)^2$ theoretical plates, if t_R is the time required for the center of the sample vapor packet to pass from the sample injection point to the detector and w_b is the time corresponding to the distance between the intersects of the peak tangents and the baseline (Fig. 1.4). If the length of the column is divided by n, the quotient h is the height equivalent to a theoretical plate. The narrowest peaks are obtained from columns with the smallest height per theoretical plate.

A. Height per Theoretical Plate

Tests concerning the theory and practice of gas chromatography are most often expressed by plotting the height h per theoretical plate against the average carrier gas velocity $\bar{u}$. Such a plot shows that with increasing gas velocity $\bar{u}$, the height h per theoretical plate first drops sharply to a minimum, then increases relatively slowly (Fig. 1.5).

Sec. 1.3 Theoretical Plates

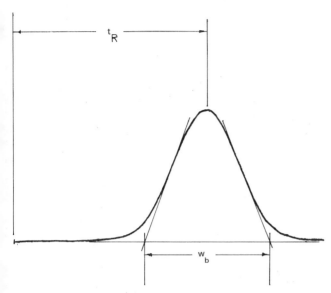

FIGURE 1.4 To calculate the number of theoretical plates that characterize a peak, (1) draw tangents to the peak at the inflection points, and extend the tangents through the baseline; (2) extend the baseline under the peak; and (3) using the same units, measure the distance t_R from sample injection to peak apex, and the distance w_b between the baseline-tangent intercepts. The number n of theoretical plates is $n = 16 \, (t_R/w_b)^2$.

For any column, an optimum carrier gas velocity exists. In practice, the carrier gas velocity may exceed this optimum value within fairly wide limits with little decrease in column performance, particularly for a very good column. Plots of h *versus* $\bar{u}$ are expressions of the van Deemter equation.

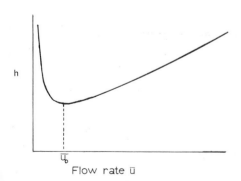

FIGURE 1.5 The height h per theoretical plate is equal to the length L of the column divided by the number n of theoretical plates produced by the column: $h = L/n$. A plot of h *versus* the flow rate $\bar{u}$ shows an initial rapid decrease to a more-or-less broad minimum, then a slower rise. The minimum shows the optimum flow rate $\bar{u}_o$.

1.4 THE VAN DEEMTER EQUATION

The van Deemter equation states that the height h per theoretical plate is

$$h = A + \frac{B}{\bar{u}} + C\bar{u}$$

where $\bar{u}$ is the carrier gas velocity. Measures to minimize A, B, and C also decrease h, result in narrower peaks, and therefore should if possible be identified and adopted.

The A term in the van Deemter equation is evaluated as

$$A = 2\lambda d_p$$

where 2λ is an *ad hoc* constant and d_p is the diameter of the particles filling the column. This is the first mention in this introduction to gas chromatography of "the particles filling the column." The first earlier sentence on gas chromatography stated, "In gas chromatography, the mixture (solute) to be separated is vaporized and swept over a relatively large absorbent or adsorbent surface inside a long narrow tube or column." The "large surface" referred to is the collective surface of the particles in the column. In gas-liquid chromatography, the collective surface is the gas-liquid interface of the liquid stationary phase supported on these particles.

Generally, these particles are highly reticulated themselves, so that the surface area is partly internal and relatively independent of the particle size. The particles, consisting of materials such as crushed firebrick, are more or less closely graded for size and size distribution. Although some very useful types of columns contain only the bare particles, almost all columns contain particles that have been coated with a solvent--the stationary phase. (Our discussion will be restricted at least for the present to columns containing a supported stationary phase.) The solvent may merely be poured over the particles and mixed. Alternatively, the stationary phase may be dissolved in a volatile solvent, the particles impregnated with the solution, and the volatile solvent then removed by evaporation. After these impregnating operations, the particles can still be poured with ease.

They are, therefore, poured with vibration, and sometimes with the help of a vacuum, into a column. Considerable care must be taken to achieve even, channel-free packing. Once packed, however, a column may be treated quite roughly without seriously decreasing its efficiency. The powdery packing is held in the column primarily by friction. It is kept from falling out of the column by a small tamp of glass wool or a screen in each end.

Thus, the stationary phase is distributed over the extensive surfaces of a ground and sized material such as crushed firebrick. In most columns, these particles have only one function: to support the widely distributed stationary phase. The particles are thus known as the support, or as the inert support.

Lack of inertness in the support is a more-or-less severe disadvantage. An adsorbent surface of a nominally inert support can cause trace components to disappear

Sec. 1.4   The van Deemter Equation   7

completely, or it can adversely affect or ruin separations by causing tailing--making a peak look like a low cliff and low following plateau rather than a proper peak. Sometimes pretreatment with the functional group to be processed is effective in countering adsorption. A more general approach is to either cover or chemically remove the active molecular groups.

The best-known procedure involves replacing the hydrogens of surface hydroxyls with silyl groups:

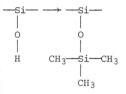

The new surface is now one of inert methyl groups. On these, adsorption is negligible.

We now return to a consideration of the van Deemter A term, assuming now that the surface of the support particles is inert and that the particles vary from one column to another primarily only in particle diameter d_p.

The A ($= 2\lambda d_p$) term of the van Deemter equation is constant with respect to carrier gas velocity, and can be so indicated on an h *versus* $\bar{u}$ plot. The A term sometimes is called the *multiple path* term, sometimes the *eddy diffusion* term. The idea is that the solute vapor molecules, in being swept around the support particles, must travel different distances and/or times to move the same distance along the column. Either way, the solute vapor molecules become less closely grouped and the peak becomes a little broader. Obviously, both effects are minimized if all the support particles have the same, minimal diameter. Thus a 70/80-mesh support gives better results--narrower peaks--than a 35/80-mesh support because the diameters do not differ so widely, and better than a 20/30-mesh support because the particles are smaller.

Peak spreading in the empty volume at the end of the column is a different effect, which in practical work accounts for much of the A term.

To summarize: The A, or *eddy diffusion*, or *multiple path* term of the van Deemter equation is not a function of carrier gas velocity, but is theoretically proportional to the support particle diameter. The A term is minimized by using an inert support consisting of minute particles--about 120 mesh--closely graded to have a narrow size distribution--about a 10 mesh cut. It is also minimized by leaving as small an unfilled volume as possible at the end of the column.

The B, or *molecular diffusion*, term of the van Deemter equation is given by

$$B = \frac{2\gamma D_G}{\bar{u}}$$

where 2γ forms an *ad hoc* constant reflecting the nature of the labyrinths within the particles of a particular inert support, D_G indicates the gas diffusivity, and $\bar{u}$ is the carrier gas velocity. The B term is minimized by using heavier rather than lighter carrier gases (and solutes), lower temperatures, and higher gas velocities. If gas velocities are kept high enough, as they are in practical work, the B term contributes little to peak broadening. (Note: Strictly, $B = 2\gamma D_G$ only.)

It is the C terms that are of practical importance. They describe forms of resistance to mass transfer. A term describing resistance to mass transfer in the liquid phase was part of the original van Deemter equation. For better description of peak broadening, Jones added terms describing resistance to mass transfer in the gas phase. However, it is the term describing resistance to mass transfer in the liquid phase that has practical importance for most work being done with gas chromatography.

The term concerning resistance to mass transfer in the liquid phase contains the expression $(d_L^2/D_L)\bar{u}$. Thus this C term varies as the square of the stationary phase film thickness d_L and, therefore, at least approximately, as the square of the weight percent of the stationary phase.

This C term varies inversely as the liquid diffusivity D_L of the stationary phase. Thus the simpler the molecular structure of the stationary phase, the lower the contribution of this C term to the height per theoretical plate. However, the stationary phase is usually chosen for a low volatility and some given, desired selectivity, with the increased convenience of a longer-lasting column that is usable at higher temperatures.

Over a very wide range of weight percent stationary phase, no price need be paid for decreasing the C term by decreasing the stationary phase film thickness. A simple experiment can demonstrate the practical value of this C term in evaluating the importance of the proportion of stationary phase in a column to peak narrowness:

A set of columns is prepared. The columns vary in length, but each column contains the same weight of (any given) stationary phase. With each column used near its optimum flow rate (at 1.0 to 1.5 times the optimum flow rate, but not less than that), results show that resolution improves sharply with decreasing weight percent stationary phase but, as postulated, constant weight of stationary phase. That is, resolution is better, for example, with a 10-ft column containing 10 wt% stationary phase than it is with a 5-ft column containing 20 wt% stationary phase. However, because the amount of stationary phase is held constant, retention time also stays roughly constant within this set of columns.

With such a set of columns, varying the injected sample size--1, 2, 5, 10, and 20 μl, for instance--shows that resolution also improves with decreasing sample size.

Within wide limits, therefore, a radical increase in column performance, evident in narrower peaks for a given retention time, is obtainable merely by decreasing the weight percent stationary phase and proportionately increasing the column length.

To summarize the most important and consistent directions for packing and using columns that can be gleaned from inspection of the van Deemter equation:

1. Use a lower rather than a higher percent stationary phase. The higher the diffusivity of the stationary phase, the better.

2. Use inert support particles of minute—about 100/120 mesh—identical—a 10 or 20 mesh cut—diameters.
3. Use as low a column temperature as possible (this also follows from other considerations; see a detailed treatment in Chap. 11).
4. Leave as small an unfilled volume as possible at the end of the column.
5. Use a carrier gas velocity somewhat higher than that indicated as optimum by a plot of h *versus* u.

1.5 MAXIMUM ALLOWABLE SAMPLE SIZE

Another experimental variable of great importance in column efficiency is sample size. Now that the number n of theoretical plates has been defined, the equation theoretically limiting sample volume can be stated.

$$V_{max} = a \frac{(V_M^o + KV_L)}{\sqrt{n}}$$

We restate this, expressing more explicitly the internal diameter d_C of the column.

$$V_{max} = a \frac{(\pi/4)d_C^2 L + KV_L}{\sqrt{n}}$$

The maximum sample (vapor) volume V'_{max} that will not cause appreciable peak broadening is inversely related to the square root of the number n of theoretical plates, but directly and linearly related to the internal volume, essentially V_M^o, of the column, the amount V_L of stationary phase in the column, and the partition coefficient K, which expresses the ratio of solute concentrations in the stationary phase and in the carrier gas. The constant a is about 0.02, but decreases with increasing column efficiency. To put it somewhat differently, the equation states that $V_{max} = a \left(V_R^o / \sqrt{n} \right)$; the maximum acceptable sample volume is directly proportional to the retention volume, for a given column.

It has already been pointed out that a lower proportion of stationary phase to inert support improves column performance. Such a lowering, however, also requires a decrease in the quantity of sample injected in the column. A practical working rule is to use the smallest sample that will satisfy the requirements of sensitivity and precision.

1.6 COLUMN PERFORMANCE

Following all the working rules laid down in the preceding sections will, with a well-designed gas chromatograph, produce a gratifyingly high number of theoretical plates. Obtaining 1,000 theoretical plates per foot is possible; 500, not uncommon. Packed columns can yield well over 100,000 theoretical plates, can easily show 15,000.

The "theoretical plate" of gas chromatography is based on the same theory as that of distillation, but in practice does not have the same meaning. Perhaps because a theoretical plate in gas chromatography is used only once per separation, whereas it is used repeatedly in distillation, the number of theoretical plates required by gas chromatography to achieve a given separation is approximately the square of that required in a distillation. But the requisite number is far more easily obtained by gas chromatography, and the accompanying information is obtained far more quickly. Yet, with the same experimental ease gas chromatography affords the potential of extractive distillation through the use of selective stationary phases.

1.7 THE SELECTIVE STATIONARY PHASE

A selective stationary phase changes the retention time of a solute from what would be expected from its boiling point. An understanding of selectivity depends on an acquaintance with molecular forces affecting solubility. These forces are (1) dispersion, London, or nonpolar forces; (2) induced dipole, Debye forces; (3) orientation, or Keesom forces; and (4) specific interaction forces, or chemical bonding.

Dispersion, London, or nonpolar forces are always present in all molecular species. They arise from synchronized variations in chance, instantaneous dipoles—charge separations—of two interacting species. They are the only source of attracting energy between two nonpolar substances. They are not temperature dependent.

Induced dipole, Debye forces result from an interaction between a permanent dipole in one molecule and an induced dipole in a neighboring molecule.

The most important forces for selectivity in gas chromatography are orientation, or Keesom, forces. Hydrogen bonding is a Keesom force of paramount importance. Keesom forces are interactions between permanent dipoles in interacting molecules. The energy of such an association depends on the sizes and positions of the interacting dipoles. For greatest selectivity in the use of Keesom forces, column temperature should be kept as low as possible. (The use of minimal weight percentages of stationary phase decreases retention times; lower temperatures become feasible as retention times decrease, and lower column temperatures aid selectivity. So minimal proportions of stationary phase indirectly make possible increased selectivity with selective stationary phases.)

The strongest forces for selectivity are not the most important, because they occur in specific interactions rather than on a general basis. These are the forces involved in chemical adducts or complexes.

Thus, the four forces causing solubility are the omnipresent dispersion forces (the only forces operating between nonpolar molecules) and the three types of forces underwriting selectivity. Weakest are the induction forces, whereby a molecule with a permanent dipole induces a dipole in another, polarizable molecule. In this way, the polarizable benzene is retained by a glycol, or any highly polar stationary phase

Sec. 1.8 Peak Movement

longer than cyclohexane, although the boiling points of the two are nearly identical. Most general are the orientation forces between permanent dipoles, such as between an alcohol and a glycol; these forces include the ubiquitous hydrogen bonds. Finally, the strongest forces are those involving specific interactions, as, for example, the selective retardation of olefins by silver ions.

Exactly what is meant by "retardation"? How do the molecules of a given solute actually move along the column at some given average rate?

1.8 PEAK MOVEMENT

A. The Basic Process

For any given column segment at any given time, the molecules of a given compound are found distributed between the stationary phase and the gas phase, in constant proportion.

When in the gas phase, the molecules are swept along in the steadily moving carrier gas. As the molecules temporarily in the gas phase enter a region just downstream from where they were, many reenter, that is, dissolve in, the stationary phase. In this downstream region the same constant distribution between the gas phase and the stationary phase is also found. Meanwhile, the fresher gas in the region upstream, from which the molecules were just swept, tends to become depleted. Therefore, in accordance with the constant and characteristic gas-stationary phase distribution, or *partition coefficient*, of these molecules, they move from the now-too-rich solvent into the gas, and then of course are swept downstream.

All the molecules of a given molecular type tend to move downstream at a rate characteristic of only this given molecular type. This rate depends only on the carrier gas velocity, the amount of stationary phase in the column, and the partition coefficient for the given molecular type-stationary phase-temperature combination. Different molecular types, each with its own partition coefficient, move along the column at different rates.

This picture of peak movement shows that the intimate physical status within the column causes an approach to equilibrium, but that equilibrium does not in fact exist. For practical purposes, however, gas chromatography constitutes a quick and powerful method for the determination of partition coefficients.

B. Temperature Programming

Successive members of a homologous series move along a column at rates that decrease logarithmically with increasing carbon number. For a given column held at a given temperature (isothermal gas chromatography), the retention times of a homologous series increase exponentially. Because peak width is proportional to retention time in isothermal gas chromatography, the peak widths also increase exponentially.

However, if after sample injection the column temperature is increased linearly with time, then retention times increase only linearly with carbon number in a homologous series, and peak width remains constant. Temperature programming the column makes the application of gas chromatography to mixtures of wide boiling range practical. The advent of temperature programming marked a profound increase in the productivity of gas chromatography (Fig. 1.6; see Ref. 3). (The expression *temperature programming* usually means a linear increase of temperature with time. When the function is not linear, this is always specified.)

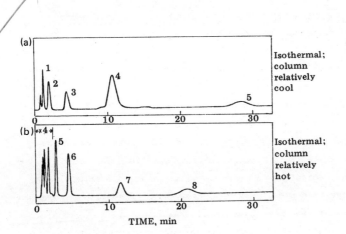

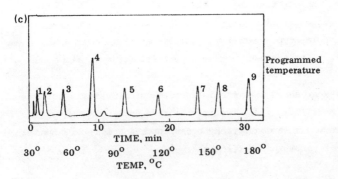

FIGURE 1.6 Temperature programming greatly increased the productivity of gas chromatography. With temperature programming, gas chromatography could be applied to mixtures of wide boiling range. The mixture shown chromatographed had a boiling range of 226°C. Peaks 1 through 6: n-paraffins, C_3 to C_8; 7, bromoform; 8, m-chlorotoluene; and 9, m-bromotoluene. (a) Isothermal, column relatively cool; (b) isothermal, column relatively hot; (c) programmed temperature. (Reprinted from Harris and Habgood, Ref. 4, *Programmed Temperature Gas Chromatography*, 1966, by permission of the authors and John Wiley and Sons, Inc. Data originally published by the authors in *Analytical Chemistry*, Ref. 3.)

1.9 RELATIVE RETENTION TIMES

The relative retention time is a means of recording a gas chromatographic datum so that gas chromatographic experimentation can become continuous, related, and cumulative. A relative retention time is a ratio referred to a standard; it is a ratio of time differences. The retention time for air is subtracted from the retention time for a given substance, and from that for a reference standard. The air-free retention time for the given substance is then divided by the air-free retention time for the reference standard. This subtraction and division frees the quotient from reflecting many experimental parameters such as weight percent stationary phase or column length. Ideally, a relative retention time reflects only the column temperature and the identities of the component chromatographed, the stationary phase, and the reference standard. Much progress has been made in this approach to data compilation, but not enough to make such data generally usable for qualitative analysis.

1.10 A RECAPITULATION AND AN OVERVIEW

Gas-liquid chromatography is a means of separating substances that are or can become gaseous. The substances separate because they dissolve to different degrees in a solvent, called the stationary phase. The stationary phase is held within a long narrow tube called the column, and is elaborately exposed within the column to the substances being separated because it is thinly spread over the surfaces of internally open, inert, support particles. The substances are swept into and through the column by a carrier gas that flows at perhaps twice the velocity that yields optimum resolution.

In day-to-day application, gas chromatography becomes easily 10 and perhaps 1,000 times more effective in yielding separations within a given and reasonable time as the following parameters, listed roughly in order of importance, are diminished: (1) the size of the sample; (2) the weight percent stationary phase; (3) the inner diameters of the column and of the auxiliary passages through which the substance gases or vapors are swept; and (4) the support particle size and size distribution.

The already phenomenally wide-ranging applicability of gas chromatography was vastly extended by two developments: temperature programming--the linear increase with time of column temperature--and the chemical elimination, most importantly by silylation, of hydroxyl groups in the sample and the support.

1.11 A PRELIMINARY GUIDE TO THE LITERATURE OF GAS CHROMATOGRAPHY

By 1965, over 2,000 publications on gas chromatography were being presented yearly [5]. The yearly number continued to increase. In 1976, over 3,000 abstracts concerned with gas chromatography were published by the Preston Technical Abstracts Company [6].

Such an outpouring is dealt with by the organization of means for retrieval and by the statement of principles.

A. Retrieval

Chemical Abstracts

Chemical Abstracts is the least specialized and least descriptive guide to the literature of gas chromatography. A great gem, nevertheless, it is highly dependable and generally available.

The Fundamental Reviews in Analytical Chemistry

A more detailed and descriptive guide to the literature is the biennial review presented in the April issue of *Analytical Chemistry* each even-numbered year (in 1976, 1,748 publications were cited). The concise summaries in these reviews enable the investigator to grasp the essence of individual papers and to decide whether to consult the original reference. Books and symposia are also listed.

Gas and Liquid Chromatography Abstracts

The most personally useful guide to the literature of and the trade names in gas chromatography is *Gas and Liquid Chromatography Abstracts*, formerly called *Gas Chromatography Abstracts* [7]. These abstracts are compiled by the members of the Chromatography Discussion Group (Trent Polytechnic, Burton Street, Nottingham NG1 4BU, United Kingdom). Edited for over two decades by C. E. H. Knapman, the abstracts had been published since 1958, first yearly as hardcover books, now quarterly as paperbacks.

Each volume of these abstracts contains a subject index so elaborate that there is also an accompanying index to the subject index. Most questions on gas chromatography have now been answered. The abstracts of the Chromatography Discussion Group are easily the quickest and surest approach available to an individual for finding answers.

B. Principles

The principles of gas chromatography are stated most succinctly in texts and they are stated in even more detail in reviews of specific areas. Finally, the original papers in which these principles were first stated also warrant close and repeated examination. There is much in them that the later condensation of knowledge omits.

Texts

The texts that I have found most useful are those of Giddings [8], Purnell [9], Harris and Habgood [4], Ettre [10], and McFadden [11].

Giddings's *Dynamics of Chromatography* provides insights that cannot be found elsewhere. Among these is a description of the fundamental interactions of flow and diffusion in the chromatographic process. Also among them is a comparison of chromatographic forms with each other and with their individual potentials. Reading *Dynamics* is a fascinating intellectual adventure.

References

Purnell's *Gas Chromatography* is a valuable reference for the physical chemistry of gas chromatography. To work straight through the book is very worthwhile, although the lack of a list of symbols makes this unduly difficult.

Harris and Habgood's *Programmed Temperature Gas Chromatography* tells and shows just about all there is to know on that subject. The writing is clear, the mathematical background is there in full, the illustrations are clear and cogent. This is a necessary reference for those who wish to understand this subject area.

Ettre's *Open Tubular Columns* presents in one place the basic theory and practice of open tubular columns, and compares them with packed columns. The eventual predominant importance among open tubular columns of the glass and fused silica varieties is not foreseen, but this is a minor fault.

McFadden's *Techniques of Combined Gas Chromatography/Mass Spectrometry* is a basic, detailed, authoritative reference for this field of rapidly growing importance. Unfortunately, the treatment of mass spectrometry and of interfacing is much better than the treatment of gas chromatography.

Journals

Reviews appear frequently in the primary journals for gas chromatography. These are the *Journal of Chromatography,* the *Journal of Chromatographic Science, Chromatographia,* and *Analytical Chemistry.* The *Journal of Chromatography* incorporates issues of *Chromatographic Reviews;* the *Journal of Chromatographic Science* and *Chromatographia* occasionally devote entire issues to reviews of specific subjects; and *Analytical Chemistry* usually carries special reviews, some of which relate to gas chromatography.

Books of Reviews

An invaluable source of insight, perspective, and understanding are the surveys that are presented in hardbound books or surveys. These surveys are invited by knowledgable editors and written by authoritative practitioners. A storehouse of such studies is the *Advances in Chromatography* series published by Marcel Dekker and edited for its first 11 volumes by J. Calvin Giddings and Roy A. Keller. Other books of such reviews have been edited by Purnell [12, 13], Domsky and Perry [14], and Grob [15].

Early Original Papers

The early studies of gas chromatography can furnish valuable insight and perspective. Magee has edited a collection of these works, and is to be thanked for that [16].

A prime source of original papers is the bound volumes of the early symposia. These were sponsored in Europe by the Institute of Petroleum [17] and in the United States by the Instrument Society of America [18].

REFERENCES

1. V. Heines, *Chem. Tech. 1,* 281 (1971).
2. M. S. Tswett, *J. Chem. Educ. 44,* 238 (1967).

3. H. W. Habgood and W. E. Harris, *Anal. Chem. 32,* 450 (1960).
4. W. E. Harris and H. W. Habgood, *Programmed Temperature Gas Chromatography,* Wiley, New York, 1966.
5. R. S. Juvet, Jr. and S. P. Cram, *Anal. Chem. 42,* 1R (April, 1970).
6. S. P. Cram and R. S. Juvet, Jr., *Anal. Chem. 48,* 411R (April, 1976).
7. C. E. H. Knapman (editor), *Gas Chromatography Abstracts,* one volume per year, bound: 1958-1962, inclusive, Butterworths, London; 1963-1969, inclusive, The Institute of Petroleum, London.
    Gas and Liquid Chromatography Abstracts, one volume per year, bound: 1970, The Institute of Petroleum; 1971-1972, Applied Science, Ripple Road, Barking, Essex, England; also, four quarterly journals per year: 1973-present, Applied Science, Essex.
    Cumulative Indexes, 1958-1963, The Institute of Petroleum, London; 1969-1972, The Institute of Petroleum, London; 1969-1973, Applied Science, Essex. (Still available from Applied Science Publishers are the volumes 1963-1967 inclusive, 1969-1972 inclusive, and the Cumulative Indexes. The *Abstracts* are available on a regular quarterly basis to members of the Chromatography Discussion Group, Trent Polytechnic, Burton Street, Nottingham NG1 4BU, England. The Group welcomes new members.)
8. J. C. Giddings, *Dynamics of Chromatography,* Dekker, New York, 1965.
9. H. Purnell, *Gas Chromatography,* Wiley, New York, 1962.
10. L. S. Ettre, *Open Tubular Columns in Gas Chromatography,* Plenum, New York, 1965.
11. W. McFadden, *Techniques of Combined Gas Chromatography/Mass Spectrometry,* Wiley-Interscience, New York, 1973.
12. J. H. Purnell (editor), *Progress in Gas Chromatography,* Interscience, New York, 1968.
13. H. Purnell (editor), *New Developments in Gas Chromatography,* Wiley, New York, 1973.
14. I. I. Domsky and J. A. Perry (editors), *Recent Advances in Gas Chromatography,* Dekker, New York, 1971.
15. R. L. Grob (editor), *Modern Practice of Gas Chromatography,* Wiley-Interscience, New York, 1977.
16. R. J. Magee (editor), *Selected Readings in Chromatography,* Pergamon, New York, 1970.
17. D. H. Desty and C. L. A. Harbourn (editors), *Vapour Phase Chromatography,* Butterworths, London, 1957; D. H. Desty (editor), *Gas Chromatography 1958,* Academic, New York, 1958; R. P. W. Scott (editor), *Gas Chromatography 1960,* Butterworths, Washington, D. C., 1960; M. Van Swaay (editor), *Gas Chromatography 1962,* Butterworths, Washington, D. C., 1962; A. Goldup (editor), *Gas Chromatography 1964,* The Institute of Petroleum, London, 1965; A. B. Littlewood (editor), *Gas Chromatography 1966,* The Institute of Petroleum, London, 1967; C. L. A. Harbourn and R. Stock (editors), *Gas Chromatography 1968,* The Institute of Petroleum, London, 1969; R. Stock and S. G. Perry (editors), *Gas Chromatography 1970,* The Institute of Petroleum, London, 1971.
18. V. J. Coates, H. J. Noebels, and I. S. Fagerson (editors), *Gas Chromatography,* Academic, New York, 1958; H. J. Noebels, R. F. Wall, and N. Brenner (editors), *Gas Chromatography,* Academic, New York, 1961; N. Brenner, J. E. Callen, and M. D. Weiss (editors), *Gas Chromatography,* Academic, New York, 1962; L. Fowler (editor), *Gas Chromatography,* Academic, New York, 1963.

Chapter 2

SOME LANGUAGE: A FEW TERMS AND SYMBOLS

Some of the terms and symbols used in descriptions of gas chromatography have become more or less standard. Learning a few of them in order to understand gas chromatography is both worthwhile and necessary.

2.1 GAS HOLDUP TIME

Consider a gas chromatogram showing an air peak and one or two other peaks (Fig. 2.1). After the sample is injected, the air peak appears (if some air was injected and the detector is air sensitive). We assume that the injected air does not dissolve detectably in the stationary phase. Two results come immediately from this first datum, t_M, called the *gas holdup time* [1].

In the first place, the gas holdup time t_M reveals the time spent in the gas phase for any solute, because they are all the same in this respect: all solutes spend the same amount of time in the gas phase. (Retention times differ only with respect to the time spent by the solutes in the stationary phase.)

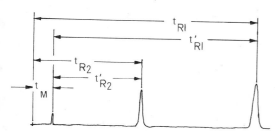

FIGURE 2.1 From a simple gas chromatogram, a number of numerical values can be calculated.

In the second place, if the flow rate F_c is known (the subscript implies that the measured flow rate has been corrected to column temperature) and the pressure gradient correction factor j--which we define in the next section--is used, then the product $jF_c t_M$ gives the gas volume of the column. With an instrument of good design and therefore of negligible extracolumn internal volume, this should approximate the gas volume of the whole gas chromatograph.

Should time be measured from the moment of sample injection or from air-peak appearance? The operator must begin waiting from the moment he injects the sample, so from his point of view retention times start at sample injection. On the other hand, appearance of the air peak marks the earliest evidence of any interaction within the gas chromatographic column. So, from the point of view of an observer of such interactions, time should be marked from the air peak.

With which point of view should the number of theoretical plates be calculated? In 1956 and again in 1958, an International Nomenclature Committee recommended [2] that the number of theoretical plates be calculated from the moment of sample injection.

So calculated, the number of theoretical plates increases with the square of the air retention time, merely by delay--which can be several minutes--and not necessarily with any resultant improvement in separation. An alternative calculation, excluding t_M and yielding smaller values, was then proposed [3]. This calculation prevailed.

The number of theoretical plates calculated as shown in Fig. 1.4 (using the total retention time) is symbolized by lower case n [1]. Capital N now symbolizes the number of *effective* theoretical plates and is calculated by excluding the air retention time (*i.e.*, gas holdup time) from the ratio numerator.

The gas holdup time t_M is used quite generally in treating retention data to make them more reproducible. A peak other than the air peak can be variously described: It may be (1) measured from sample injection or air peak; (2) expressed as retention time or retention volume; (3) corrected for carrier gas compressibility or temperature, or for the weight of the stationary phase; or (4) compared to the similar behavior of standards such as the series of normal paraffins. Most of the alternatives must be learned--although not all in this chapter--because they are met so frequently.

What the individual does in his own laboratory about recording retention times depends on his need to communicate. Retention data that are to be reproduced or recognized at a later time or a different place must be *adjusted* (as defined in the next section) and referred to standards, usually *via* the Kovats Retention Index (see Sec. 13.2), but that is all. Neither *correction* (also defined in the next section) nor full reduction to the specific retention volume are often done. (An elaborately worked-out example calculation of a specific retention volume is shown in the first edition of Littlewood, Ref. (4), a much-shortened example is found in the second edition, Ref. (5).)

2.2 HANDLING RETENTION DATA: SYMBOLS AND LANGUAGE

These procedures, the language for which must be learned, begin with the pressure gradient correction factor j [1, 6]. This is expressed in terms of the inlet/outlet pressure ratio P_i/P_o

$$j = \frac{3[(P_i/P_o)^2 - 1]}{2[(P_i/P_o)^3 - 1]}$$

Although the pressure gradient correction is appreciable enough--given that the outlet pressure is simply atmospheric, then for 14 psig inlet pressure, j is already 0.64; and for 28, 0.46--it is rarely applied in practical work. In the practical matter of keeping track of retention times, ratios are quite generally used. Because the pressure gradient correction factor j cancels out of these ratios--see the note and proof at the end of this section--it is not applied numerically.

In gas chromatography, the pressure gradient correction factor j is rarely applied in calculations on retention times, but is always applied in the development of theory. The gas chromatographer learns about the pressure gradient correction factor j not in order to use it numerically, ever, on retention data, but rather to understand the language of gas chromatography.

Gas is compressible. If the equations of gas chromatography are to be kept as simple and linear as possible, then, to begin with, the pressure gradient correction factor must be included. In addition the flow-independent retention volume is used in the development of theory, rather than the flow-dependent retention time.

In the language describing retention, the retention volumes that are multiplied by the pressure gradient correction factor j are spoken of as being *corrected* and are indicated by the superscript "o", unless they are also adjusted. The retention volumes from which the gas holdup volume has been subtracted are spoken of as being *adjusted* and are indicated by a prime, unless they are also corrected. A retention volume that has been both adjusted and corrected is called a *net* retention volume and is indicated by a subscript capital N: V_N. The following six equations recapitulate this material [1].

The observed retention time t_R, multiplied by the column temperature-corrected exit flow rate F_c, gives the observed retention volume V_R:

$$V_R = F_c t_R \qquad (2.1)$$

If the combined volume of the injection port, connecting lines, and detector can be neglected, then we also have the special case of Eq. (2.1) for the gas holdup time t_M and volume V_M:

$$V_M = F_c t_M \qquad (2.2)$$

Multiplying the observed retention volume V_R by the pressure gradient correction factor j gives the *corrected* retention volume V_R^o:

$$V_R^o = j V_R \qquad (2.3)$$

Note the corresponding, corrected gas holdup volume V_M^o

$$V_M^o = jV_M \qquad (2.4)$$

where the capital subscript M refers to the mobile phase.

Alternatively, the observed retention volume can have the gas holdup volume subtracted from it, giving the *adjusted* retention volume V_R'

$$V_R' = V_R - V_M \qquad (2.5)$$

If the observed retention volume is both corrected and adjusted, it becomes the *net* retention volume V_N

$$V_N = V_R^o - V_M^o = jV_R' \qquad (2.6)$$

The concept of the net retention volume is desirable because it can be so concisely expressed and used in equations relating gas chromatographic variables.

We can now see how the pressure gradient correction factor j cancels out of ratios of adjusted retention volumes and of the corresponding retention times. The added subscripts 1 and 2 refer to two solutes.

$$\frac{V_{N,2}}{V_{N,1}} = \frac{jV_{R,2}'}{jV_{R,1}'} = \frac{jF_c t_{R,2}'}{jF_c t_{R,1}'}$$

Therefore

$$\frac{V_{N,2}}{V_{N,1}} = \frac{V_{R,2}'}{V_{R,1}'} = \frac{t_{R,2}'}{t_{R,1}'}$$

Note too that the more elaborate and precise calculation methods for compiling retention data use ratios and thus do not call for numerical use of the pressure gradient correction factor. The most important of these methods, the Kovats Retention Index, is considered in detail in Chaps. 5 and 13, where the stationary phase and qualitative analysis are discussed.

2.3 THE PARTITION COEFFICIENT, PARTITION RATIO, AND PHASE RATIO

The net retention volume V_N is directly and linearly related to the volume V_L of the stationary phase by the partition coefficient K [1]:

$$V_N = KV_L \qquad (2.7)$$

In other words, for a given solute and stationary phase used at a given temperature, doubling the amount of stationary phase doubles the net retention volume. For a given flow rate, this amounts to doubling the observed retention times, because the gas holdup volume is proportionately so small in most cases.

The partition coefficient succinctly describes the solute/solvent interaction for a given temperature. It gives the ratio of the weight-per-volume concentration C_L of

Sec. 2.3 The Partition Coefficient, Partition Ratio, and Phase Ratio

the solute in the stationary liquid phase to the corresponding concentration C_M of the solute in the mobile gas phase [3]:

$$K = \frac{C_L}{C_M} \tag{2.8}$$

One of the three basic assumptions of gas-liquid chromatography is that the sample is initially distributed throughout the first theoretical plate. The other two assumptions—that equilibrium is attained instantaneously in the column, and that the partition coefficient is constant throughout the column—are implied by Eq. (2.8).

Equilibrium is not actually attained instantaneously as the peak moves along the column. Indeed Giddings [7] has shown that gas chromatography can be more precisely described if it is viewed as a nonequilibrium process. However, as the sample size approaches zero or at least does not overload the column; as the column construction improves column efficiency by minimizing the local depth of the stationary phase, supporting the stationary phase on supports of minimal and uniform particle diameter and optimal and uniform pore size, namely, Type I supports (see Chap. 8), and minimizing the column internal diameter; as the carrier gas flow rate is at least not increased beyond the optimum flow rate: as all these conditions obtain, the degree of nonequilibrium lessens—peaks tend to remain as sharp as they were when they entered the column—and equilibrium is more nearly attained. In short, if gas chromatography is performed well experimentally, then the assumption that equilibrium is attained "instantaneously" is not too far off.

Equation (2.8) implies, in accord with the basic assumption that the partition coefficient is constant throughout the column, that if the solute, the stationary phase, and the temperature are given, then a definite, finite number exists that describes the distribution of the solute between the solvent and the carrier gas. That same distribution exists throughout the column. It exists at the local center of the solute peak, and equally at locations far from there. The same just-mentioned experimental conditions that tend to allow the attainment of equilibrium also tend to permit a spatially constant partition coefficent.

Whether or not the partition coefficient is exactly constant, it is certainly always finite and, at any rate, nearly constant. The solute concentration in the gas phase is always related to that in the liquid phase, by a constant. This means that the solute concentration in the gas phase is never zero. Thus, when after injection the solute enters the column and dissolves in the stationary phase, its vapors do not vanish from the gas phase. No matter how soluble the solute, some of the solute molecules are always in the gas phase in the local concentration described by Eq. (2.8).

The astonishing realization follows that at least some molecules of every injected solute emerge simultaneously with the air peak. Further, at least some molecules of every sample ever injected into a gas chromatographic column *are still there*. Thus, there is no such thing as a "clean" gas chromatographic column, once a sample has been injected into it.

Equation (2.8) can be expanded and rearranged to yield two more concepts and symbols that are frequently met in papers on gas chromatography. We express the weight-per-volume concentrations explicitly, then rearrange to form a weight fraction and a volume fraction:

$$K = \frac{C_L}{C_M} = \frac{w_L/V_L}{w_M/V_M^o} = \frac{w_L}{w_M} \frac{V_M^o}{V_L} = k\beta \qquad (2.9)$$

In Eq. (2.9), two new symbols, k and β, have been introduced. The partition ratio k (also called the capacity ratio and the capacity factor) gives the ratio of the weights w_L and w_M of solute in the stationary (L) and mobile (M) phases, respectively, in the column [3]:

$$k = \frac{w_L}{w_M} \qquad (2.10)$$

The partition ratio also expresses the ratio of the fractions of the solute in the two phases. To show this, we divide the numerator and denominator of Eq. (2.10) by $(w_L + w_M)$:

$$k = \frac{w_L/(w_L + w_M)}{w_M/(w_L + w_M)}$$

Thus, if p expresses the fraction $w_L/(w_L + w_M)$ of solute in the stationary phase, and q expresses the fraction $w_M/(w_L + w_M)$ of solute in the mobile phase, then

$$k = \frac{p}{q} \qquad (2.11)$$

In one sense, the partition coefficient K is more fundamental than the partition ratio k. The partition coefficient K is a function only of the solute, the stationary phase, and the temperature at which the solute and the stationary phase are brought together. The partition ratio k, however, depends not only on these but also on the ratio of the volumes V_M^o and V_L that is the *phase ratio* β [1]:

$$\beta = \frac{V_M^o}{V_L} \qquad (2.12)$$

(Another name for the phase ratio β is the column characteristic.)

If less fundamental than the partition coefficient K, the partition ratio k, as K/β, embodies more meaning: The partition coefficient K expresses solubility and is solely chemical in meaning. The phase ratio β expresses a space relationship and is solely physical in meaning. Thus, the partition ratio can be considered to express the ratio of the chemical to the physical aspects of the column. Further comment on this concept is made in Chap. 7.

The partition ratio can be calculated directly from the chromatogram itself. As we shall see in a moment, this calculation is not approximate but exact, because in

Sec. 2.3 The Partition Coefficient, Partition Ratio, and Phase Ratio

it the pressure gradient correction factor j cancels out. We begin with Eq. (2.9), which we rearrange:

$$K = k\beta \qquad (2.9)$$

so

$$k = \frac{K}{\beta} \qquad (2.13)$$

But

$$V_N = KV_L \qquad (2.7)$$

and

$$\beta = \frac{V_M^\circ}{V_L} \qquad (2.12)$$

We use Eqs. (2.7) and (2.12) to replace K and β in Eq. (2.13):

$$k = \frac{V_N/V_L}{V_M^\circ/V_L} = \frac{V_N}{V_M^\circ} \qquad (2.14)$$

Both V_N and V_M° are corrected retention volumes that can be reexpressed as products of the flow rate and the corresponding retention times. Thus,

$$V_N = jV_R' \qquad (2.6)$$

and

$$V_R' = F_c t_R' \qquad (2.15)$$

so

$$V_N = jF_c t_R' \qquad (2.16)$$

Also,

$$V_M^\circ = jV_M \qquad (2.4)$$

and

$$V_M = F_c t_M \qquad (2.2)$$

so

$$V_M^\circ = jF_c t_M \qquad (2.17)$$

We can now substitute the expressions of Eqs. (2.16) and (2.17) for V_N and V_M° into Eq. (2.14):

$$k = \frac{V_N}{V_M^\circ} \qquad (2.14)$$

$$k = \frac{jF_c t_R'}{jF_c t_M}$$

therefore

$$k = \frac{t_R'}{t_M} \qquad (2.18)$$

Equation (2.18) shows that the partition ratio k can be calculated exactly as the ratio of the adjusted retention time t'_R--that is, the observed retention time t_R minus the observed gas holdup time t_M--to the observed gas holdup time t_M.

2.4 RELATIVE RETENTION

If two solutes are to be separated by a gas chromatographic column, then no matter how good the column or the gas chromatograph, the two solutes must have different solubilities in the stationary phase. If solute 1 has partition coefficient K_1 and solute 2, K_2, then the separation is possible only if the ratio K_1/K_2 is different from unity. The ratio of partition coefficients is called the *relative retention* α [1].

$$\alpha = \frac{K_1}{K_2} \qquad (2.19)$$

It is always expressed as equal to or larger than unity. The larger the relative retention α, the easier the separation.

For a given column and column temperature, the relative retention can be calculated from the ratio of adjusted retention times:

$$K = k\beta \qquad (2.9)$$

and

$$k = \frac{t'_R}{t_M} \qquad (2.18)$$

therefore,

$$\frac{K_1}{K_2} = \frac{(t'_{R1}/t_M)\beta}{(t'_{R2}/t_M)\beta}$$

and

$$\alpha = \frac{K_1}{K_2} = \frac{t'_{R1}}{t'_{R2}} \qquad (2.20)$$

2.5 CALCULATING SOME NUMERICAL VALUES RELEVANT TO THESE CONCEPTS

The stationary phase within a given column is usually measured by weight. Therefore, the volume of stationary phase within a column can be calculated by dividing the weight W_L of stationary phase in the column by the stationary phase density ρ_L. (Stationary phase densities can usually be found in reference handbooks or suppliers' descriptions.)

$$V_L = \frac{W_L}{\rho_L} \qquad (2.21)$$

Sec. 2.5 Calculating Some Numerical Values Relevant to These Concepts

Thus, if we measure only the flow rate and the weight of stationary phase used in a given column, and look up the stationary phase density for the temperature at which the column is to be used, we can find the partition coefficient K and the partition ratio k for a given solute, and the phase ratio β for the column, from a gas chromatogram made with that column at that temperature.

For example, suppose we hold at 100°C a column that contains 3.75 g stationary phase quoted to have a density of 0.75 g/ml at that temperature. At room temperature, 22°C, we measure the column exit flow rate as 79.3 ml/min. We find retention times of 2 min for air and 32 min for the solute peak of interest. The partition ratio k can be immediately calculated from the retention times by Eq. (2.18):

$$k = \frac{t'_R}{t_M} = \frac{32 - 2}{2} = 15$$

The phase ratio β can be calculated once the volume V_L of the stationary phase is known. This, by Eq. (2.21), is

$$V_L = \frac{W_L}{\rho_L} = \frac{3.75 \text{ g}}{0.75 \text{ g/ml}} = 5 \text{ ml}$$

The volume V_M^o of the mobile phase is, by Eqs. (2.2) and (2.4)

$$V_M^o = jF_c t_M$$

Let us assume that the pressure gradient correction factor for this example is about 0.7. Then

$$V_M^o = (0.7)[(79.3)(373/295) \text{ ml/min}](2 \text{ min})$$

so

$$V_M^o = 140 \text{ ml}$$

Therefore, by Eq. (2.12), the phase ratio β is

$$\beta = \frac{V_M^o}{V_L} = \frac{140 \text{ ml}}{5 \text{ ml}} = 28$$

Finally, the partition coefficient K can be calculated by Eq. (2.9):

$$K = k\beta = (15)(28) = 420$$

Thus, in this example the gas volume is 28 times the stationary phase volume. Yet 15 times as much solute is found in the stationary phase as in the mobile phase. Or, when expressed with respect to concentration, the weight-per-volume solute concentration in the stationary phase is 420 times that in the mobile phase.

Consider a second solute peak emerging after 35 min. The relative retention α of the two peaks is

$$\alpha = \frac{35 - 2}{32 - 2} = \frac{33}{30} = 1.10$$

REFERENCES

1. ASTM Committee E-19 on Chromatography, E 355-77, *Gas Chromatography Terms and Relationships*, American Society of Testing and Materials, 1916 Race Street, Philadelphia, Pa., 1977; Analytical Chemistry Division, International Union of Pure and Applied Chemistry, *Compendium of Analytical Nomenclature* (H. M. N. H. Irving, H. Freiser, and T. S. West, preparers), Pergamon, Elmsford, N. Y., 1978, Chap. 13.
2. Nomenclature Committee Recommendations, in *Vapour Phase Chromatography* (D. H. Desty, editor), Academic, New York, 1957, pp. xi-xiii; and in *Gas Chromatography 1958* (D. H. Desty, editor), Academic, New York, 1958, p. xi.
3. J. H. Purnell, *J. Chem. Soc.*, 1268 (1960).
4. A. B. Littlewood, *Gas Chromatography*, Academic, New York, 1962, pp. 30-35.
5. A. B. Littlewood, *Gas Chromatography*, 2nd ed., Academic, New York, 1970, pp. 33-34
6. A. T. James and A. J. P. Martin, *Biochem. J. 50*, 679 (1952).
7. J. C. Giddings, *Dynamics of Chromatography*, Dekker, New York, 1965.

Chapter 3

SAMPLE INTRODUCTION

One of the assumptions of gas chromatographic theory is that at the start of the gas chromatographic separation, the sample is uniformly distributed throughout the first theoretical plate.

3.1 SAMPLE SIZE

In gas chromatography, as technique improves, sample size decreases. The amounts of sample required on the grounds of detector linear dynamic range (described in Sec. 9.1C) and column overloading are as far removed from ordinary experience as distances on the atomic scale, and about as hard to comprehend. We can start by considering the size of a normal drop.

A. Overloading the Detector

A drop weighs about 0.05 g--about 50 mg. Calculations shown in Chap. 14 indicate that the very largest amount of a solute charged to a flame ionization detector (FID) and appearing at the detector after about 8 min in a fairly sharp peak, without exceeding the linear dynamic range of the FID, is about 10^{-5} g. One drop would be about 5,000 times too much for even the maximum permissible charge. However, this amount represents the very top of the FID linear dynamic range, which is 10 million. As a sample charge that would exploit to within a factor of 10 the full linear dynamic range of the FID, a drop would be five billion times too large.

It thus becomes more understandable that a detector-suitable amount of solute might be injected as 0.1 µl of a 0.01% solution. This is less than a drop of the neat solute by a factor of five million, and is only 1,000 times too large to exploit the full linear dynamic range of the flame ionization detector.

B. Overloading the Column

Another reason for using minute samples is to avoid overloading the column. A column is said to be overloaded when the sample size has caused a decrease of 10% or more in the number n of theoretical plates from the maximum number observed with any smaller sample. No given size of sample can be recommended for all columns and situations in gas chromatography. The closest approach to specifying the maximum permissible sample size for a given solute that will not overload the column is $V_{max} = aV_R^o/\sqrt{n}$, where V_{max} is the volume of sample vapor, a is a constant (about 0.02; see Ref. 1), and V_R^o is the corrected retention volume for that solute. The prediction of such a calculation can and should be experimentally checked as follows:

Sample size should be decreased until no further increase in the number n of theoretical plates is found for the column at hand. The necessary factor of decrease may run into three orders of magnitude for certain low-loaded or open tubular columns, and for early peaks. As a start for selecting sample size, work toward smaller sizes from 0.10 µl liquid until no further increase in n is found. It is not probable with modern equipment that detector sensitivity will be inadequate and therefore limiting to the decrease of sample size required for maximum peak sharpness.

C. Adequate Temperature for Introduction of Liquids

In tests for the variation of the number of theoretical plates with sample size, and indeed generally, the sample introduction port temperature must be high enough that slow vaporization of injected liquids does not broaden peaks. The injection temperature should be perhaps 50°C higher than the highest boiling point of any component in the mixture being injected. Increasing the injection port temperature 20°C should not decrease peak width or increase n if that temperature was already high enough.

D. Never Use an Eyepiece to Measure Peak Width

Measuring the peak width for calculating the number n of theoretical plates in these trials may be difficult if the peak emerges early and the column is efficient. The error in n increases as the square of the error in the measurement of peak width. In such cases, decrease the measurement error by using a much higher chart speed during the tests, rather than by trying to apply a micrometer eyepiece to the existing narrow peak.

3.2 PLUG INTRODUCTION

Unless special techniques such as discussed in Sec. 11.2,B are used, the sample must be not only small, but also concentrated.

Ideally, we wish to achieve plug introduction. Given plug introduction, an instant after the sample has been introduced a hypothetical plot of the sample

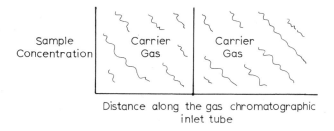

FIGURE 3.1 Ideal plug introduction: vertical plug sides; no mixing with the carrier gas; no plug width, zero sample size.

concentration *versus* distance along the tube would show a spike, as in Fig. 3.1. Ideally, the spike has vertical sides: the sample vapors do not mix with the carrier gas. Ideally, the spike has no width: the sample size should approach zero, and sample introduction should be instantaneous. Perhaps "spike introduction" would be more descriptive of the desired effect.

Plug introduction favors separability and sensitivity. Peaks of minimum width are easiest to separate. Also, peak area is directly and usually linearly proportional to the quantity of the component or components producing the peak, and to the product of the peak height and half-width. Consequently, if this product is to produce the requisite area, then for the ideal--zero peak width at introduction--peak height becomes maximized insofar as sample introduction can affect it. Therefore, the sensitivity of detection, which for a given detector and sensitivity setting is the distance from the peak apex to the baseline, is also maximized for the peak of minimum width.

In summary, two restrictions should govern sample introduction in gas chromatography: the sample must not overload either the column or the detector, and the sample must be introduced within a period that is at most no longer than one-tenth the duration of the narrowest peak to be eluted. Both restrictions are discussed in greater detail in the treatments of detectors (Chap. 9), open tubular columns (Chap. 10), and quantitative analysis (Chap. 14).

3.3 EXPERIMENTAL METHODS

A. Gas Samples

In considering the means for sample introduction, let us start with gas samples. If the sample is not already a gas, it must be made one before it can be handled by gas chromatography. If it is already a gas, it must be kept gaseous and discrete: it must neither condense nor mix with the carrier gas.

Gases may be injected directly from a gas syringe of the type shown in the diagram in Fig. 3.2 and by the photograph in Fig. 3.3.

A gas may also be injected by means of a gas sampling valve like the one illustrated in Fig. 3.4. (The gas sample must not come into contact with any solvent other

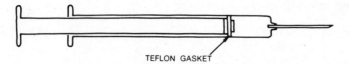

FIGURE 3.2 A diagram of a syringe for gas injection. (Courtesy of the Hamilton Company.)

FIGURE 3.3 Photograph of a gas-injection syringe. (Courtesy of the Hamilton Company.)

than the stationary phase properly displayed within the column. Therefore the gas sampling valve must operate lubricant free.) By the action of the gas sampling valve, the gas is really being placed into the carrier gas stream by the sudden insertion into the carrier gas line of the sample gas container, which is a long, narrow tube. Again, the purpose of this design and action is to prevent the carrier gas from mixing with the sample, thus to introduce the sample as a plug. Therefore, if the gas sample volume is to be increased, this is accomplished primarily by lengthening the sample-containing tube rather than by increasing its radius.

We point out, in passing: of the three phases of matter, only gases impose no fundamental limit on the rapidity of gas chromatographic separations. With liquids or solids, the time-consuming heat transfer into the sample for vaporization becomes limiting.

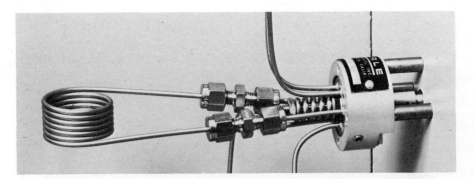

FIGURE 3.4 In a gas sampling valve, the gas sample is held in the long, narrow, coiled tube. When the valve shaft is turned from the sampling position to the injection position, the gas sample is inserted into the carrier gas without mixing with it. (Photograph courtesy of the Carle Instrument Company.)

Sec. 3.3 Experimental Methods

FIGURE 3.5 A 10-µl, plunger-in-barrel syringe. (Photograph courtesy of the Hamilton Company.)

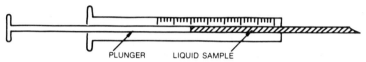

FIGURE 3.6 The plunger-in-barrel syringe, showing the measurement of a liquid sample prior to injection. After injection, the needle remains filled and can cause further unwanted injection. (Diagram courtesy of the Hamilton Company.)

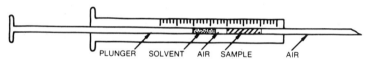

FIGURE 3.7 With this use of the plunger-in-barrel syringe, the solvent seals the syringe and washes out any remaining sample. (Diagram courtesy of the Hamilton Company.)

FIGURE 3.8 Volumes smaller than a microliter can be precisely measured with a plunger-in-needle syringe. (Photograph courtesy of the Hamilton Company.)

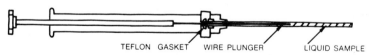

FIGURE 3.9 This diagram of the plunger-in-needle syringe shows that the less than 1-µl sample can be precisely measured, is contained only in the needle, and is expelled completely on injection. (Drawing courtesy of the Hamilton Company.)

B. Liquid Samples

Normal Injection

A liquid may be injected into the carrier gas stream with a syringe. But the injection must be instantaneous and so, also, must volatilization within the instrument. *Instantaneous* here means negligibly brief compared to the duration of the earliest and therefore narrowest of any resultant peaks. (This requirement, however, is eased if the column is much cooler during sample injection than later, when it is warmed during elution. See the discussions on trace detection in Chaps. 11 and 14.) Compared to a peak 60 sec wide, for instance, an injection time of 0.6 sec is negligibly brief.

Two types of syringes are used for injecting liquids. These are shown in Figs. 3.5 and 3.8. In one, the plunger is in the barrel of the syringe (Figs. 3.5 to 3.7); in the other, the plunger is in the needle (Figs. 3.8 and 3.9).

When the plunger is in the barrel, it can be seen and so can the contents--the user can see what he is doing. This is comforting. By pumping the air out of the tip of the needle when it is held in the sample, withdrawing the syringe, and inspecting the liquid in front of the plunger, the analyst can ascertain that the liquid is free of bubbles. He can then move the plunger forward until it reaches a desired volume calibration. Because the bubble-free sample can be seen emerging from the needle during this forward movement, the volume ahead of the plunger must be filled by the sample (this stage is shown in Fig. 3.6). When the plunger is pressed home, a known volume of sample will be ejected from the needle tip. Meanwhile, the plunger can be retracted, drawing the measured volume back into the barrel, where it will neither leak out nor be changed by some inadvertent moving of the plunger.

With this combination of syringe and technique, the needle is filled during injection of the sample. At the same time, because the needle is being held at least temporarily inside the very hot sample injection port, the needle is becoming rapidly hotter. Thus, whatever liquid sample may be in the needle expands and partially or perhaps completely evaporates. The actual volume injected becomes uncertain, and reflects technique (bad: a good analytical method cancels out technique). Using a high-boiling solvent [2] only alleviates the problem.

Another technique (Fig. 3.7) for using the plunger-in-barrel syringe calls for drawing a microliter or so of solvent into the syringe first, next to the plunger. This is followed by a volume of air, and then by approximately the desired volume of sample. The sample volume is measured by reading the front and back limits of the trapped sample. With this technique, the solvent provides a seal between the plunger and barrel. If some of the solvent is lost, no matter. On ejection, the solvent washes out the surface in front of it, tending to ensure quantitative injection.

The disadvantages of the plunger-in-barrel syringe are unfortunately preponderant. The minimum volume precisely measurable and injectable is far too large for the capacities of either a good column or a flame detector. Therefore, for most work, certainly for most work with the FID, the plunger-in-needle type of syringe is required. In this

Sec. 3.3 Experimental Methods   33

design, as shown in Figs. 3.8 and 3.9, the plunger is a fine wire that fills the needle. The inner end of the needle cavity is a chemically inert Teflon bushing through which the plunger slides. Inside the glass part of the syringe, the wire moves inside a narrow tube that supports the wire. Another tube that the operator can see slides outside the inner tube. The end of the larger tube can be seen against the graduations on the glass and reflects the position of the end of the wire inside the needle.

When the sample is injected from the plunger-in-needle syringe, only an acceptably small fraction of the volume injected remains inside the needle. The range of the volume injected is set by the internal diameter of the needle rather than by its length, so that much smaller volumes can be read with much higher precision than with the plunger-in-barrel type of syringe.

Despite this improved precision, the prudent analyst does not depend on injection reproducibility in quantitative analysis. Instead, he injects a solution that incorporates an internal standard. Then, the ratio of the unknown peak area to the internal standard peak area will not reflect the inevitable variations in injected volume.

On-Column Injection

At the beginning of the separation, ideally, the sample is uniformly distributed throughout the first theoretical plate. Injecting the liquid sample directly into the column has been said to satisfy this requirement. The practice is called *on-column injection*. Let us examine it.

On-column injection is done only with packed columns. (The 0.01 to 0.03 in. inner diameters of open tubular columns do not permit the entry of a syringe needle.) The packing of a packed column is a stationary phase on an inert support. The support is usually a diatomaceous earth. Rarely, the support may be glass beads. Both supports are excellent thermal insulators.

The stationary phase, typically only 2 or 3% by weight of the light and fluffy powder that is the support, comprises only a small spatial part of the packing. Thus the packing is also an excellent thermal insulator.

It is into a spot in this powdery insulation that the cool sample is injected.

The diatomaceous earth is an excellent support and insulator because it has a highly extended microscopic structure (see Figs. 8.1 to 8.9) made of a material that is itself nonconductive. During the instant that the injected sample is supposed to be vaporizing, the insulator packing cannot receive heat from the column wall by conduction, but can receive heat only from the carrier gas, and then only by convection. However, the carrier gas has negligible heat capacity compared to the solid wall of an injection port from which the sample would otherwise evaporate. Neither is the carrier gas moving rapidly, as a gas must if it is to carry heat effectively.

Thus the sample injected on column can evaporate only by cooling the fluffy insulator and the low-heat capacity gas around it, and by spreading, through capillary action, into the surrounding insulation, there to evaporate a bit more.

Chemically, on-column injection is surely the mildest possible injection method. It is ideally suited to the injection of thermally labile substances, such as biochemicals. To the separation of these, gas chromatography is often just marginally applicable, being limited by sample lability and by adsorption. So the chemical mildness of on-column injection very probably helps to extend the advantages of gas chromatography to the analysis of such materials.

Physically, on-column injection could hardly be worse. It could not possibly provide an environment less well suited to the rapid sample volatilization required for plug introduction and high column efficiency. At the start of the separation, the sample can be uniformly distributed throughout the first theoretical plate by introduction of the vaporized sample as a vapor plug, but not by the deposition of the sample within the column as a thermally insulated, cool, concentrated liquid mass.

The poor physical characteristics of on-column injection could be alleviated, while yet preserving the possibly necessary and beneficially mild conditions. The operator could stop the flow of carrier gas from perhaps 30 sec before injection until perhaps 30 sec after. This would allow the carrier gas to come nearly to a halt at the head of the column, and then give the injected sample a little time in which to evaporate at the point of injection. Resumption of carrier gas flow would then present the column with the more nearly ideal plug introduction of the sample vapors. (The flow should be interrupted not by a hurried and futile turning and resetting of the flow controller needle valve, but by the throwing of a toggle valve installed downstream from the flow controller.)

Further: The column should be made as narrow as possible in the sample-injection region, to maximize the thermal contact of the injected sample and the heat-exchanging carrier gas with the column wall. The gas chromatograph should afford separate temperature control of the sample-injection region. This temperature should be set at the maximum temperature allowable for the stationary phase (see Sec. 14.10A, Sample Injection)[3]. The sample injection part of the column should then be separately packed with a packing having a high-temperature stationary phase, such as preconditioned SE-30. This can be easily done, and will not appreciably affect retention behavior.

C. Solid Samples

Solids can be injected by a device such as that shown in Fig. 3.10. Here the sample sits within the needle and is pushed out and deposited within the sample injection port inside the instrument. The gas chromatographic demand for instantaneous injection does not lapse if solids are injected, so performance deteriorates instead: wider peaks result. As mentioned earlier, the techniques of programmed temperature gas chromatography can alleviate the requirement for instantaneity of sample introduction.

Sec. 3.4 The Sample Introduction Port   35

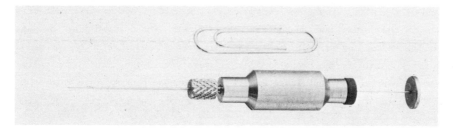

FIGURE 3.10 A solid sample. The solid is ejected from the needle into the hot injection port. (Photograph courtesy of the Hamilton Company.)

3.4 THE SAMPLE INTRODUCTION PORT

Figure 3.11 shows a modern sample introduction port. Certain elements of the design--the carrier gas concentric flow pattern, and the vaporizer tube removability, replaceability, material (glass), and inner diameter (minimal)--have become more or less standard. They can be legitimately looked for as minimal evidence of good design in a new instrument. In addition, the septum may well be shielded from the hot carrier gas by an internal ring with only a small center hole for the needle. This minimizes pickup by the gas of detectable organic vapors from the septum: septum "bleed" that is all too detectable at modern detector sensitivities.

In the modern port, the carrier gas enters the side from a narrow tube, flows through a narrow but uniform annulus outside the liner--heating it in the process--to

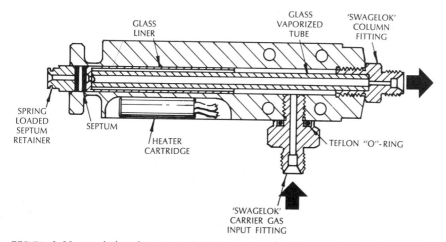

FIGURE 3.11 An injection port should be dynamically swept throughout and have a removable and replaceable glass vaporizer tube of minimal inner diameter. (Drawing courtesy of the Hamilton Company.)

the outer end of the port, sweeps the septum or septum holder, and reverses direction and flows toward the column down the narrow center liner. The carrier gas flows as an annulus over the end of any syringe needle that has been pushed through the septum into the port. Not only is the proper injection space within the liner dynamically swept out by the carrier gas, but there are also no upstream pockets into which the suddenly injected and isotropically expanding sample vapors can penetrate, thus ruining resolution.

The vaporizer tube is glass, removable, cleanable, and replaceable. Therefore, a sample containing nonvolatiles may nevertheless be injected without ruining a column or leaving a permanent residue in the injection port to affect a subsequent sample injection or furnish "echo" injections.

REFERENCES

1. A. I. M. Keulemans, *Gas Chromatography*, 2nd ed., Reinhold, New York, 1959, p. 194.
2. M. M. E. Metwally, C. H. Amundson, and T. Richardson, *Anal. Chem. 39*, 551 (1969).
3. D. W. Grant and A. Clarke, *J. Chromatog. 97*, 115, 129 (1974).

Chapter 4

RESOLUTION: SEPARABILITY *VERSUS* PLATE NUMBER

4.1 INTRODUCTION

This chapter offers perspective for the study of the succeeding four.
    These four chapters deal with two aspects of the gas chromatographic column: Separability and plate number. Separability depends on the stationary phase, to which Chapter 5 is devoted. Plate number depends on the manner in which the stationary phase is used--Chapters 6, 7, and 8.
    Which is more important, separability or plate number? Placing the peaks far apart, or keeping the peaks narrow?
    To examine this question, we first derive and study an expression for resolution. Then, from examples in the history of gas chromatography (GC), we seek and find a consistent direction in GC development and use.
    This consistent direction leads to an answer that transcends the original question, and *that* is the perspective to be gleaned from this chapter.

4.2 TWO AUXILIARY METHODS

Temperature programming and derivatization have enormously increased the productivity of gas chromatography. Each is considered in a later chapter (see Chaps. 11 and 12, respectively). However, before we discuss resolution, let us establish that neither method affects the problem of emphasis inherent in evaluating the two aspects of resolution.

A. Temperature Programming

Two problems arise if column temperature is held constant during a separation. First, the peak widths increase linearly with time. Second, the center-to-center distance, delay, and wait between peaks increase exponentially with carbon number (in a homologous series). These problems, illustrated in Fig. 1.6, are solved by temperature programming.

"Temperature programming" means raising the column temperature during the separation, usually at a constant rate. With temperature programming, also as shown in Fig. 1.6, all the peaks of a homologous series have the same width and arrive at roughly equal intervals. However, although it appears to do so, and is most persuasive to watch, temperature programming does not increase resolution.

For a given flow rate, the resolution achievable with a given column approaches a maximum as the rate of temperature increase approaches zero [1, 2]. For each separation of a close pair, there is one best temperature [3]. But the very definition of resolution refers to the separation of just such a close pair. Thus temperature programming does not affect the theoretical nature of resolution nor the problem of emphasis in considering resolution. Nor, on the practical side, does temperature programming make any given degree of resolution more attainable.

Each mixture usually contains numerous pairs. Temperature programming affords to each successive pair the optimum temperature for best resolution--but fleetingly. For any given pair, resolution is maximized as the rate of temperature increase approaches zero.

B. Derivatization

Derivatization--making gas chromatographically separable derivatives of substances otherwise less or not at all amenable to gas chromatography--greatly widened the applicability of gas chromatography.

Sugar, for instance, merely chars under heat, normally. Vaporizing sugar would seem to be out of the question. But sugar can be silylated in seconds, and the vaporizable silyl derivatives of sugars are nicely amenable to separation by gas chromatography.

Like other substances, however, derivatives may be more or less difficult to separate. In any case, once made, they remain to be separated successfully. Although there are derivatives that are made solely because they are more easily separable than the parent compounds, still derivatization as such does not affect the nature of resolution nor its inherent problem: separability *versus* plate number.

4.3 RESOLUTION

A. A Derivation

The two chromatographic peaks shown in Fig. 4.1 represent that pair of solutes most difficult to separate in any group of solutes. Thus, any separation reduces to the consideration of this figure.

Sec. 4.3 Resolution

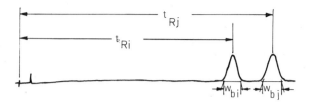

FIGURE 4.1 The resolution of these two peaks is given by the peak-to-peak separation divided by the average peak width.

The resolution R_s of solute i from solute j is given by the peak-to-peak separation ($t_{Rj} - t_{Ri}$) divided by the average peak width ($w_{bj} + w_{bi}$)/2. Because the peaks are close to each other, the width w_b of each is nearly the same, and is thus taken for the average. We start, therefore, with

$$R_s = \frac{t_{Rj} - t_{Ri}}{w_b} \qquad (4.1)$$

From this basic expression for resolution [4] we are going to derive an equation that will express resolution as a function of the relative retention α and the number n of theoretical plates. We will then be able to examine the relationship of separability (α) to plate number (n).

Into Eq. (4.1), we introduce the number n of theoretical plates. From the definition

$$n = 16 \left(\frac{t_{Rj}}{w_b} \right)^2 \qquad (4.2)$$

we divide by 16 and take the square root:

$$\frac{\sqrt{n}}{4} = \frac{t_{Rj}}{w_b} \qquad (4.3)$$

We now multiply Eq. (4.1) by t_{Rj}/t_{Rj}, namely, unity, switch numerators, and substitute $\sqrt{n}/4$ for t_{Rj}/w_b:

$$R_s = \frac{t_{Rj} - t_{Ri}}{w_b} \frac{t_{Rj}}{t_{Rj}} = \frac{t_{Rj}}{w_b} \frac{t_{Rj} - t_{Ri}}{t_{Rj}}$$

$$R_s = \frac{\sqrt{n}}{4} \frac{t_{Rj} - t_{Ri}}{t_{Rj}} \qquad (4.4)$$

The techniques of the rest of this derivation are similarly voluminous, but simple enough--multiplying by unity (fractions having the same numerator and denominator), adding and subtracting a given quantity at the same time, and rearranging fractions. When the desired expression for R_s arrives, we simplify it a little more by the same techniques. Now we return to the derivation.

We divide both numerator and denominator of Eq. (4.4) by t_M, multiply by unity expressed as $(t_{Ri} - t_M)/(t_{Ri} - t_M)$, and interchange the top denominators:

$$R_s = \frac{\sqrt{n}}{4} \frac{(t_{Rj} - t_{Ri})/t_M}{t_{Rj}/t_M} \frac{t_{Ri} - t_M}{t_{Ri} - t_M}$$

$$= \frac{\sqrt{n}}{4} \frac{(t_{Rj} - t_{Ri})/(t_{Ri} - t_M)}{t_{Rj}/t_M} \frac{t_{Ri} - t_M}{t_M}$$

We rearrange the right fraction into a ratio of reciprocals, and also add and subtract t_M from both numerators of the left fraction:

$$R_s = \frac{\sqrt{n}}{4} \frac{[(t_{Rj} - t_M) - (t_{Ri} - t_M)]/(t_{Ri} - t_M)}{(t_M + t_{Rj} - t_M)/t_M} \frac{1/t_M}{1/(t_{Ri} - t_M)}$$

We reexpress the larger fraction by dividing the numerators by the denominators, and multiply the right fraction by unity expressed as $(t_{Rj} - t_M)/(t_{Rj} - t_M)$:

$$R_s = \frac{\sqrt{n}}{4} \frac{[(t_{Rj} - t_M)/(t_{Ri} - t_M)] - 1}{1 + [(t_{Rj} - t_M)/t_M]} \frac{(t_{Rj} - t_M)/t_M}{(t_{Rj} - t_M)/(t_{Ri} - t_M)}$$

Interchanging denominators yields the desired result, which we shall then reexpress in terms of the relative retention α and the partition ratio k:

$$R_s = \frac{\sqrt{n}}{4} \frac{[(t_{Rj} - t_M)/(t_{Ri} - t_M)] - 1}{(t_{Rj} - t_M)/(t_{Ri} - t_M)} \frac{(t_{Rj} - t_M)/t_M}{1 + [(t_{Rj} - t_M)/t_M]} \quad (4.5)$$

As we saw in Eqs. (2.19), (2.7), (2.16), (2.6), and (2.5),

$$\alpha = \frac{K_j}{K_i} = \frac{V_{Nj}/V_L}{V_{Ni}/V_L} = \frac{jF_c(t_{Rj} - t_M)}{jF_c(t_{Ri} - t_M)}$$

so that the relative retention α is given exactly by

$$\alpha = \frac{t_{Rj} - t_M}{t_{Ri} - t_M} \quad (4.6)$$

Also, the partition ratio k is given exactly--see Eq. (2.18)--by

$$k = \frac{t_{Rj} - t_M}{t_M} \quad (4.7)$$

With Eqs. (4.6) and (4.7), we can now reexpress Eq. (4.5):

$$\boxed{R_s = \frac{\sqrt{n}}{4} \frac{\alpha - 1}{\alpha} \frac{k}{1 + k}} \quad (4.8)$$

Equation (4.8) could be our goal, except that J. H. Purnell [5] objected to using the number n (it was his objection that caused the eventual demotion of the symbol

Sec. 4.3 Resolution

from capital to lowercase), which increases merely as the square of the delay t_M in arrival of an air peak or its equivalent, as an index of column performance. A better index would exclude t_M, yielding a number N of *effective* plates:

$$N = 16 \left(\frac{t_R - t_M}{w_b} \right)^2 \tag{4.9}$$

It follows, from Eqs. (4.2) and (4.9), that

$$\frac{\sqrt{N}}{\sqrt{n}} = \frac{t_R - t_M}{t_R}$$

or that

$$\sqrt{N} = \sqrt{n} \, \frac{t_R - t_M}{t_R} \tag{4.10}$$

We can now quickly show that the fraction $(t_R - t_M)/t_R$ in Eq. (4.10) is equal to $k/(1 + k)$. Divide both $(t_R - t_M)$ and t_R by t_M, and simultaneously add and subtract t_M to and from the denominator t_R:

$$\frac{t_R - t_M}{t_R} = \frac{(t_R - t_M)/t_M}{(t_M + t_R - t_M)/t_M}$$

But by Eq. (4.7),

$$k = \frac{t_R - t_M}{t_M} \tag{4.7}$$

so

$$\frac{t_R - t_M}{t_R} = \frac{k}{1 + k} \tag{4.11}$$

Therefore, we can substitute Eq. (4.11) into Eq. (4.10) to get

$$\sqrt{N} = \sqrt{n} \, \frac{k}{1 + k} \tag{4.12}$$

and substitute Eq. (4.12) into Eq. (4.8) for a simpler (and final) expression of resolution R_s:

$$\boxed{R_s = \frac{\sqrt{N}}{4} \frac{\alpha - 1}{\alpha}} \tag{4.13}$$

B. Chemical Separability *versus* Physical Efficiency

We can now see from Eq. (4.13) that the resolution R_s is a function of just two variables: The relative retention α and the number N of effective theoretical plates. The relative retention expresses the chemical capability for separation. The number of effective theoretical plates expresses the physical exploitation of this capability.

The relative retention α is a ratio of distribution coefficients. Each distribution coefficient is a thermodynamic quantity, independent of time and form. The relative retention can be evaluated equally well from either a sealed system or an ideal chromatographic column (one in which equilibrium is instantaneously attained).

The number N of effective theoretical plates, on the other hand, is completely determined by dynamics and form. Carrier gas flow rate, duration of sample introduction, column diameter and length, and stationary phase thickness are examples of the many time- and form-dependent factors that determine column efficiency.

We can also see from Eq. (4.13) that a trade-off exists between the number N of effective theoretical plates and the relative retention α. Although each must be usably large, nevertheless as one is the larger the other may be smaller. Which merits more attention? Which yields the greater return for experimental effort?

In 1955, Glueckauf published a graph that shows how separability, the physical efficiency of separation, and peak overlap are related [6]. We examine the segment of that graph for 1% peak overlap, in Fig. 4.2. (Fig. 4.2 shows separation factor SF versus n. The separation factor SF is the ratio of observed rather than of adjusted retention times. Similarly, n refers to observed retention times.)

Consider Fig. 4.2. If the separation factor SF approaches unity even only gradually, the number n of theoretical plates required for the separation of a given pair rises very sharply: At SF = 1.2, for example, about 700 theoretical plates are required. At SF = 1.05, 15,000 are needed; at 1.02, over 50,000.

Obviously, increasing separability radically decreases the number of theoretical plates required to achieve a given separation. Thus, to separate a given pair, first choose that stationary phase most favorable to the separation.

For instance, after they had been selected from a number of stationary phases already cited for the purpose, five stationary phases were tested for the separation of menthol-menthone stereoisomers [7]. Of the five, only Carbowax 400 separated neoisomenthol from both neomenthol and menthol.

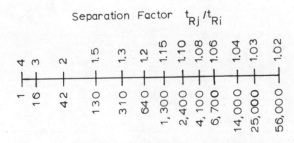

FIGURE 4.2 As the ratio of the observed retention times of two peaks approaches unity, the number of theoretical plates required to separate those peaks increases rapidly. (The numbers shown correspond to a 1% peak overlap. The relationship is based on the work of Glueckauf, Ref. 6.)

Sec. 4.4   Studies of Three Independent Areas   43

However, such mere citing of examples will not yield the insight needed to decide which aspect of resolution merits more attention.

For the decision, we can instead examine areas in chromatography that have been pursued for years. There we can look for the direction in change of emphasis in achieving resolution. Has separability been increasingly emphasized, or has the physical efficiency of separation?

Moreover, if we do find independent areas that do exhibit clear direction in change of emphasis, does this direction alter from one area to another, or is this direction always the same?

Is there one consistent direction that indicates the emphasis in resolution that is ultimately the more profitable?

4.4 STUDIES OF THREE INDEPENDENT AREAS

A. The Changing Direction

Petroleum and Petrochemicals

We shall consider three independent areas: Petroleum and petrochemicals, flavors, and urinary steroids. Gas chromatography was first applied industrially to petroleum and petrochemicals.

In 1956, Whitham [8] presented gas chromatograms such as that of Fig. 4.3 for petrochemical solvent analysis. These emphasized the selectivity of fluorene picrate.

Similarly, in 1958 Taylor and Dunlop [9] reported that "one of the most useful stationary liquids employed is dimethylformamide...especially where a separation between isobutylene and n-butene-1 is required." Characteristically for that time,

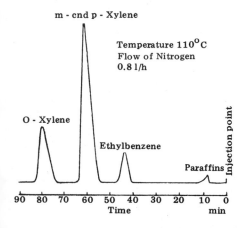

FIGURE 4.3 This early chromatogram from a 1957 paper showed in its caption a typical emphasis on stationary phase selectivity. The caption read, "Analysis of industrial xylenes on fluorene picrate column." (Reprinted from Whitham, Ref. 8, by permission of the author and Academic Press.)

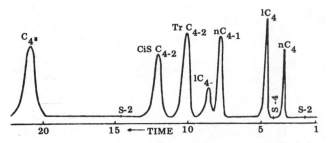

FIGURE 4.4 Another early (1958) gas chromatogram, showing that a number of C_4 hydrocarbons can be separated by dimethylformamide at 0°C. (Reprinted from Taylor and Dunlop, Ref. 9, by permission of the authors and Academic Press.)

Taylor and Dunlop used a "weight ratio of 40:100 of stationary liquid to solid support," thus inadvertently guaranteeing poor column efficiency. The separation achieved is shown in Fig. 4.4. It required a column temperature of 0°C and depended almost exclusively on the selective interactions of the solute-stationary phase system.

An imminent and radical change of emphasis came in a paper presented at the Instrument Society of America's Second International Symposium on Gas Chromatography (Lansing, Michigan; June, 1959). Ostensibly, the paper [10] reemphasized separability--"specially prepared liquid modified adsorbents." Yet the thrust of the paper was toward column efficiency--the new, more efficient, longer packed columns carrying lower loadings of stationary phase. But, in the next-to-the-last sentence, the authors leaped into the future, setting aside not only packed columns but also selectivity itself: "Capillary columns of 100,000 plate efficiencies coated with nonspecific liquid phases can effect...complex olefin-paraffin...separations." The final sentence offered mitigation: "Considerably less efficient columns may be used with supporting liquids such as [the highly selective] dimethylsulfolane."

Amos and Hurrell [11] confirmed in 1962 that column efficiency is greatly enhanced by (1) stationary phase loadings of 5 wt% or less, and (2) narrower column diameters, reduced to 1/8 in. O.D. (1.5 mm I.D.) from 1/4 in. O.D. (5 mm I.D.). (These two improvements became standard with packed columns.) Amos and Hurrell moved deliberately with all their columns and stationary phases toward higher column efficiencies, as just indicated, and found "in all instances, greater precision...and analysis time...reduced by at least a factor of two."

In 1961, reporting in the United States, the English team of D. H. Desty, A. Goldup, and W. T. Swanton [12], showed the shape of things to come in gas chromatography. They displayed gas chromatograms embodying such a remarkable efficiency of column and instrument that it has never since been equalled, let alone surpassed. Figure 4.5 shows their "chromatogram of the light fraction in Ponca crude." The stationary phase was the nonselective squalane. This paper defined the reward potentially available to those who emphasize column efficiency.

Sec. 4.4 Studies of Three Independent Areas   45

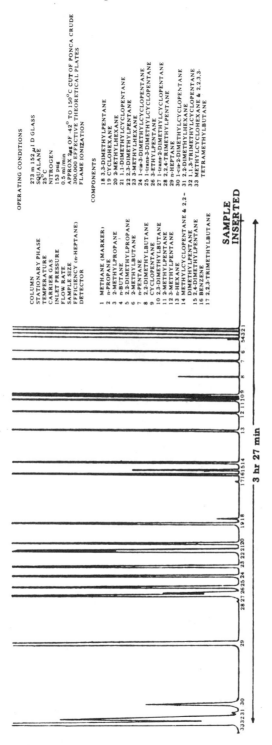

FIGURE 4.5 The chromatogram of the light fraction in Ponca crude. The column in which this separation was made exhibited almost a million effective plates, an unprecedented performance. (From Desty, Goldup, and Swanton, Ref. 12. Reprinted by permission of the authors and Academic Press.)

Flavors

The many difficulties of separating flavor components by gas chromatography are described in the book, *Flavor Research* [13]. Both separability (evidenced as selectivity) and column efficiency are needed:

> For the GC separations of complex mixtures a variety of columns and various stationary liquid phases with a range of polarity and selectivity are needed... Compounds that cannot be well resolved on one phase may be separable on another phase at different polarity.
>
> The efficiency of a chromatographic column is a measure of peak broadening as the solutes pass through the column. The more efficient the column, the less broadening occurs. The greater the efficiency of the column, the better the resolution of compounds [the] partition coefficients [of which] may differ only slightly [14].

These two paragraphs acknowledge only that both separability and column efficiency are necessary. But which is to be emphasized?

> If we have a mixture of a few compounds, changes in selectivity can be used to improve resolution. In aroma research, we often have mixtures containing hundreds of compounds. In such cases, different selectivities would only change retention times such that different mixtures would be found in any peak...Because the selectivities of [certain] closely related compounds do not change much with different stationary liquids, it is necessary in aroma research to use columns with large numbers of theoretical plates [15].

In practical flavor analysis such as quality control, certain key components must be monitored. Specific selectivity for these is sought, because in any event many peaks overlap. Figure 4.6 [16] brings home the irrelevance of high selectivity with highly complex mixtures, and the sharply increasing rewards of high and still higher column efficiencies.

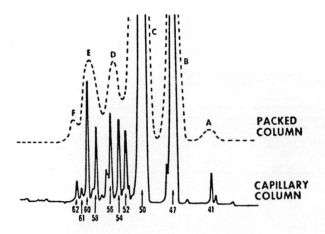

FIGURE 4.6 With mixtures of high complexity, high selectivity tends to become irrelevant. High column efficiency becomes the prerequisite. This figure shows a separation of sesquiterpenes by packed and capillary columns. (Reprinted with permission from Buttery, Lundin, and Ling, Ref. 16. Copyright by the American Chemical Society.

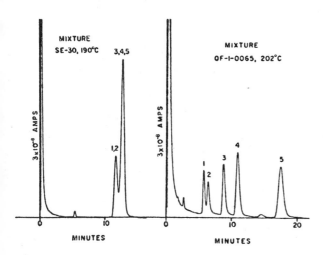

FIGURE 4.7 A very early paper on the separation of steroids emphasized the selectivity of the stationary phase. QF-1 is selective, SE-30 is not. The steroids: (1) 5α-pregnane-3β,20β-diol, (2) 5α-pregnane-3β,20α-diol, (3) 5α-pregnane-3β-ol-20-one, (4) 5α-pregnane-20β-ol-3-one, and (5) 5α-pregnane-3,20-dione. (From Horning, Vandenheuvel, and Haahti, Ref. 17. Reprinted by permission of the authors and Academic Press.)

The Steroids

Applying gas chromatography to the separation of steroids was thought to be impossible. Steroids were too large to vaporize, too unstable to high temperature, too slow to elute. However, since the late 1950s, E. C. Horning and his colleagues have

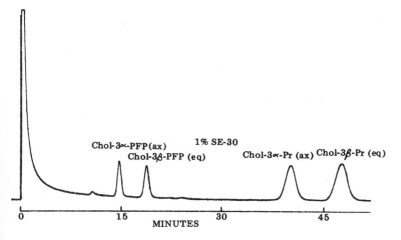

FIGURE 4.8 Separability can be increased through either higher stationary phase selectivity or derivatization. Through derivatization, two pairs of closely related steroids could be separated by the nonselective SE-30. Horning was perhaps the first to realize and state the full potential of derivatization for gas chromatography. (Reproduced by permission of Marcel Dekker, Inc., from Horning and Vandenheuvel, Ref. 18).

demonstrated that it can be done. As early as 1959, this group pioneered in using very low proportions of stationary phase on specially deactivated supports. They developed thermostable, nonadsorptive, separable derivatives from unstable, or adsorptive, or inseparable parent steroids. From the first they published repeatedly on aspects of stationary phase selectivity relevant to steroid separations. Their first emphasis was decidedly on separability, on selectivity.

For instance, in their first general report to American gas chromatographers, they showed (see Fig. 4.7) that the selective phase QF-1 would separate five closely related steroids, but that the nonselective SE-30 would not [17]. They found their relatively low column efficiency adequate enough: The "efficiency required for the urinary 17-ketosteroid separations is 4000-5000 theoretical plates, and this was attained with relatively rapid (15-20 min) elution of the steroids." This was in 1961.

Reviewing their work in 1965 [18], they again emphasized separability, particularly the separability that can be achieved over SE-30 by means of derivatives. See, for example, Fig. 4.8. But by this time, they were speaking wistfully of the potentially high efficiency of open tubular columns, an efficiency as yet unrealized for steroid separations. Most of their steroid separations were still being done with 6-ft packed columns yielding perhaps 2,500 theoretical plates.

By 1971, their language and thinking had changed markedly [19]. They knew by this time that they could make steroid derivatives that could be separated over SE-30. This assured, they could proceed to demonstrate their concept of the "metabolic profile," obtainable from a *single* gas chromatographic separation. The metabolic profile would reflect not only the concentration of many individual components, but also the important "pattern of concentration relationships... For example, if a com-

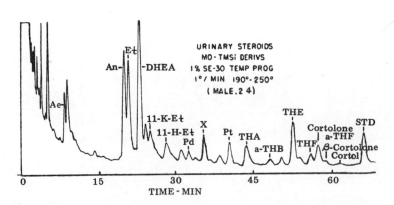

FIGURE 4.9 This is an early metabolic profile. It was made with resolution that was higher than normal for the time in that field, but lower than needed for the task. I shows the drive toward and the advantages of treating the sample at an earlier stage of separation, in a more complex state. Such an advance requires a shift in emphasis toward higher column efficiencies. (Reproduced with permission of Marcel Dekker, Inc from Horning and Horning, Ref. 19.)

Sec. 4.4 Studies of Three Independent Areas

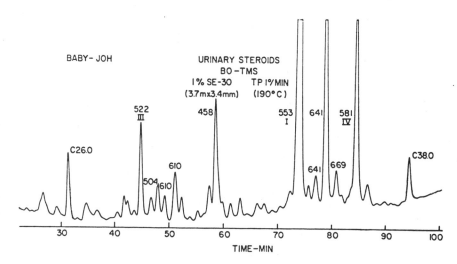

FIGURE 4.10a Metabolic profile of urinary steroids for a new-born human, obtained by separation of TMS and BO-TMS derivatives with a 12 ft × 3.4 mm 1% SE-30 column (temperature programmed at 1°C/min from 190°C). The molecular weights of major steroid derivatives are indicated. Compounds I, III, and IV are androst-5-en-3β, 16α-diol-17-one (I); androst-5-en-3β, 16α, 17β-triol (III); and pregn-5-en-3β, 16α-diol-20-one (IV). Reference compounds, C_{26} and C_{38} n-alkanes.

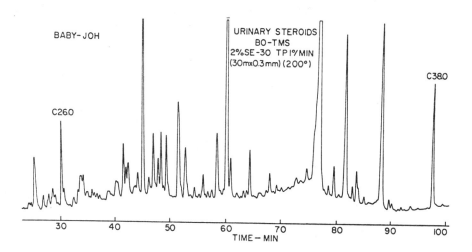

FIGURE 4.10b Metabolic profile of urinary steroids for a newborn human, obtained with a 30 m, 0.3 mm i.d., capillary column (temperature programmed at 1°C/min from 200°C). The sample was the same as that used for Fig. 4.10a. (Reproduced by permission of Preston Publications, Inc., from German and Horning, Ref. 20 in the Journal of Chromatographic Science.)

pound normally present in very small amount is encountered in high concentration because of pathologic circumstances, this fact is immediately apparent." Fig. 4.9 shows an example of a urinary steroid metabolic profile, based on the methoxime (MO) and trimethylsilyl ether (TMS) derivatives.

Still unable to use open tubular columns, they preferred to take more time per separation in order to gain more theoretical plates. Twelve-foot packed columns were used, yielding 6,000 to 7,000 theoretical plates.

However, in 1973 they reported on "Thermostable Open Tube Capillary Columns for the High Resolution Gas Chromatography of Human Urinary Steroids" [20]. These columns yielded 42,000 to 48,000 theoretical plates.

The improvement in the metabolic profile afforded by the higher column efficiency is shown in Figs. 4.10a and b. Figure 4.10a shows a profile from a 12-ft packed column yielding about 6,500 plates; Fig. 4.10b shows the improved profile from the same sample but obtained with the open tubular column, yielding about 45,000 plates.

The higher column efficiency brings out not only far more detail but also the need for still higher column efficiency. Such a higher efficiency was immediately sought, gained (Fig. 4.11), and reported [21], just 6 weeks after its predecessor [20]. Figure 4.11 shows a urinary steroid profile made with a 57-m column yielding 100,000 theoretical plates [21].

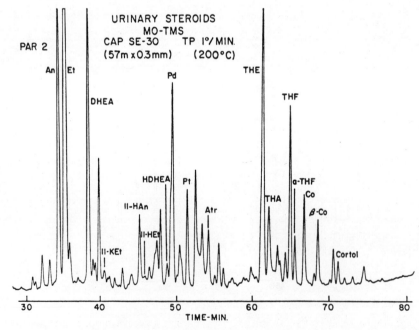

FIGURE 4.11 This metabolic profile shows over 100,000 theoretical plates. In it, a complex mixture of derivatized steroids is separated over the nonselective SE-30. Obviously, resolution was increased by increasing the number of theoretical plates rather than by increasing the selectivity of the stationary phase. (Reprinted with permission from German *et al.*, Ref. 21. Copyright by the American Chemical Society.)

B. Why the Direction Changed

The Three Independent Areas

Petroleum and petrochemicals. In this case, first gas chromatography itself and then the higher column efficiencies increasingly replaced piecemeal methods of analysis.

Before gas chromatography made it unnecessary, samples would be fractionally distilled. The fractions would then be further analyzed by such methods as mass spectrometry and infrared spectrophotometry. Eventually, all the various numbers would be correlated to make up the total analysis, if everything had gone well.

Often everything had not gone well.

The truly elaborate variety of miscarriages can only be suggested. Samples and fractions could, during repeated handling, be lost, mislabeled, dropped, stored until useless, or rerun until used up without result. (That list is not complete.) Correlated analyses became meaningless. At best, analyses took weeks.

Development of the ability to carry out one complete analysis on one sample vastly improved matters. To err was still human but no longer catastrophic. Delays became nominal.

The complementary effects of temperature programming and very high plate numbers were responsible for the new efficiency--and for the introduction of computers to handle the resultant deluge of data.

As for separability *versus* column efficiency: With extremely high plate numbers routinely available, almost any degree of stationary phase selectivity will do for most analyses of complex mixtures. Selectivity has been replaced by high operating temperature as the most important characteristic of a stationary phase.

Flavors. As we have seen, both flavor analysis and research deal with such complex mixtures that extra selectivity merely changes the composition of the many unresolved peaks. Rather, more and more theoretical plates are needed.

In flavor research, the precursor of understanding and flavor synthesis must be ultimate analysis--although a good guess may work very well [22]. The analysis itself, however, good guess or no, must not create artifacts. Therefore extra steps are to be avoided. Yet avoiding intermediate steps presents the ultimate method of separation--gas chromatography--with mixtures of higher complexity. Hence avoiding artifacts decreases the emphasis on selectivity still further and increases the need for more theoretical plates.

The brightest hope for flavor analysis is the gas chromatograph-mass spectrometer. With this, as we shall describe, even more theoretical plates are needed.

The human nose is more sensitive than the mass spectrometer. Furthermore, the overall reaction of the nose is frequently determined by trace components. Hence no nose-detectable trace must be overlooked by the far less sensitive mass spectrometer. To alleviate this mismatch, flavor researchers inject oversize samples. Such sample sizes reduce column efficiency. The flavor researchers counter with longer (1,000 ft) open tubular columns that are also wider and can therefore accommodate larger samples [13].

In sum, flavor research moves steadily toward high column efficiency. There is also an attempt to combine high column efficiency with larger sample sizes.

The metabolic profile. The very concept of the metabolic profile hinges directly on high plate numbers. A certain effective selectivity is gained by steroid derivatization. But beyond this, the concept reiterates the practical necessity of dealing with the whole sample. Attempting to get the profile by repeated fractionation combined with later correlation would not be practical.

The Basic Reasons

The reasons why the change in emphasis toward more theoretical plates always occurs are implicit in the examples just reviewed. The reasons are quite general.

The basic reasons why the number of components to be determined in a given mixture will always tend to increase are these: Mixtures are naturally complex, humans naturally make mistakes, and people always want more information.

All mixtures are originally highly complex. (Every substance is a highly complex mixture. A "pure" substance merely has a low concentration of impurities, not a low number of them.) To separate a very few components from any mixture requires work that introduces quantitative and qualitative errors.

Quantitative errors are brought about by operations that precede the final separation and determination of composition. These errors are expected and unexpected. The expected errors are the type described in textbooks. The unexpected errors are the ones that so often render all but the most carefully scrutinized analyses meaningless.

Qualitative errors are caused by inadequate information on which to base interpretation and decision. The more complex analysis is sought because it is the more informative.

For these reasons--the natural complexity of mixtures, the errors introduced by mutually dependent sets of operations, and the continuing need for better information--all analyses of mixtures always tend toward fewer steps per analysis, and the mixtures in each step tend to be more complex. The ideal analysis deals with the original mixture in a single step.

4.5 THE DIRECTION OF GAS CHROMATOGRAPHY

We have now reviewed three independent areas with respect to a change in emphasis regarding the two aspects of resolution. In each case, the progress of the change could be and was traced through a number of years. In the direction of this change in emphasis, the areas are representative, and we have seen why.

The direction of gas chromatography has always been away from an early emphasis on separability toward an ultimate emphasis on high plate number. The direction within a given area is never reversed; high plate number is never abandoned for higher selectivity.

From this chapter, we were to gain orientation and perspective by seeking a consistent direction in GC development and use. This we found. Having found this direction, we can now reach a further conclusion: The two complementary aspects of resolution are not mutually opposed. The more efficient column makes possible a better use of any stationary phase.

Through the more efficient use of stationary phases, highly efficient columns have made unnecessary the tremendous range of column lengths and loadings and variety of stationary phases that characterized the early years of gas chromatography. In a good instrument, high plate number achieved in a highly efficient column makes most selectivities adequate to separate most solute pairs. The easier separations are produced more quickly; the time and expense of the more difficult or elaborate separations are minimized [23, 24].

REFERENCES

1. W. E. Harris and H. W. Habgood, *Programmed Temperature Gas Chromatography*, Wiley, New York, 1966, p. 9.
2. *Ibid.*, Section 5.10, P. 138.
3. *Ibid.*, Section 4.05, p. 104.
4. ASTM Committee E-19 on Chromatography, E 355-77, *Gas Chromatography Terms and Relationships*, American Society of Testing and Materials, 1916 Race Street, Philadelphia, Pa., 19103, 1977; Analytical Chemistry Division, International Union of Pure and Applied Chemistry, *Compendium of Analytical Nomenclature* (H. M. N. H. Irving, H. Freiser, and T. S. West, preparers), Pergamon, Elmsford, N. Y. 10523, 1978, Chap. 13.
5. J. H. Purnell, *J. Chem. Soc.* 1268 (1960).
6. E. Glueckauf, *Trans. Far. Soc. 51*, 34 (1955).
7. D. G. Gillen and J. T. Scanlon, *J. Chromatog. Sci. 10*, 729 (1972).
8. B. T. Whitham, in *Vapour Phase Chromatography* (D. H. Desty, editor), Butterworths, London, 1957, pp. 395-412, and also Figure 10.
9. G. W. Taylor and A. S. Dunlop, in *Gas Chromatography* (V. J. Coates, H. J. Noebels, and I. S. Fagerson, editors), Academic, New York, 1958, pp. 73-85, and also Figure 4.
10. A. Zlatkis and H. R. Kaufman, in *Gas Chromatography* (H. J. Noebels, R. F. Wall, and N. Brenner, editors), Academic, New York, 1961, pp. 339-342.
11. R. Amos and R. A. Hurrell, in *Gas Chromatography 1962* (M. van Swaay, editor), Butterworths, Washington, D. C., 1962, pp. 162-177, and also Figure 2.
12. D. H. Desty, A. Goldup, and W. T. Swanton, in *Gas Chromatography* (N. Brenner, J. E. Callen, and M. D. Weiss, editors), Academic, New York, 1962, pp. 105-135, and also Figure 7.
13. R. Teranishi, I. Hornstein, P. Issenberg, and E. Wick, *Flavor Research*, Dekker, New York, 1971.
14. *Ibid.*, p. 24.
15. *Ibid.*, p. 82.
16. R. G. Buttery, R. E. Lundin, and L. Ling, *J. Agr. Food Chem. 15*, 58 (1967).

17. E. C. Horning, W. J. A. Vandenheuvel, and E. O. A. Haahti, in *Gas Chromatography* (N. Brenner, J. E. Callen, and M. D. Weiss, editors), Academic, New York, 1962, pp. 507-518, and also Figure 6.
18. E. C. Horning and W. J. A. Vandenheuvel, in *Advances in Chromatography*, Vol. 1 (J. C. Giddings and R. A. Keller, editors), Dekker, New York, 1965, pp. 153-198, and also Figure 5.14.
19. E. C. Horning and M. G. Horning, in *Recent Advances in Gas Chromatography* (I. I. Domsky and J. A. Perry, editors), Dekker, New York, 1971, pp. 341-376, and also Figure 3.
20. A. L. German and E. C. Horning, *J. Chromatog. Sci.* 11, 76 (1973); see also Figures 6 and 7.
21. A. L. German, C. D. Pfaffenburger, J. P. Thenot, M. G. Horning, and E. C. Horning, *Anal. Chem.* 45, 930 (1973); see also Figure 3.
22. R. G. Buttery, in *Techniques of Combined Gas Chromatography/Mass Spectrometry* (W. H. McFadden, editor), Wiley, New York, 1973, pp. 330-334.
23. B. O. Ayres, R. J. Loyd, and D. D. DeFord, *Anal. Chem.* 33, 968 (1961).
24. D. D. DeFord, in *Gas Chromatography* (L. Fowler, editor), Academic, New York, 1963, pp. 23-31.

Chapter 5

THE STATIONARY PHASE

5.1 INTRODUCTION

The stationary phase is the liquid in gas-liquid chromatography.

This chapter discusses the stationary phase--bases of its action, ways of describing it, lists of preferred stationary phases, and criteria for selecting given stationary phases for given separations.

The stationary phase dissolves the components of the sample. In the resulting (column-long) solution, the components spend different amounts of time. Separation depends on these differences in solubility.

Sections 5.2 and 5.3 provide an orientation to the chief intermolecular forces that affect gas chromatographic retention of substances by solubility and adsorption.

Any solvent can be a stationary phase. Hundreds have been. Yet only a very few--perhaps five, no more than ten--are now commonly used. How did this happen, and where are we now?

Sections 5.4 and 5.5 discuss stationary phase characterization and the lists of preferred stationary phases that resulted from that successful characterization.

Finally, criteria for selecting stationary phases for specific separations are presented, and illustrative examples are cited.

5.2 SOLUBILITY

"Intermolecular Forces," Chapter IV of *The Solubility of Nonelectrolytes*, by Hildebrand and Scott [1], is a good starting guide to this subject.

A. Coulombic Forces

These forces exist between ions. They are mentioned here only because molten salts are used in gas chromatography as stationary phases for a very few, high-temperature applications. However, these forces in general do not affect solubility in gas chromatography, so they will not be discussed further here.

B. Orientation Forces (Dipole-Dipole)

Keesom Forces

Given the assumption of a free molecular rotation that is possible only in a gas, the average potential energy of interaction between two molecules having dipole moments u_1 and u_2 and separated by distance r can be expressed as follows:

$$E = -\left(\frac{1}{(3/2)k_B T}\right)\left(\frac{u_1 u_2}{r^3}\right)^2$$

Here k_B is the Boltzmann constant, which expresses the thermal energy per molecule per degree T, Kelvin. This expression shows that the average dipole-dipole energy is directly proportional to the square of the dipole product and inversely proportional to the thermal energy per molecule and to the *sixth* power of the intermolecular separation.

This expression is not directly applicable to gas-liquid chromatography because in the derivation of it free rotation of molecules is assumed. This is impossible in liquids. As Hildebrand and Scott point out, "The attraction between polar molecules does not fall off so rapidly as this formulation would require" [1].

Nevertheless, the Keesom expression suggests that orientation forces grow stronger at lower temperatures. This agrees with experience. It is one of several reasons for decreasing column temperatures.

The Keesom expression also strongly suggests that the smaller the stationary phase molecules, the stronger the orientation force. Hence the more selective is that stationary phase. Given the same dipole, a smaller stationary phase molecule that permits twice as close an approach of a given dipole will, in theory, generate an orientation force 64 times as large. Dimethylformamide, an unusually small polar molecule, makes a highly selective stationary phase.

Hydrogen Bonds

The hydrogen bond provides a special type of orientation force. The smallness of the hydrogen allows such a close approach of the dipoles of which hydrogen is a part that the orientation force becomes very strong, indeed essentially electrostatic. And if several or many hydrogen bonds are involved per molecule, then the behavior or even the shape of that molecule is strongly influenced or even completely controlled by its hydrogen bonding:

"The strengths of hydrogen bonds fall off roughly in the order FHF, OHO, OHN, NHN, CHO, but they are very dependent upon the geometry of particular combinations, upon the nature of neighboring atoms and upon resonance, and upon acid-base characters" [1]. "Basicity is usually considered only in terms of the hydrogen bonding of alcohols, and these are the most important, but carbonyls and ethers can hydrogen bond with stationary phases containing -OH groups" [2].

Sec. 5.2 Solubility

The Keesom equation is even less accurate in predicting behavior here, but it is still useful in directing experimental effort: An inverse dependence on temperature and an inverse and exponential dependence on molecular size can be expected. So the column temperature should be kept as low as possible for greatest selectivity and the stationary phase molecule should be as small as possible.

C. Induction Forces (Dipole-Induced Dipole)

Induction forces depend on polarizability. A molecule with no permanent dipole moment may have a dipole moment induced by an electric field. Its electrons become shifted by the field. It acquires a temporary polar character that is proportional to its polarizability α_p and to the strength of the polarizing field.

"If the molecule, A, has a moment u_A and the molecule, B, a polarizability $\alpha_{p,B}$, there is a reciprocal interaction energy

$$E_i = -\frac{u_A \alpha_{p,B}}{r^6}$$

where r is the distance separating their centers" [1].

The more loosely the electrons of a molecule are held, the greater its polarizability. Thus the more double bonds a molecule contains, or the greater its resonance energy, or the larger its component atoms, the greater is its polarizability.

So benzene is more polarizable than cyclohexane, and carbon tetrachloride is more polarizable than methane. Thus the retention of benzene (b.p., 80°C) relative to cyclohexane (b.p., 81°C) increases when the stationary phase is changed from a paraffin oil to a polyglycol. The polyglycol is a polarizing solvent but the paraffin oil is not. The polyglycol dissolves benzene more strongly than the essentially nonpolarizable cyclohexane.

D. Dispersion Forces (Interaction of Nonpolar Molecules)

Dispersion forces exist among all molecules. They result from the attraction of induced dipoles for the instantaneous and varying dipoles among molecules. Dispersion interactions increase with the sizes and unsaturation of solute molecules [2].

Between a molecule A of polarizability $\alpha_{p,A}$ and ionization energy I_A and a similarly characterized molecule B, the dispersion energy of interaction is, according to London,

$$E_d = -\frac{3}{2} \frac{\alpha_{p,A} \alpha_{p,B}}{r^6} \frac{I_A I_B}{I_A + I_B}$$

where r is, as before, the separation of the interacting molecular centers.

Neither induction nor dispersion forces are temperature dependent.

E. Specific Interaction

By specific interaction we mean a strong but reversible interaction between a metal ion and a polarizable molecule. An example is the silver ion-olefin complex, where the silver ion exists as part of the stationary phase and the olefin is the solute [3-6].

This type of solubility force is strong, highly selective, and useful where applicable--but it is rarely applicable.

F. Chemical-Mechanical Constraint: Liquid Crystals

"Liquid crystals...are formed by certain compounds with elongated, relatively polar molecules whose mutual attraction tends to orient them with their long axes parallel... If a material whose molecules are rod-shaped is dissolved in a liquid crystal, we may expect the solute molecules to be oriented parallel to the molecules of the solvent... [The] relative affinities [of liquid crystals] for a series of isomers [should] depend on the shapes of the latter" [7].

Thus, here is another highly specialized type of solubility force. It is the basis of one of the few ways to separate m-xylene from p-xylene (they are eluted in that order, the more rod-like p-xylene being more strongly retained).

5.3 ADSORPTION ON THE GAS-LIQUID SURFACE

In 1961, R. L. Martin [8] proved that adsorption at the gas-liquid interface effectively retards solutes. The importance of this effect has grown steadily.

Adsorption is proportional to surface area. The gas-liquid surface area increases with decreased loading of the stationary phase on the support. Column efficiency also increases with decreased stationary phase loading, over a wide range of loading. Therefore, maximizing column efficiency also maximizes gas-liquid adsorption.

Gas-liquid adsorption increases more quickly than solubility with decreasing column temperature. This makes changing the column temperature additionally effective in varying and enhancing selectivity.

The relative importance of gas-liquid adsorption *versus* solubility increases with stationary phase polarity. This increase holds for both polarizable solutes and for the nominally unpolarizable saturated hydrocarbons. The solubility of saturated hydrocarbons decreases sharply, by one or two orders of magnitude, with the increase of stationary phase polarity from hexadecane to β,β'-thiodipropionitrile [8].

The adsorbability--defined below--of saturated hydrocarbons on β,β'-thiodipropionitrile differs little from that of polarizable solutes, but is only a fraction of that of polar solutes:

Hydrocarbon	Adsorbability
n-Pentane	15.4
n-Heptane	10.9
Cyclohexane	7.3
Cyclohexene	9.1
1-Hexene	18.9
Benzene	16.6
Isopropyl alcohol	31
Ethyl alcohol	54
Acetone	85
Methyl alcohol	119

(Adsorbability is given by the product of the vapor pressure with an adsorption coefficient K_a analogous to the partition coefficient K. When the coefficient K_a has by definition the dimensions of centimeters, then adsorbability has the dimensions, torr-cm $\times 10^{-3}$.)

5.4 STATIONARY PHASE CHARACTERIZATION

A. History

The history of any chromatographic technique shows first an emphasis on existence and only later one on efficiency. So it is with gas chromatography. Its first emphasis was on the stationary phase, whereby the separations would occur if they were to occur at all. Early gas chromatography saw a flood of stationary phases, each new one eagerly sought and tried.

As gas chromatography has matured, an increasingly successful effort has been made to cope with this flood. A necessary part of this effort has been the development of a method to tell reproducibly, accurately, and precisely whether one stationary phase is different from another, and if so, in what way.

What follows is a partial history of this slow and painstaking development, and of its success.

B. Polarity

In part, stationary phases proliferated because they could not be accurately described, related, or ranked: They could not be characterized.

The first tool tried for characterizing stationary phases was polarity--but polarity itself had first to be defined, or at least adequately expressed by a polarity index. An adequate polarity index would express solute-solvent interaction on a single numerical scale, without exceptions or regard to experimental conditions.

Reichardt reviewed a number of empirical parameters devised for use as polarity indices, but found none adequate: "All these empirical parameters constitute a more comprehensive measure of the polarity of a solvent than...any single physical characteristic, since they reflect more faithfully the complete picture of all the intermolecular forces...[Coulomb, directional, inductive, dispersion, and charge-transfer forces, as well as hydrogen-bonding forces]...acting in solution...[However, the] use of these parameters for characterizing the solvating action of any solvent and [the] conclusions as to the solvation of any organic molecules appear to be very restricted" [9].

The gas chromatographer Kovats also thought that "polarity" was not going to help stationary phase characterization:

> In order to define the separating character of a stationary phase, *i.e.*, the separation properties towards different groups of substances, one would like to avoid the rather vague and ill-defined term polarity. Though one understands well enough what is implied by the terms weakly polar, moderately polar, and highly polar, and although such terms are adequate to describe gross effects, it is not possible to express this concept with one simple parameter. The separating power can be expressed in more detail by comparing a given stationary phase with a nonpolar stationary phase by means of [the numerical] values [of the retention index differences, ΔI][10].

C. The Retention Index Difference ΔI

As we shall be explaining, a retention index difference ΔI is the difference between two values of the Kovats Retention Index I. The Kovats Retention Index I is determined as follows:

For any solute X eluted from a given column at a given temperature, two successive normal paraffins $n\text{-}C_z$ and $n\text{-}C_{z+1}$ exist that are eluted before and after X. From the corresponding adjusted retention times, the Kovats Retention Index I for solute X is determined:

$$I = 100 \frac{\log t_R'(X) - \log t_R'(n\text{-}C_z)}{\log t_R'(n\text{-}C_{z+1}) - \log t_R'(n\text{-}C_z)} + 100z \tag{5.1}$$

(The retention index for the solute X should properly bear a superscript indicating the stationary phase and a subscript indicating the column temperature in degrees centigrade. Thus, for example, $I_{100}^{squalane}$. Also, although the notation log indicates common logarithms, because of the properties of logarithms it does not matter whether common or natural logarithms are used in this expression; the 2.3 factor of difference cancels out.)

It can be seen that the retention index for any normal paraffin $n\text{-}C_z$ is $100z$. The retention index for n-hexane, for instance, is 600, by definition.

Also, from a given column at a given temperature, any substance eluted after n-hexane but before n-heptane will have a retention index between 600 and 700. Dioxane, for instance, eluted from a squalane column at 100°C, is found to have a retention index of 651 [11].

Note again that the retention index for a normal paraffin is fixed--600 for n-hexane, for example--for any stationary phase at any temperature. However, the retention index for any solute but a normal paraffin may vary with the stationary phase. For a polar solute, it varies a great deal.

Sec. 5.4 *Stationary Phase Characterization*

Consider dioxane again, eluted at 100°C. With the nonpolar stationary phase, squalane, dioxane shows a retention index of 651. But with diisodecylphthalate, dioxane shows a retention index of 779; with neopentylglycolsuccinate, 1,080; and with diethyleneglycolsuccinate, 1,363 [11]. The more polar the stationary phase, the higher the retention index of a polar solute.

The increase in retention index of a polar solute could measure stationary phase polarity, but such a scale of retention indices would not start at zero, as a proper scale should. Enter the retention index difference.

For a given column temperature, let the retention indices of a given polar solute be determined over polar stationary phases of interest. Then subtract from each index the corresponding index determined over a given nonpolar stationary phase. The resultant index differences now would not only rank the stationary phase with respect to polarity, but also equal zero for zero stationary phase polarity.

For dioxane, its retention index 651 for the nonpolar squalane would be subtracted from each of its other retention indices. The resultant index differences would be 779 − 651 = 128 for diisodecylphthalate, 429 for neopentylglycolsuccinate, and 712 for diethyleneglycolsuccinate. For the polarity "probe" dioxane, these stationary phases rank in the numerical order just given, with respect to polarity, and the nonpolar phase ranks zero, as desired.

This approach might be used to rank all stationary phases with respect to polarity. Standards would first have to be defined and agreed on: A standard test method; a standard expression for the results; a standard, completely nonpolar stationary phase; a standard, 100% polar stationary phase; standard operating conditions for the tests; and a standard polar solute.

These standards came largely from Chovin and Lebbe [12]. They suggested using successive normal paraffins for characterizing stationary phase polarity; to obtain a linear relationship, using the logarithm of the ratio of the adjusted retention times from these paraffins; using squalane as a standard, completely nonpolar stationary phase; and using β,β'-oxydipropionitrile as the standard, 100% polar stationary phase, with the important proviso that values above 100% polarity not be excluded.

Rohrschneider [11] took up the matter of the standard polar solute, which was still to be resolved.

D. The Rohrschneider Constants

As we have already seen, a given polarity involves a certain balance among several intermolecular forces. That balance changes if the temperature changes, but for simplification a given temperature can be specified. However, no given polar solute can adequately reveal the components of this balance.

How many independent intermolecular polar attractions must be accounted for? Rohrschneider found five:

> ...besides the intermolecular forces which for instance retain benzene stronger than the alkanes, a separate factor had to be considered for the alcohols (*e.g.*,

ethanol) and [another for] the carbonyl compounds (e.g., methyl ethyl ketone). In the course of these investigations it was found that the retention behavior of nitromethane, chloroform, and phenylacetylene cannot be completely defined by the forces which result in the retention of benzene, ethanol, and methyl ethyl ketone, and that a fourth value had to be introduced which is characterized by the retention of nitromethane... [for] pyridine and dioxane still a fifth factor had to be considered [11].

Note the introduction of a radically new operational principle: for each independent intermolecular polar force, a given solute is selected that then *defines* that force.

Consider, for example, ethanol. Representing alcohol-polarity, ethanol responds by Rohrschneider's definition to just that one intermolecular polar force, no other. In mathematical terms, ethanol is assigned a coefficient of unity for the alcohol-type polar force and coefficients of zero for all the other polar forces. Similarly for each of the other standard solutes--five in all.

In this way, Rohrschneider characterized 22 liquid phases by the "five index differences of benzene (x_p), ethanol (y_p), methyl ethyl ketone (z_p), nitromethane (u_p), and pyridine (s_p)" [11]. (Rohrschneider chose not to multiply by 100. For example, he would express the dioxane retention index differences we mentioned earlier not as 128, 429, and 712, but as 1.28, 4.29, and 7.12.)

Rohrschneider used generally accepted test conditions, including the following: Loading 20% by weight; 100°C column temperature; 0.1 to 0.3 μl sample.

The real value of Rohrschneider's study lay not in his characterization of 22 liquid phases nor in his choice of solutes but in his approach to stationary phase characterization: Pure solutes used as probes, retention index differences used as expressions of the data, and several solutes employed as mutually independent probes with the results from each separately expressed.

Four pragmatic assertions can be seen in Rohrschneider's approach:

1. The number of distinct intermolecular polar forces is without a *priori* limitation.
2. The various intermolecular polar forces are mutually independent.
3. The effect of a given intermolecular polar force is identical to the retention behavior of a given gas chromatographic solute selected as a probe.
4. The intermolecular polar forces effective in gas chromatography are adequately describable by tests that are wholly gas chromatographic.

By the last two assertions, probe retention index differences may be used without change as the elements of stationary phase characterization. By the first three assertions, as many dissimilar probes as needed may be used.

The Rohrschneider experimental method was sound, explicit, precise, reproducible, and usable by any gas chromatographer using only standard gas chromatographic equipment.

Rohrschneider solved stationary phase characterization. The later alterations of others merely elaborated Rohrschneider's principles.

Sec. 5.5 Preferred Stationary Phases

McReynolds [13] made Rohrschneider's method somewhat easier to use. Mainly, to obtain probes the bracketing hydrocarbons of which are liquid at room temperature, he substituted butanol, 2-pentanone, and nitropropane for ethanol, methyl ethyl ketone, and nitromethane, respectively.

Both the Rohrschneider and the McReynolds constants are Kovats Retention Index differences for selected solutes used as polarity probes. Both sets of probes and the corresponding symbols for the retention index differences measured for these probes are shown in Table 5.1.

McReynolds characterized over 200 stationary phases. He then listed these in order of increasing polarity, along with their constants. As he intended, this polarity listing showed clearly that any given stationary phase differed little from its neighbor. With this study, stationary phase proliferation began to come under control.

TABLE 5.1 Rohrschneider and McReynolds Constants

Rohrschneider constants		McReynolds constants	
Probe	ΔI Symbol	Probe	ΔI Symbol
Benzene	x_p	Benzene	X'
Ethanol	y_p	Butanol	Y'
Methyl ethyl ketone	z_p	2-Pentanone	Z'
Nitromethane	u_p	Nitropropane	U'
Pyridine	s_p	Pyridine	S'
		2-Methyl-2-pentanol[a]	H

a The constant H, indicating the polarity of a stationary phase for a branched-chain alcohol, is sometimes quoted along with the other McReynolds constants.

5.5 PREFERRED STATIONARY PHASES

A. A List of Twelve

In 1973, Leary *et al.* [14] reported applying to the 226 stationary phases of the McReynolds survey [13] a statistical "nearest-neighbor" technique suitable for determining "groups and similarities" among them. They suggested that just 12 stationary phases could supplant the 226.

These 12 stationary phases were selected as well tested, readily available, generally thermally stable, and wide ranging in polarity. They are listed in Table 5.2. With each phase is listed its maximum operating temperature, polarity distance from squalane (the square root of the sum of the squares of the McReynolds constants), chemical nature, and McReynolds constants. A scale at the bottom of the table shows the 12 stationary phases ranked along the "nearest-neighbor" polarity scale.

TABLE 5.2 Twelve Stationary Phases Suggested As "Preferred"

Phase Number	Phase	Max. °C	Polarity distance	Chemical nature	McReynolds constants[a]				
					X	Y	Z	U	S
1	Squalane	150	0	Cycloparaffin	0	0	0	0	0
2	SE-30	350	100	Dimethyl-siloxane	15	53	44	64	41
3	OV-3	350	194	Phenylmethyl-dimethyl-siloxane	44	86	81	124	88
4	OV-7	350	271	Phenylmethyl-dimethyl-siloxane	69	113	111	171	128
5	DC-710	300	377	Phenylmethyl-siloxane	107	149	153	228	190
6	OV-22	350	488	Phenylmethyl-diphenyl-siloxane	160	188	191	283	253
7	QF-1	250	709	Trifluoropropylmethyl-siloxane	144	233	355	463	305
8	XE-60	275	821	Cyanoethylmethyl-siloxane	204	381	340	493	367
9	Carbowax-20M	250	1052	Polyethyleneglycol	322	536	368	572	510
10	DEG-Adipate	200	1269	Diethylene glycol adipate	378	603	460	665	658
11	DEG-Succinate	225	1612	Diethylene glycol succinate	492	733	581	833	791
12	TCEP	175	1885	1,2,3-Tris(2-cyanoethoxy)propane	593	857	752	1028	915

```
                     STATIONARY PHASE NUMBER
 1   2   3   4   5   6   7   8   9   10   11   12
 0   200    400    600    800   1000  1200  1400  1600  1800  2000
           NEAREST-NEIGHBOR POLARITY DISTANCE
```

[a] McReynolds probes: Benzene, X; butanol, Y; 2-pentanone, Z; nitropropane, U; pyridine, S.
Source: Data from Ref. 14; McReynolds constants from Ref. 13.

Sec. 5.5 Preferred Stationary Phases

Winnowed from the stationary phases surveyed by McReynolds, the list of Table 5.2 could not reflect any introduced after early 1970. We use the list as a valuable basis for study and for a still better list--shorter if possible.

The stationary phases shown in Table 5.2 fall into groups: saturated hydrocarbons (one), silicones (seven), polyethylene glycols (one), polyesters (two), and highly polar compounds (one). We take these up in turn.

Saturated Hydrocarbons

Squalane, usable only to 100°C, now ranks only as a scientific standard for non-polarity, not as a stationary phase of wide practical use. In routine practice, its place is taken by either Apiezon L, usable to 250°C, or, more probable, the silicone SE-30, usable to 350°C to 400°C. Both have low polarity:

	X	Y	Z	U	S
Apiezon L	32	22	15	32	42
SE-30	15	53	44	64	41

Silicones

Haken [15] contributed a 30-page review of the silicones used in gas-liquid chromatography in 1972. This extensive study included the silicones' chemistry, tradenames, replications and substitutes, viscosities, information as available on sundry other properties, and polarities. In 1973, Trash [16], of the Ohio Valley Specialty Company, and Coleman [17], of Silar Laboratories, also contributed incisive though shorter reviews.

Increasingly, silicones have become the stationary phases of choice in gas-liquid chromatography. They permit higher column temperatures--witness SE-30 *versus* Apiezon L, for example. The newer silicones offer the adequately high polarities that the earlier silicones lacked. Working from the list in Table 5.2, we now examine the silicones in order of increasing polarity.

The silicones of lowest polarity are the methyl silicones. These have a siloxane chain, $-Si-O-(Si-O)_n-Si-$, with methyl substituents on all the otherwise unconnected bonds of the tetravalent silicon atoms. The polarity of the silicones increases as the methyl groups are increasingly replaced by phenyl or other polar groups. Of the 12 "preferred" stationary phases shown in Table 5.2, the polarities of the first four silicone phases derive from the degree of methyl replacement by phenyl:

"Twelve" number	Name	Max °C	Substituent	% Phenyl
2	SE-30	350	Methyl	0
3	OV-3	350	Methylphenyl	10
4	OV-7	350	Methylphenyl	20
5	DC-710	300	Methylphenyl	50
6	OV-22	350	Methylphenyldiphenyl	65

Trash [16], commenting on the methylsilicones, such as SE-30, warned that samples such as strong acids or bases injected onto silicone columns can cause depolymerization of the silicone stationary phases, as can traces in the silicones of equilibration catalysts such as ferric chloride. The conscientious manufacturer of stationary phases will have tried to produce a stable silicone by removing the catalyst. If

> the stability and efficiency [of a given methylsilicone column] are good, there is no catalyst present.
>
> The only opportunity [of the gas chromatographer] for improving the liquid phase would be if it contained low molecular weight species which could be removed by heating...while sparging with inert gas. This is essentially what chromatographers have been doing for years as they condition the column for several hours while purging with inert carrier gas...The methylsilicones are considerably more stable in the absence of oxygen.
>
> A wide variety of methylsilicone gums has been...used...for GC... Of this list, only three are prepared and marketed exclusively as GC liquid phases. GC Grade SE-30...a [closely controlled] product of General Electric...[and] OV-1 and OV-101, prepared by the Ohio Valley Specialty Company...represent the ultimate in methylsilicone stability and performance...[16].

Is a silicone fluid preferable to a silicone gum, given the same polarity and maximum operating temperature? SE-30 and OV-1 are gums, whereas OV-101 is a fluid, yet equally usable to 350°C [16]. The choice therefore depends on efficiency.

Stationary phase efficiency varies directly with diffusivity. Diffusivity is inversely related to viscosity, albeit not simply [18, 19]. Nevertheless, one might expect a lower diffusivity with a more viscous stationary phase--and thus a lower efficiency with a gum than with a fluid. This was tested.

Butler and Hawkes [20] tested diffusivity as a function of viscosity. They used two pairs of polymeric stationary phases--a pair of methylsilicones and a pair of polyethylene glycols. The molecular weights and the viscosities varied widely within each pair.

> Diffusivities in the two methyl silicones SE-30 (M. WT. 400,000) and SF-96-200 (M. Wt. 11,000) are so different as to suggest a different mechanism of diffusion or liquid structure. Surprisingly, the diffusivities in SE-30 [the gum] are higher than in the light oil... As better diffusivities apparently go with the longer chain and better thermal stability, the long-chain elastomer seems to be preferable...
>
> In complete contrast the difference [between]...the two polyethylene glycols (PEG 400 and PEG 20M) [is]...insignificant. Hence choice of a polyethyleneglycol

Sec. 5.5 Preferred Stationary Phases

apparently needs to be based only on its partition properties as shown by its Rohrschneider-McReynolds indices.

The use of the viscosity to predict diffusivity in polymers is completely erroneous...it can no longer be believed that the viscosity of a polymeric stationary phase is an index of its general usefulness [20].

Later, Millen and Hawkes found that SE-30 causes less peak dispersion with the lighter ($<C_{11}$) n-alkanes, "while for the longer chain n-alkanes, SF-96-200 appears to be a better choice" [21]. At 200°C or higher, the difference between the two may become negligible [21].

We must mention a polymer of singular thermal stability, Dexsil. Dexsil, introduced in 1970 and not included in the McReynolds study, could not become a "preferred" stationary phase in 1973. However, because its high-temperature limit is unprecedented, it needs no such assistance. Dexsil can be used from 50 to 450°C routinely and to 500°C briefly. Both high temperatures are beyond the limit of most gas chromatographs. Chemically, Dexsil [22] was something new. The Dexsil polymer chain incorporates both siloxane and carborane units: $[Si(CH_3)_2-CB_{10}H_{10}C(Si(CH_3)_2-O)_3]_n$. The polarity of Dexsil, cited below in Rohrschneider constants, varies from that of OV-3 to that of OV-7 (McReynolds constants).

	X	Y	Z	U	S
OV-3	44	86	81	124	80
Dexsil	.43	.64	1.11	1.51	1.01
OV-7	69	86	111	171	128

Each of the first five silicones of Table 5.2 has a 350°C maximum operating temperature except DC-710 with 300°C. The decreased high-temperature limit of DC-710 bears out the comments of Coleman [17]: "...[these] Dow Corning fluids...[are] low molecular weight polymers [that] contain relatively large amounts of volatiles, which can be removed by vacuum stripping at high temperatures, and residual catalyst, which can be removed by water washing, to produce polymers more suitable as GC liquid phases."

The lower high-temperature limit of DC-710 could be avoided by using OV-11 instead. OV-11 is only 19 of the 111 polarity distance units from the fifth "preferred" phase, DC-710, to the sixth, OV-22. OV-11 has a high-temperature limit of 350°C. OV-11 is a methylphenyl silicone having 35% phenyl substitution [16].

The seventh "preferred" stationary phase is QF-1, a (3,3,3-trifluoropropyl)methyl silicone manufactured by Dow Chemical not primarily for gas chromatography and not specially purified. As with the methylsilicone, vacuum stripping and water washing would purify QF-1 [17]. Polymerized by base catalysts, QF-1 should not be subjected to samples that are strong bases. For QF-1, Coleman suggested using either OV-210 or SP-2401, both "clean and well devolatilized" made-for-GC versions of this silicone [17].

[In order of increasing polarity, carbonyl/alcohol selectivity reverses after QF-1 in the list of "preferred" stationary phases. The carbonyl polarity (Z) is higher than the alcohol polarity (Y) in OV-7, DC-710, OV-22, and QF-1, but lower thereafter, in XE-60, Carbowax 20M, DEGA, DEGS, and TCEP.]

The eighth "preferred" phase is XE-60, in which 25% of the substituents on the silicone chain are β-cyanoethyl rather than methyl. Made by General Electric as an experimental polymer and not specially for GC, it shows batch-to-batch variability and an "unexpectedly high volatility" [23]. It cannot be readily purified [17]. On the other hand, OV-225 (25% γ-cyanopropyl-25% phenyl) has the same polarity as XE-60 in all respects [14, 15], is made specially for gas chromatography, and has a 25°C higher temperature limit of 275°C. OV-225 is clearly the choice over XE-60.

Since 1972, silicones even more polar than OV-225 have been introduced. The first of these was SILAR-5CP [24]. "SILAR-5CP...is a (γ-cyanopropyl) phenyl homopolymer (50% γ-cyanopropyl-50% phenyl)...has no residual catalyst and has been effectively devolatilized" [17]. The polarity [15] of SILAR-5CP is comparable with the ninth and tenth "preferred" stationary phases, and its maximum operating temperature is higher:

"Twelve" number	Name	X	Y	Z	U	S	Max. °C
9	Carbowax-20M	322	536	368	572	510	250
	SILAR-5CP	319	495	446	637	531	275
10	DEGA	378	603	460	665	658	200

The still more-polar silicones, SILAR-10C and OV-275, shall be discussed along with the polyesters (DEGA, DEGS), and with TCEP, the twelfth "preferred" phase.

Polyethylene Glycols

Carbowax-20M, usable from 60 to 250°C, is "synthesized by joining polyethylene glycol 6000 molecules [structure, $HOCH_2-CH_2-(-O-CH_2-CH_2-)_n-O-CH_2-CH_2OH$, Ref. 25] with a diepoxide" [26]. At highest temperatures, trace oxygen in the carrier gas causes trace decomposition into acetaldehyde and acetic acid, especially if catalyzed by acid samples or supports. Columns containing "a per cent or so of potassium hydroxide" will be more stable [26].

Polyesters

The tenth and eleventh "preferred" stationary phases are the diethyleneglycol polyesters of adipic and succinic acids, respectively. Because of the high polarities and reasonably good temperature stability of these polyesters, they have long been standards, especially DEGS. The high-temperature limits of these polyesters *should be very carefully observed*, however, and they must be very well protected from contact with oxygen or water [27].

Sec. 5.5 Preferred Stationary Phases

As seen below, the silicone Silar-10C [28] offers a higher polarity and maximum operating temperature than the polyesters. Like other silicones, Silar-10C can briefly be taken some 25°C above its recommended maximum temperature limit without--as would be the case with the polyesters--disaster.

"Twelve" number	Name	X	Y	Z	U	S	Max. °C
10	DEGA	378	603	460	665	658	200
11	DEGS	492	733	581	833	791	225
	Silar-10C	523	757	659	942	801	250

TCEP

The twelfth and most polar of the 12 "preferred" stationary phases is TCEP: 1,2,3-tris(2-cyanoethoxy)propane. TCEP offers a higher maximum operating temperature than the more polar cyanoethyl sucrose. However, the cyano-silicone OV-275 shows a higher operating temperature and polarity than either TCEP or cyanoethyl sucrose.

"Twelve" number	Name	X	Y	Z	U	S	Max. °C
12	TCEP	593	857	752	1028	915	175
	Cyanoethyl sucrose	647	919	797	1043	976	100
	OV-275	629	872	763	1106	849	250

Note that tolerance to high temperature is more important than high polarity. The proposed standard for 100% polarity, β,β'-oxydipropionitrile [12], was not even included in the McReynolds study, presumably because its maximum operating temperature is 75°C. Similarly, each of the early, volatile, highly polar stationary phases--safrole, propylene carbonate, acetylacetone, benzyl cyanide--finds diminishingly little use today. For the same reason, replacements operable to 350°C will eventually be found for such silicones as Silar-10C and OV-275, which are also highly polar but too limited thermally.

B. A List of Six

In 1975, a further attempt to reduce the number of stationary phases was reported by a committee chaired by S. Hawkes [29]. The committee presented three lists of stationary phases: A primary list of 6, to cover the great bulk of separations; a secondary list of 18 more, to fill in the polarity range in more detail; and a third list of 13, to perform special separations.

We present the primary list without further comment (we have just commented on each type in it, in the preceding pages). In the primary list, both dimethylsilicones and 50%-phenylmethyl silicones were suggested. The committee commented that if one did not work, the other probably would not either. Examples of dimethylsilicones are SE-30, OV-101, SP-2100, and SF-96; examples of 50%-phenylmethyl silicones are OV-17 and SP-2250.

Polyethyleneglycols (the Carbowaxes) were next. These differ little from each other in retention properties. Of the polyesters, diethyleneglycol succinate made the primary list, butanediol succinate the secondary list, but diethyleneglycol adipate was not mentioned. The highly polar 3-cyanopropyl silicones such as Silar-10C, Apolar 10C, and SP-2340 were on the primary list. These were described by the committee as chromatographically similar to the polyethyleneglycols and to DEGS.

The sixth type of stationary phase suggested as primary is the trifluoropropylmethyl silicone, examples of which are OV-210 and SP-2401.

5.6 SOLVENT SELECTION

A. Principles

To separate solutes, seek and emphasize the differences among them.

A solvent functionally different from the solutes tends to separate them more efficiently than one functionally similar to them. It tends to be more selective and to yield quicker separations.

The physical chemistry of this approach was first presented and discussed by Langer [30] and Langer and Sheehan [31]: "...favorable selective interactions will reinforce each other and are desirable. However, if it is desired to utilize specific interaction for a selective separation, it is appropriate to minimize other solvating interactions which may be counter to the desired separation and time of column residence if nonselective..." [30].

The Hildebrand-Scatchard solubility parameter (which is about to be mentioned) is defined as the square root of a ratio, the ratio of the energy of vaporization to the molal volume [32]. Now we continue with Langer:

> Where solutes with no functional groups are involved, such as aliphatic and aromatic solutes, ...the effect of any small difference in [the Hildebrand-Scatchard solubility parameter] δ between two [such] solutes is enhanced by choosing a solvent with a δ which is considerably different from that of the solutes. [If this approach cannot be used indiscriminately, nevertheless]...the model is useful as an indicator of how to accentuate differences. Of course, this approach differs greatly from the frequent recommendation to 'use a stationary phase of similar nature to the solute being separated' [30].

If the solutes differ in heats of solution, then the differences between the corresponding activity coefficients is enhanced by operating the column at the lowest possible temperature. This source of selectivity is most effective when the heat of

Sec. 5.6 Solvent Selection

solution is small, that is, when solubility is low. Here again, a stationary phase should be chosen that differs from the solute and dissolves it poorly. "Nonselective strong polar interactions should also be avoided, if possible" [30].

A solvent functionally different from the solutes dissolves them less strongly than one functionally similar. Such solutes are eluted more quickly, giving quicker separations, but the sample size must be reduced. The maximum allowable sample size is directly and linearly proportional to the net retention volume, and thus to the partition coefficient, which is the measure of solubility. The sample size must decrease as the partition coefficient decreases.

B. Practical Examples

A Computer Search

The American Society for Testing and Materials publishes the massive *Gas Chromatographic Data Compilation* [33, 34]. Nominally, one need only find in the compilation those phases on which two components to be separated have already been run. One then calculates the corresponding relative retentions for the two components. Indeed, from this, the corresponding number of theoretical plates needed for the separation can also be calculated, once the desired resolution is decided on. In practice, however, finding the desired data is almost impossible.

Recognizing this, J. S. Lewis, an editor of *Compilation*, wrote a computer program, available from him upon request. This program directs a computer to perform the search, identify the relevant stationary phases, and calculate the requisite number of theoretical plates required for the separation with each phase. If this approach is relevant--if the gas chromatographer has access to the data compilation and to a suitably programmable computer, and if the relevant measurements have already been inserted into the data compilation--then it provides an easy answer to this type of problem.

Separating Ethyl Alcohol from Ethyl Acetate

This case is presented in the introduction to the first issue of the ASTM retention data compilation [33]: Find the best stationary phase for separating ethyl alcohol, b.p. 78°C, from ethyl acetate, b.p. 77°C. Conventional wisdom suggests, "Like dissolves like," and thus the stationary phase should probably be an alcohol or an ester. Such a stationary phase is Emulphor-O. We can compare Emulphor-O with the ester didecyl phthalate and the polyethyleneglycol PEG 4000 (and with OV-225, for our own instruction):

	X	Y	Z	U	S
Didecyl phthalate	136	255	213	320	235
Emulphor ON-270	202	396	251	395	345
OV-225	228	369	338	492	386
PEG 4000	325	551	375	582	520

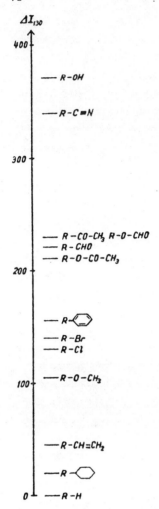

Figure 5.1 The "retention dispersion" of Wehrli and Kovats, for Emulphor-O. The "R" indicates a n-paraffin chain of more than six carbons. (Reproduced by permission of Verlag *Helvetica Chimica Acta* and the authors, from Ref. 35.)

The "retention dispersion" of Emulphor-O, presented by Wehrli and Kovats [35], is also helpful and is shown in Figure 5.1. It shows a wide difference between esters and alcohols. Conventional wisdom would seem to be supported.

However, the two components, ethyl alcohol and ethyl acetate, have markedly different chain lengths, two carbons *versus* four. A nonpolar phase would not respond to the polar alcohol and ester groups, but would differentiate such different numbers of carbons. In accordance, the data from the ASTM compilation show that the nonpolar hexadecane would require only 53 theoretical plates for this separation [33]. The column would need to be only about one inch long.

Sec. 5.6 Solvent Selection

Separating Some Hydrocarbons

"Like dissolves like" suggests that a paraffin or olefin would be a good first choice for separating paraffins and olefins from each other. We now cite several examples to the contrary.

Figure 5.2 shows that hexadecane would not separate n-hexane well, if at all, from 1-hexene, but that the separation would be easy with β,β'-thiodipropionitrile, a stationary phase of extreme polarity. Similarly, a benzene-cyclohexene-1-hexene separation would be difficult with hexadecane, but facile with the nitrile--note that the nitrile retention of the benzene is off-scale in Fig. 5.2. Note also from the ordinate values how very much faster the nitrile separation would be. We reemphasize: Sample sizes must be reduced with this approach. Here, the solubilities of these hydrocarbons in the nitrile are 1/7 to 1/50 of those in hexadecane [8]; the sample sizes must be reduced proportionately.

Lekova and Gerasimoff [36] devised a computer program for selecting the optimum three stationary phases for separating a given mixture. The search and comparison were based on experimental data. These data included, principally, the relative retention times of the mixture components over each phase.

The results of using the program were illustrated by searching among thirty-seven stationary phases for the best three to separate each of six mixtures of hydrocarbons.

Three of the mixtures were made up of saturated hydrocarbons. For all three mixtures, the computer comparison of retention times showed that polar stationary phases were indicated over the nonpolar squalane, which was the stationary phase only of second choice. For two of the mixtures, dichlornaphthalene was first choice; for the third mixture, butilsulfide acetic acid was first choice.

One of the other three mixtures was made up of aromatics. For this mixture, the highly polar diethyleneglycol succinate was selected as first choice over the nonpolar Apiezon L as second choice.

The other two mixtures were made up of saturated hydrocarbons and olefins. For each of these, only highly polar stationary phases were selected by the computer. For one mixture, Carbowax 1500, dimethyl sulfolane, and a 1:1 mixture of propyleneglycol and dimethylsulfolane were chosen in that order. For the other mixture, dimethylsulfolane, a mixture of ethylbenzoate and dimethylformamide, and a mixture of n-propylsulfone and dimethylsulfolane were selected in that order.

Separating Ethanol from Isopropanol

"The initial attempt to separate these on propylene glycol, the stationary phase most widely recommended for alcohols, gives almost coincident peaks... A nonpolar stationary phase such as dimethylsilicone gives $t'_{isoPr}/t'_{Et} = 1.4$. Retention [by the dimethylsilicone] is solely by dispersion interaction so that lower temperatures must be used than is suggested by their boiling points, which are high due to hydrogen bonding" [2]. The low solubility of a dimethylsilicone for these alcohols would also give very fast elution, but it would require a proportionately decreased sample size.

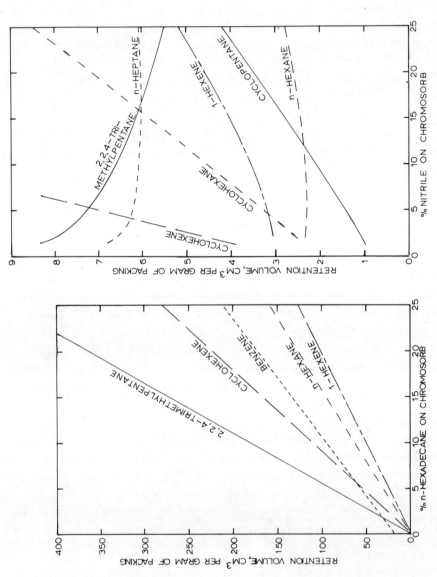

FIGURE 5.2 Unlike often separates better—and faster. The highly polar nitrile (right) would separate these "unlike" nonpolar and polarizable hydrocarbons far more easily and quickly than the similar, "like dissolves like" hexadecane (left). Note that the ordinates differ by two orders of magnitude, and that benzene is off the scale of the nitrile plot—making the separation just that much easier. (Reprinted with permission from Martin, Ref. 8. Copyright by the American Chemical Society.)

References

Separating Butyl Ether from n-Heptane

"This pair has a retention ratio, t'_{BuEt}/t'_{C_7}, of 0.9 on squalane. Better separation could be obtained by using the weak dipole or basicity of the ether" [2]. With polyethyleneglycol, the retention ratio increases to 2.0; with diglycerol, to 2.7. Again, n-heptane is much less soluble in either polyethyleneglycol or diglycerol than in squalane; therefore, elution would be much faster and the maximum allowable sample size, much smaller.

REFERENCES

1. J. H. Hildebrand and R. L. Scott, *The Solubility of Nonelectrolytes*, 3rd ed., Dover, New York, 1964.
2. W. Butler and S. J. Hawkes, *J. Chromatog. Sci.* **15**, 185 (1977).
3. B. W. Bradford, D. Harvey, and D. E. Chalkley, *J. Inst. Petrol.* **41**, 80 (1955).
4. M. E. Bednas and D. S. Russell, *Can. J. Chem.* **36**, 1272 (1958).
5. J. Shabtai, *J. Chromatog.* **18**, 302 (1965).
6. A. Zlatkis, G. S. Chao, and H. R. Kaufman, *Anal. Chem.* **36**, 2354 (1964).
7. M. J. S. Dewar and J. P. Schroeder, *J. Am. Chem. Soc.* **86**, 5235 (1964).
8. R. L. Martin, *Anal. Chem.* **33**, 347 (1961); see also Figures 2 and 3.
9. C. Reichardt, *Angew. Chem. (Int.)* **4**, 29 (1965).
10. E. Kovats in *Advances in Chromatography*, Vol. 1 (J. C. Giddings and R. A. Keller, editors), Dekker, New York, 1965, pp. 229-247.
11. L. Rohrschneider, *J. Chromatog.* **22**, 6 (1966).
12. P. Chovin and J. Lebbe, *J. Gas Chromatog.* **4**, 37 (1966).
13. W. O. McReynolds, *J. Chromatog. Sci.* **8**, 685 (1970).
14. J. J. Leary, J. B. Justice, S. Tsuge, S. R. Lowry, and T. L. Isenhour, *J. Chromatog. Sci.* **11**, 201 (1973).
15. J. K. Haken, *J. Chromatog.* **73**, 419 (1972).
16. C. R. Trash, *J. Chromatog. Sci.* **11**, 196 (1973).
17. A. E. Coleman, *J. Chromatog. Sci.* **11**, 198 (1973).
18. S. J. Hawkes and E. F. Mooney, *Anal. Chem.* **36**, 1473 (1964).
19. R. C. Reid, J. M. Prausnitz, and T. K. Sherwood, *The Properties of Gases and Liquids*, 3rd ed., McGraw-Hill, New York, 1977, p. 576.
20. L. Butler and S. Hawkes, *J. Chromatog. Sci.* **10**, 518 (1972).
21. W. Millen and S. Hawkes, *J. Chromatog. Sci.* **15**, 148 (1977).
22. *Analabs Research Notes* **10**, No. 3 (July, 1970), and **12**, No. 1 (February, 1972), Analabs, North Haven, Conn.
23. R. W. McKinney, J. F. Light, and R. L. Jordan, *J. Gas Chromatog.* **6**, 97 (1968).
24. *Gas-Chrom Newsletter* **13**, No. 4 (July/Aug., 1972); **13**, No. 6 (Nov./Dec., 1972); **14**, No. 3 (May/June, 1973); and **14**, No. 4 (July/Aug., 1973), Applied Science Laboratories, P. O. Box 440, State College, Pa. 16801.
25. H. Rotzche and M. Hofmann, in *Handbuch der Gas-Chromatographie* (E. Leibnitz and H. G. Struppe, editors), Verlag, Leipzig, 1967, p.364.

26. H. E. Persinger and J. T. Shank, *J. Chromatog. Sci. 11*, 190 (1973).

27. D. G. Anderson and R. E. Ansel, *J. Chromatog. Sci. 11*, 192 (1973).

28. *Gas-Chrom Newsletter 14*, No. 5 (Sept. Oct., 1973); *14*, No. 6 (Nov./Dec., 1973); and *15*, No. 1 (Jan./Feb., 1974), Applied Science Laboratories, P. O. Box 440, State College, Pa. 16801.

29. S. Hawkes, D. Grossman, A. Hartkopf, T. Isenhour, J. Leary, and J. Parcher, *J. Chromatog. Sci. 13*, 115 (1975).

30. S. H. Langer, *Anal. Chem. 39*, 524 (1967).

31. S. H. Langer and R. J. Sheehan, in *Progress in Gas Chromatography* (J. H. Purnell, editor), Interscience, New York, 1968, pp. 289-324.

32. J. H. Hildebrand and R. L. Scott, *The Solubility of Nonelectrolytes*, 3rd. ed., Dover, New York, 1964, Chap.23.

33. O. E. Schupp, III, and J. S. Lewis (editors), *Gas Chromatographic Data Compilation* (ASTM DS 25 A), 2nd ed., American Society for Testing and Materials, 1916 Race Street, Philadelphia, Pa. 19103, 1967.

34. O. E. Schupp, III, and J. S. Lewis (editors), *Gas Chromatographic Data Compilation* (Suppl. 1, ASTM AMD 25A S-1), American Society for Testing and Materials, 1916 Race Street, Philadelphia, Pa. 19103, 1971.

35. A. Wehrli and E. Kovats, *Helv. Chim. Acta 42*, 2709 (1959).

36. K. Lekova and M. Gerasimoff, *Chromatographia 7*, 595 (1974).

Chapter 6

THE JONES-VAN DEEMTER EQUATION: DERIVATIONS

The Jones-van Deemter equation provides one of the keys to understanding gas-liquid chromatography. It describes the physical parameters within the column that contribute to peak broadening. In preparation for the study of this equation, we shall define diffusion, arrive at the Einstein diffusion law, and relate the broadening of gas chromatographic peaks to statistical theory. We shall then derive and discuss the Jones-van Deemter theory.

Considerable insight can come from this. We shall understand better, for instance, why separation reflects the rate of carrier gas flow, the support particle size, and the amount of stationary phase in the column. Also, much of the further language and forms of data presentation in gas chromatography will become more familiar, providing an improved base for further study.

6.1 DIFFUSION

The succinct derivation of Einstein's law of diffusion as presented by Moelwyn-Hughes [1] is reproduced here more or less intact. A fairly elaborate but less easily grasped treatment is given by Purnell [2]. Giddings has provided a highly instructive chapter, "Diffusion and Kinetics in Gas Chromatography," in his *Dynamics of Chromatography* [3].

Webster's Third [4] defines diffusion: "3a: the process whereby particles (as molecules and ions) of liquids, gases, or solids intermingle as the result of their spontaneous movement caused by thermal agitation and in dissolved substances move from a region of higher concentration to one of lower concentration." Diffusion is a random movement of molecules that tends to eliminate any concentration gradient. In a pure substance and in a mixture, self-diffusion and interdiffusion occur, respectively. The laws of diffusion describe mixing on a larger scale also.

Diffusion is important for gas chromatography. If it is the fundamental mechanism by means of which solute molecules move into and out of the gas and liquid phases, it is also a fundamental reason why peaks broaden. Because diffusional peak broadening occurs when everything else--operator performance, instrument design, support surface--is perfect, it merits our first attention. For this, we start by deriving Einstein's law for linear diffusion from Fick's first law of diffusion.

Fick's first law has the form

$$\frac{dN}{dt} = -DO\frac{dn}{dx} \tag{6.1}$$

(We shall define the symbols in a moment.) From this law, we wish to derive an expression for the diffusion coefficient D in terms of distance diffused per unit time. This expression, which is Einstein's diffusion law, is used repeatedly in the derivations of the Jones-van Deemter equation.

For this desired definition of D, first postulate a planar area of magnitude O with axis X normal to it, as indicated in Fig. 6.1. Project onto this axis the distance Δ that a molecule will travel by random motion in time t.

So on one side of the area O, and within distance Δ from it, we will have n_1 molecules, and similarly, n_2 molecules on the other side. Molecular motion, being

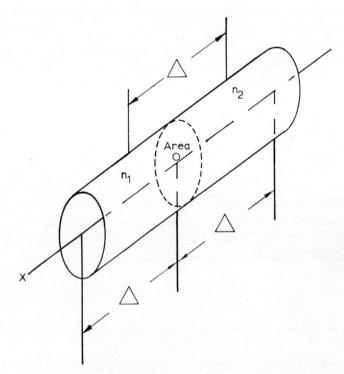

FIGURE 6.1 Within a distance Δ normal to area O are n_1 molecules on one side and n_2 on the other. By random motion, such a molecule travels distance Δ in time t.

Sec. 6.2 Gaussian Peak Shape

random, is directionally equally probable. Therefore, in unit time, half the n_1 molecules will move through the O-sized area to the n_2 side, and conversely.

The net number N of molecules crossing the area will be equal to the difference $(n_1 - n_2)$. Quite generally,

$$\frac{dN}{dt} = \frac{1}{2} \frac{O\Delta(n_1 - n_2)}{t} \tag{6.2}$$

The net differential molecular increment dN per differential time increment dt is directly and linearly proportional to the area O, the distance Δ projected onto the X-axis and likely to be traversed by the molecule in time t, and the net difference in numbers of molecules crossing from left to right. The coefficient 1/2 expresses equal directional probability for this random molecular motion.

A difference in concentration along the axis X can be expressed in the basic terms of these groups of n_1 and n_2 molecules and the random-motion distance Δ separating the centers of these groups. We can thus express the concentration gradient dn/dx:

$$\frac{dn}{dx} = \frac{n_2 - n_1}{\Delta} \tag{6.3}$$

We now have what is needed to derive Einstein's law of diffusion from Fick's: We rearrange Eq. (6.1) explicitly for D, then substitute Eqs. (6.2) and (6.3) into the rearranged Eq. (6.1):

$$\frac{dN}{dt} = -DO\frac{dn}{dx} \tag{6.1}$$

$$D = -\frac{dN/dt}{O\,(dn/dx)} = -\frac{(1/2)[O\Delta(n_1 - n_2)/t]}{O[(n_2 - n_1)/\Delta]} = \frac{\Delta^2}{2t} \tag{6.4}$$

6.2 GAUSSIAN PEAK SHAPE

By more or less elaborate mathematics, the reasons why a gas chromatographic peak tends eventually to have a Gaussian shape have been reviewed in texts [5,6] on gas chromatography. Purnell [5], for example, showed that an equation representing a Gaussian distribution is derivable from Fick's second law of diffusion. Dal Nogare and Juvet [6] showed how in theory the distribution of solute molecules along the column changes as the peak moves down the column. From a plug shape as the solute enters the column, the distribution becomes successively describable as binomial, Poisson, and finally Gaussian (Fig. 6.2).

Three cardinal assumptions of gas-liquid chromatography underlie the reasoning that leads to the conclusion regarding Gaussian peak shape. They are (1) that the first theoretical plate initially contains all the solute, (2) that the partition

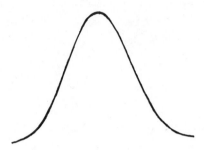

FIGURE 6.2 Ideally, peak shapes eventually become Gaussian.

coefficient remains constant at all points in the column at the same temperature, and (3) that equilibrium is instantaneously attained.

The first assumption tends to be warranted, given plug introduction of a small enough sample. The second assumption requires, in the absence of ideal solutions, such minute samples that the stationary phase-dissolved solute molecules do not influence each other even at highest concentration--the peak maximum. Thus the first two assumptions are experimentally attainable. But we can only ease the attainment of equilibrium, in ways we shall be describing. Equilibrium is not attained instantaneously; indeed, peak broadening can be accounted for on the basis that gas chromatography is a nonequilibrium process [7]. We return to the Gaussian distribution of the solute along the column.

A Gaussian distribution is asymptotic with the baseline. This has two implications: At least some molecules from every solute injected emerge without retention, that is, with the air peak. Also, some molecules from every solute injected remain in the column. Thus "separation" and "cleanliness" are matters only of degree in gas chromatography. True separations never occur. Columns never become "clean" once used.

But much more important, the properties of a normal frequency of Gaussian distribution allow the separate consideration of independently operating causes for peak broadening [8, 9; or see any basic reference on statistics]. The symmetrical, bell-shaped Gaussian distribution curve has a characteristic distance called the *standard deviation* σ. This is the distance, measured parallel to the baseline, from the apex to either of the inflection points. This distance, the standard deviation, is the measure of peak broadening that we shall be using.

The square of the standard deviation σ is the *variance* σ^2. The total variance is the sum of the variances arising from independent causes:

$$\sigma_T^2 = \sigma_1^2 + \sigma_2^2 + \sigma_3^2 + \ldots + \sigma_n^2 \tag{6.5}$$

By this means the reasons why peaks broaden can be considered one cause at a time. Conversely, taking one cause at a time, we can learn how to keep peaks narrow.

Sec. 6.2 Gaussian Peak Shape

The broadening of a peak, as caused by the whole column, when divided by the length L of that column, gives a unit measure of this broadening. This unit measure is called the *height per theoretical plate*, h:

$$h = \frac{\sigma_T^2}{L} \quad (6.6)$$

From Eq. 6.5, we can see that h represents the sum of a number of independent reasons for peak broadening:

$$h = \frac{\sigma_1^2}{L} + \frac{\sigma_2^2}{L} + \ldots + \frac{\sigma_n^2}{L} \quad (6.7)$$

In the rest of this chapter we consider the Jones-van Deemter equation, which is a detailed expression of Eq. (6.7) for gas-liquid chromatography.

Before proceeding to this study, however, we can now understand that, and pause to consider why, the equation that gives the number n of theoretical plates contains the constant 16 and a ratio that is squared. The height h per theoretical plate is expressed as both L/n and, as we have just seen, the total variance σ^2, i.e., spreading, per column length L:

$$h = \frac{\sigma^2}{L} = \frac{L}{n}$$

We solve for n:

$$n = \frac{L^2}{\sigma^2} \quad (6.8)$$

It is the quality of additivity of the variance σ^2--see Eq. (6.5)--that causes it to be used rather than the nonadditive standard deviation σ, which expresses the distance from the Gaussian peak center--the mean--to the inflection point of the Gaussian curve. Arbitrarily, the peak area is taken as that included within the two lines that are parallel to the mean and each 2σ distant from it. This includes 95.45% of the peak area, a property of the Gaussian distribution [10]. Therefore, the distance w_b between these two lines is 4σ,

$$w_b = 4\sigma$$

and so

$$\sigma = \frac{w_b}{4} \quad (6.9)$$

We substitute Eq. (6.9) into Eq. (6.8):

$$n = \frac{L^2}{(w_b/4)^2}$$

or,

$$n = 16 \left(\frac{L}{w_b}\right)^2 \quad (6.10)$$

Eq. (6.10) gives the dimensionless number n of theoretical plates in terms of distance along the column. The numerator L is the distance the center of the solute band has moved since injection, and w_b is the corresponding 4σ measure taken to express the total spreading accrued from the movement of the peak through the n theoretical plates, each producing unit spreading.

The equation usually seen for n expresses the ratio L/w_b in the corresponding, directly, and linearly proportionate terms of total retention time t_R and base width w_b. If the retention is measured in distance along the chart, then

$$n = 16 \left(\frac{t_R}{w_b}\right)^2 \tag{6.11}$$

We can now also see that, because

$$\frac{\sigma^2}{L} = \frac{L}{n}$$

and

$$\sigma = \frac{L}{\sqrt{n}} = \frac{hn}{\sqrt{n}} = h\sqrt{n} \tag{6.12}$$

Thus, given constant h, the peak width 4σ increases as the square root $\sqrt{n}$ of the number n of theoretical plates traversed. However, as we can see by dividing both sides of Eq. (6.12) by n, the ratio σ/n, which expresses relative peak width, decreases with the same square root $\sqrt{n}$ of the number n of theoretical plates traversed:

$$\frac{\sigma}{n} = \frac{h}{\sqrt{n}} \tag{6.13}$$

Thus chromatography becomes a feasible method of separating molecular species: As peaks move, they separate from each other faster than they spread into each other.

6.3 THE JONES-VAN DEEMTER EQUATION

The equation we are about to study suggests that the independent reasons why peaks broaden can be isolated, studied separately, and acted on to keep peaks narrow. Each of these reasons is expressible as a separate term in Eq. (6.7). The work of van Deemter and his colleagues [11, 12] resulted in an equation of three terms, called the A, B, and C terms.

It is possible to point out defects in the original van Deemter equation, and we shall. Nevertheless, the language of gas-liquid chromatography is based on this understandable and instructive equation. The principles by which columns can be made more efficient are deducible from a study of its terms. It was a most worthwhile contribution.

Sec. 6.3 The Jones-van Deemter Equation

Therefore we shall derive and study first the simpler, three-term equation, then the expression of it by four more C terms. Then we shall review the further development of the A term. Finally, we shall review the complete equation and the derivation of it.

A. The Eddy Diffusion Term A

The variance (or peak broadening) caused by the particles in the column is proportional to the number n of opportunities the solute molecules have to experience this type of broadening, and to the square of the distance ℓ, the travel through which causes this type of broadening:

$$\sigma^2 = n\ell^2 \tag{6.14}$$

This form of equation will be seen to recur frequently in these derivations.

Note that the peak spreads along the column, so that the measure of standard deviation σ has the dimensions of distance. Thus σ^2 has the dimensions of distance^2. Hence each side of Eq. (6.14) has the dimensions of distance^2.

We now eliminate n and then ℓ. The number n of opportunities for broadening that are caused by the carrier gas's having to flow around the particles is taken as the number n of particle layers in the column. This number is given by the ratio, L/d_p, where L is the column length and d_p is the diameter of a particle:

$$n = \frac{L}{d_p} \tag{6.15}$$

The distance through which any two solute molecules must travel to experience broadening by going at different effective rates around a particle of diameter d_p, that is, to become displaced from each other in projection along the column axis, is directly proportional to the particle diameter. We introduce proportionality constant λ, called the packing constant, and express ℓ^2:

$$\ell^2 = 2\lambda d_p^2 \tag{6.16}$$

We substitute Eqs. (6.15) and (6.16) into Eq. (6.14):

$$\sigma^2 = \frac{L}{d_p} \, 2\lambda d_p^2 \tag{6.17}$$

In the rearrangement of this, we want to find the variance per unit distance along the column, σ^2/L. Every term of the van Deemter equation assumes this form.

The A term, where

$$A = \left(\frac{\sigma^2}{L}\right)_A \tag{6.18}$$

is therefore

$$\left(\frac{\sigma^2}{L}\right)_A = 2\lambda d_p \tag{6.19}$$

so

$$A = 2\lambda d_p \tag{6.20}$$

B. The Molecular Diffusion Term B

Here is the first use in these derivations of the Einstein diffusion law.

Solute molecules tend to move away from each other while they are in the gas phase: Ordinary gaseous diffusion is an independently acting cause of peak broadening.

The variance--dimension (distance)^2--due to this cause is given by the relevant restatement of Einstein's law,

$$D = \frac{\Delta^2}{2t} \tag{6.4}$$

namely,

$$D_G = \frac{\sigma^2}{2t_G} \tag{6.21}$$

Let us express this explicitly for the variance:

$$\sigma^2 = 2t_G D_G \tag{6.22}$$

where D_G is the gas diffusivity and t_G is the time spent by the solute molecule in the gas phase. (Remember that all solute molecules--and all carrier gas molecules, for that matter--spend the same amount of time in the gas phase.)

We must evaluate t_G in terms of experimental parameters of the column. This can be done by the relationship,

$$\text{Velocity} = \frac{\text{distance}}{\text{time}}$$

or,

$$\text{Time} = \frac{\text{distance}}{\text{velocity}}$$

The distance the solute molecule must travel is the length L of the column. However, the path through the column but around the particles is tortuous. Therefore the actual gas velocity within the column is higher than that calculated from measuring the flow at the exit. We introduce a smaller-than-unity tortuosity constant γ as a divisor; the actual velocity u/γ is greater than the measured velocity u:

$$t_G = \frac{L}{u/\gamma} \tag{6.23}$$

Now Eq. (6.23) can be substituted into Eq. (6.22):

$$\sigma^2 = \frac{2D_G \gamma L}{u} \tag{6.24}$$

Sec. 6.3 The Jones-van Deemter Equation

As before, we express the variance per unit length of column:

$$\frac{\sigma^2}{L} = \frac{2\gamma D_G}{u}$$

For the first time in this set of derivations, the gas velocity u has appeared in a final result. We wish to have a final expression for h as a function of u. Therefore we shall, whenever possible, express each term as a coefficient of u. In this case,

$$\left(\frac{\sigma^2}{L}\right)_B = 2\gamma D_G \frac{1}{u} \qquad (6.25)$$

Therefore the B term, without the u, is

$$B = 2\gamma D_G \qquad (6.26)$$

C. The Resistance to Mass Transfer in the Liquid-Phase Term C_{LP}

The C terms have to do with "resistance to mass transfer." These words, resistance to mass transfer, indicate the tendency of molecules to remain in a given phase. A solute molecule, once dissolved in the stationary phase, moves in a random direction. Only by chance and after some delay does a typical solute molecule arrive at the stationary phase-carrier gas interface and move into the gas phase.

Similarly, to dissolve, a solute molecule in the carrier gas must somehow move from where it is over to a surface of the stationary phase.

In each case, the solute molecules do not move instantaneously back and forth from one phase into the other. This slowness, this wandering of each molecule within a given phase, is called *resistance to mass transfer*. The "mass" is the collective group of all the solute molecules. The "transfer" is the movement of these molecules (one at a time) from one phase into the other.

In the derivation of the C_{LP} term, we start with the approach used in the A-term derivation. The variance from this cause is given by the product of the number of opportunities for the solute molecules to experience this type of broadening--that is, to become mutually displaced along the column axis--and the square of the distance through which the molecules travel in this process:

$$\sigma^2 = n\ell^2 \qquad (6.27)$$

In this case, n is the number of times a solute molecule moves from within the stationary phase to the stationary phase-carrier gas interface (more simply, the gas-liquid interface), during its passage through the column. This number n is given by a ratio of durations: The total time t_L spent by a typical solute molecule dissolved in the stationary (liquid) phase, divided by the unit time τ_L spent diffusing once through the liquid phase to the gas-liquid interface. So,

$$n = t_L/\tau_L \qquad (6.28)$$

We evaluate τ_L, the time for an average solute molecules to diffuse out of the liquid phase once, according to Einstein's diffusion law:

$$d_L^2 = 2D_L\tau_L \tag{6.29}$$

Here D_L is the liquid phase diffusivity and d_L is the depth of the liquid phase. (Th actual configuration of the stationary phase varies with support loading and type. For our purposes, we shall use d_L as the effective depth of the liquid phase.)

We rearrange explicitly for τ_L and substitute c_L for 1/2:

$$\tau_L = \frac{c_L d_L^2}{D_L} \tag{6.30}$$

Having evaluated the denominator τ_L of Eq. (6.28), we now evaluate the numerator t_L. Because

$$k = \frac{t_L}{t_G}$$

and

$$t_G = \frac{L}{u}$$

$$t_L = k\frac{L}{u} \tag{6.31}$$

We can now evaluate Eq. (6.28) using Eqs. (6.30) and (6.31):

$$n = \frac{k(L/u)}{c_L d_L^2/D_L}$$

which is

$$n = \frac{kLD_L}{uc_L d_L^2} \tag{6.32}$$

Having now evaluated n from Eq. (6.27), we wish also to evaluate ℓ. In Eq. (6.27), ℓ indicates the unit distance through which the solute molecules travel while unit C_{LP}-type spreading is occurring; we shall evaluate it as the product, velocity times time. First, velocity:

The overall velocity along the column of all the solute molecules of a given injected species is given by the velocity u_L of the peak maximum. This velocity u_L is equal to the length of the column divided by the time needed for the peak maximum to emerge from the column, namely, the retention time, t_R:

$$u_L = \frac{L}{t_R} \tag{6.33}$$

[Note that Eq. (6.33) does not imply that molecules diffuse through the stationary

Sec. 6.3 The Jones-van Deemter Equation

phase at velocity u_L, or that the solute molecules reach the end of the column by diffusing there through the stationary phase.]

The velocity of an unretained gas, which would have a velocity equal to that of the carrier gas, is given by the quotient of the length L of the column divided by the time t_G spent in the gas phase:

$$u = \frac{L}{t_G} \tag{6.34}$$

To express u_L explicitly as a function of u, we must first reexpress Eqs. (6.33) and (6.34) in order to eliminate L:

$$L = u_L t_R = u t_G$$

$$u_L = \frac{t_G}{t_R} u$$

But

$$t_R = t_G + t_L$$

so

$$u_L = \frac{t_G}{t_G + t_L} u$$

We now divide by t_G, substitute k for t_L/t_G, and obtain the desired expression for u_L in terms of u and k:

$$u_L = \frac{u}{1 + k} \tag{6.35}$$

The distance traveled by all the solute molecules during this type of spreading is given by the product $u_L t_L$. The unit distance traveled is thus $u_L t_L$ divided by the number of opportunities for C_{LP}-type broadening:

$$\ell = \frac{u_L t_L}{n} \tag{6.36}$$

We can now put together what has been developed. The variance from resistance to mass transfer in the liquid phase was first expressed by Eq. (6.27):

$$\sigma^2 = n\ell^2 \tag{6.27}$$

We first substitute for unit distance ℓ, using Eq. (6.36):

$$\sigma^2 = n \left(\frac{u_L t_L}{n} \right)^2$$

$$\sigma^2 = \frac{u_L^2 t_L^2}{n} \tag{6.37}$$

Into Eq. (6.37), we substitute Eqs. (6.35), (6.31), and (6.32), for u_L, t_L, and n, respectively:

$$\sigma^2 = \frac{(u/1+k)^2 \, k^2 \, (L/u)^2}{kLD_L/uc_L d_L^2}$$

$$\sigma^2 = c_L \frac{k}{(1+k)^2} \frac{d_L^2}{D_L} Lu \tag{6.38}$$

Equation (6.38) gives the C_{LP} variance over the whole column, but we wish to express the variance per unit length of column, that is, the height per theoretical plate. Therefore we divide Eq. (6.38) by the length L of the column:

$$\left(\frac{\sigma^2}{L}\right)_{C_{LP}} = c_L \frac{k}{(1+k)^2} \frac{d_L^2}{D_L} u \tag{6.39}$$

From Eq. (6.39), we can express the C_{LP} term as a coefficient of the carrier gas velocity u, concluding the derivation of the C_{LP} term:

$$C_{LP} = c_L \frac{k}{(1+k)^2} \frac{d_L^2}{D_L} \tag{6.40}$$

D. The Resistance to Mass Transfer in the Gas-Phase Term C_G

This term, the velocity distribution term to follow, and the correlation term are due to Jones [13]. In the material that follows all of these derivations, we comment in considerable detail on the background and meaning of these terms.

The C_G term we are about to consider is completely analogous in concept to the C_{LP} term, except that the phenomena take place solely in the gas phase rather than solely in the liquid phase.

The derivation begins as those for the A and C_{LP} terms do:

$$\sigma^2 = n\ell^2 \tag{6.41}$$

The variance or spreading from this independently acting cause is given by the product of the number n of opportunities for this spreading to occur and the unit distance ℓ traveled during unit spreading of this type. We first evaluate n:

The number n of opportunities is, analogously with Eq. (6.28), given by a ratio of times:

$$n = \frac{t_G}{\tau_G} \tag{6.42}$$

The numerator of this ratio of times is the total time t_G spent in the gas phase,

Sec. 6.3   The Jones-van Deemter Equation

measurable as

$$t_G = \frac{L}{u} \tag{6.43}$$

where t_G is the time required to travel the length L of the column at the gas velocity u. The denominator τ_G of this ratio of times is the unit time required to diffuse once through distance ℓ.

The distance ℓ diffused through is taken as the depth d_G of the gas phase, so that in this C_G term

$$\sigma^2 = nd_G^2 \tag{6.44}$$

By Einstein's diffusion law, which has by now been cited or used in Eqs. (6.4), (6.21), (6.22), and (6.29),

$$d_G^2 = 2D_G\tau_G \tag{6.45}$$

where D_G is the gas diffusivity.

In order to evaluate the gas-phase depth d_G, we shall now develop two different expressions for the velocity of the group of solute molecules relative to the velocity of the carrier gas. [We just went through this for the liquid phase, starting at Eq. (6.33) and culminating in Eq. (6.35). Note that the argument here is analogous but not identical.]

The total time spent in the column by all the solute molecules of a given molecular type, that is, the total time required for the peak maximum to appear at the exit of the column, is the sum $t_G + t_L$ of the times spent in the gas and liquid phases. The carrier gas moves at velocity u through distance L in time t_G:

$$u = \frac{L}{t_G} \tag{6.46}$$

The peak maximum travels at a velocity less than the carrier gas velocity only if the solute molecules spend time t_L in the liquid phase. The ratio, the peak-maximum velocity u_S through the carrier gas, divided by the carrier gas velocity u, thus equals the ratio of the time t_L spent in the liquid phase to the total time $t_G + t_L$ spent in the column. (Notice that the peak velocity u_S is the velocity, not along the column, but through the carrier gas. If the solute is a gas that does not dissolve in the stationary phase, then it will spend zero time t_L in the stationary phase and the resultant relative velocity u_S will also be zero: The solute will not move through the carrier gas but will stay right with it. On the other hand, if the solute is highly soluble in the stationary phase, it will tend to move very slowly, if at all, through the column. Then t_L will almost equal $t_G + t_L$, and u_S will almost equal u: The nearly stationary peak maximum will be moving "through" the carrier gas at a velocity nearly equal to that of the carrier gas through the column.)

$$\frac{u_S}{u} = \frac{t_L}{t_G + t_L} \tag{6.47}$$

The ratio of Eq. (6.47) can be expressed in terms of the partition ratio k, where $k = t_L/t_G$, by dividing both t_L and t_G by t_G:

$$\frac{t_L/t_G}{(t_G/t_G) + (t_L/t_G)} = \frac{k}{1 + k} \tag{6.48}$$

By substituting Eq. (6.48) into Eq. (6.47), we can get an explicit expression for u_S in terms of the measurable u and k:

$$u_S = u \frac{k}{1 + k} \tag{6.49}$$

The solute molecules also move through the carrier gas by diffusion, moving at velocity u_S through depth d_G in unit time τ_G:

$$u_S = \frac{d_G}{\tau_G} \tag{6.50}$$

We can now evaluate the gas depth d_G from the two different expressions, Eqs. (6.49) and (6.50), that express the velocity of a group of solute molecules through the carrier gas. We equate them and eliminate u_S:

$$d_G = \tau_G u \frac{k}{1 + k} \tag{6.51}$$

In Einstein's diffusion law, the distance d_G is squared. We wish to develop an expression for $1/\tau_G$, to be used presently. We do this by squaring Eq. (6.51), equating the result to Eq. (6.45), eliminating d_G, and expressing that result explicitly for $1/\tau_G$:

$$d_G^2 = \tau_G^2 u^2 \left(\frac{k}{1 + k}\right)^2 = 2D_G \tau_G$$

Divide through by τ_G^2 and $2D_G$:

$$\frac{1}{\tau_G} = \frac{u^2}{2D_G} \left(\frac{k}{1 + k}\right)^2 \tag{6.52}$$

We can now evaluate the C_G term. We had

$$\sigma^2 = n\ell^2 \tag{6.41}$$

and saw that this could be taken as

$$\sigma^2 = nd_G^2 \tag{6.44}$$

and that

Sec. 6.3 The Jones-van Deemter Equation

$$n = \frac{L/u}{\tau_G} \qquad (6.42, 43)$$

Also, we have just developed Eq. (6.52) for $1/\tau_G$. Combining these, we get

$$\sigma^2 = \frac{L}{u} d_G^2 \frac{1}{\tau_G} = \frac{L}{u} d_G^2 \frac{u^2}{2D_G} \left(\frac{k}{1+k}\right)^2$$

Therefore

$$\left(\frac{\sigma^2}{L}\right)_{C_G} = c_G \left(\frac{k}{1+k}\right)^2 \frac{d_G^2}{D_G} u \qquad (6.53)$$

and the C_G term is

$$C_G = c_G \left(\frac{k}{1+k}\right)^2 \frac{d_G^2}{D_G} \qquad (6.54)$$

E. The Velocity Distribution Term C_2

The gas passages through the packing differ in size. As a result, a solute molecule may move locally in one stream faster or slower than its neighbor in another, and so become displaced from it along the column axis: The peak spreads. This local spreading caused by gas streams of differing velocity is described by the velocity distribution term C_2.

As is more or less usual for us by now, we state that the variance σ^2 is given as a product—the square of the distance ℓ through which the molecules travel during a unit spreading of this type, multiplied by the number n of times this type of spreading can occur during passage through the column:

$$\sigma^2 = n\ell^2 \qquad (6.55)$$

The number n is evaluated as a ratio of durations: The numerator of the ratio

$$n = \frac{L/u}{\tau_2} \qquad (6.56)$$

is the length L of the column divided by the gas velocity u; L/u is the time spent in the gas phase.

The denominator τ_2 is the time during which unit spreading of this type takes place. Thus

$$d_p = \tau_2 u \qquad (6.57)$$

where d_p is the diameter of a particle in the column and u is the carrier gas velocity.

We also have Einstein's diffusion law, as expressed for this case:

$$d_p^2 = 2D_G \tau_2 \qquad (6.58)$$

We combine Eq. (6.58) and the square of Eq. (6.57), eliminate d_p^2, and express the result explicitly for $1/\tau_2$, so that it can be substituted into Eq. (6.56):

$$d_p^2 = \tau_2^2 u^2 = 2D_G \tau_2$$

$$\frac{1}{\tau_2} = \frac{u^2}{2D_G} \tag{6.59}$$

The distance ℓ is taken as equal to the particle diameter; while unit spreading of this type is taking place, the two solute molecules exemplifying it are moving away from one another by going around different sides of a particle. So,

$$\ell = d_p \tag{6.60}$$

and

$$\ell^2 = d_p^2 \tag{6.61}$$

We now substitute Eqs. (6.56), (6.59), and (6.61) into Eq. (6.55), finding the variance from this cause:

$$\sigma^2 = \frac{L}{u} \frac{1}{\tau_2} \ell^2$$

and

$$\sigma^2 = \frac{L}{u} \frac{u^2}{2D_G} d_p^2 \tag{6.62}$$

The variance per unit column length, that is, the height per theoretical plate caused by velocity distribution, we get by dividing both sides of Eq. (6.62) by the column length L:

$$\left(\frac{\sigma^2}{L}\right)_{C_2} = c_2 \frac{d_p^2}{D_G} u \tag{6.63}$$

where c_2 is a constant that includes the $1/2$ in Eq. (6.62).

The C_2 term, therefore, is

$$C_2 = c_2 \frac{d_p^2}{D_G} \tag{6.64}$$

The complete Jones-van Deemter equation includes a correlation term C_3 because the effects described by the nominally independent C_G and C_2 terms interact. However, this is not important to spell out, nor instructive to consider, so we shall do neither.

6.4 A REVIEW, WITHOUT COMMENT, OF THE DERIVATIONS

A

$$\sigma^2 = n\ell^2$$

$$n = \frac{L}{d_p}$$

$$\ell^2 = 2\lambda d_p^2$$

$$\sigma^2 = \frac{L}{d_p} 2\lambda d_p^2$$

$$\left(\frac{\sigma^2}{L}\right)_A = 2\lambda d_p$$

B

$$\sigma^2 = 2D_G t_G$$

$$t_G = \frac{L}{u/\gamma}$$

$$\sigma^2 = \frac{2D_G \gamma L}{u}$$

$$\left(\frac{\sigma^2}{L}\right)_B = 2\gamma D_G / u$$

C_{LP}

$$\sigma^2 = n\ell^2$$

$$n = \frac{t_L}{\tau_L}$$

$$\underline{t_L}$$

$$k = \frac{t_L}{t_G}$$

$$t_G = \frac{L}{u}$$

$$t_L = k \frac{L}{u}$$

$$\underline{\tau_L}$$

$$d_L^2 = 2D_L \tau_L$$

$$\tau_L = \frac{c_L d_L^2}{D_L}$$

$$n = \frac{kL/u}{c_L d_L^2 / D_L} = \frac{kLD_L}{u c_L d_L^2}$$

$\underline{C_{LP}}$ (continued)

    σ^2 (continued)

        $\underline{\ell}$

$$u_L = \frac{L}{t_R}$$

$$u = \frac{L}{t_G}$$

$$L = u_L t_R = u t_G$$

$$u_L = \frac{t_G}{t_R} u$$

$$t_R = t_G + t_L$$

$$u_L = \frac{t_G}{t_G + t_L} u = \frac{u}{(t_G/t_G) + (t_L/t_G)} = \frac{u}{1 + k}$$

$$\ell = \frac{u_L t_L}{n} = \frac{u}{1 + k} k \frac{L}{u} \frac{1}{n}$$

$$\sigma^2 = n\ell^2 = n \left(\frac{u_L t_L}{n}\right)^2 = \frac{t_L^2 u_L^2}{n} = \frac{k^2 (L^2/u^2)[u^2/(1+k)^2]}{kLD_L/uc_L d_L^2} = c_L \frac{k}{(1+k)^2} \frac{d_L^2}{D_L} Lu$$

$$\left[\frac{\sigma^2}{L}\right]_{C_{LP}} = c_L \frac{k}{(1+k)^2} \frac{d_L^2}{D_L} u$$

$$C_{LP} = c_L \frac{k}{(1+k)^2} \frac{d_L^2}{D_L}$$

$\underline{C_G}$

    $\sigma^2 = n\ell^2$

    $\sigma^2 = nd_G^2$

        $\underline{n}$

$$n = \frac{t_G}{\tau_G}$$

$$t_G = \frac{L}{u}$$

$$\left(\frac{1}{\tau_G}\right)$$

$$d_G^2 = 2D_G \tau_G$$

$$d_G = \tau_G u_S$$

Sec. 6.4 A Review, Without Comment, of the Derivations

$$\frac{u_S}{u} = \frac{t_L}{t_G + t_L} = \frac{t_L/t_G}{(t_G/t_G) + (t_L/t_G)} = \frac{k}{1+k}$$

$$u_S = u \frac{k}{1+k}$$

$$d_G = \tau_G u \frac{k}{1+k}$$

$$d_G^2 = 2 D_G \tau_G = \tau_G^2 u^2 \left(\frac{k}{1+k}\right)^2$$

$$\frac{1}{\tau_G} = \frac{u^2}{2 D_G} \left(\frac{k}{1+k}\right)^2$$

$$\sigma^2 = n\ell^2 = n d_G^2 = \frac{L/u}{\tau_G} d_G^2 = \frac{L}{u} d_G^2 \frac{u^2}{2 D_G} \left(\frac{k}{1+k}\right)^2$$

$$\left(\frac{\sigma^2}{L}\right)_{C_G} = c_G \left(\frac{k}{1+k}\right)^2 \frac{d_G^2}{D_G} u$$

$$C_G = c_G \left(\frac{k}{1+k}\right)^2 \frac{d_G^2}{D_G}$$

C_2

$$\sigma^2 = n\ell^2$$

$\underline{n}$

$$n = \frac{L/u}{\tau_2}$$

$$\frac{1}{\tau_2}$$

$$d_p = \tau_2 u$$

$$d_p^2 = \tau_2^2 u^2 = 2 D_G \tau_2$$

$$\frac{1}{\tau_2} = \frac{u^2}{2 D_G}$$

$\underline{\ell}$

$$\ell = d_p$$

$$\ell^2 = d_p^2$$

$$\sigma^2 = n\ell^2 = \frac{L/u}{\tau_2} d_p^2 = \frac{L}{u} \frac{u^2}{2 D_G} d_p^2$$

$$\left(\frac{\sigma^2}{L}\right)_{C_2} = c_2 \frac{d_p^2}{D_G} u$$

$$C_2 = c_2 \frac{d_p^2}{D_G}$$

REFERENCES

1. E. A. Moelwyn-Hughes, *Physical Chemistry*, 2nd rev. ed., Pergamon, New York, 1961, pp. 14-16.
2. H. Purnell, *Gas Chromatography*, Wiley, New York, 1962, pp. 46-54.
3. J. C. Giddings, *Dynamics of Chromatography, Part I: Principles and Theory*, Dekker, New York, 1965, pp. 227-264.
4. P. B. Gove (editor), *Webster's Third New International Dictionary of the English Language, Unabridged*, Merriam, Springfield, Mass., 1961.
5. H. Purnell, *Gas Chromatography*, Wiley, New York, 1962, Chapter 7.
6. S. Dal Nogare and R. S. Juvet, Jr., *Gas-Liquid Chromatography*, Interscience, New York, 1962, pp. 57-64.
7. J. C. Giddings, *Dynamics of Chromatography. Part I: Principles and Theory*, Dekker, New York, 1965, Chapters 3 and 4.
8. A. G. Worthing and J. Geffner, *Treatment of Experimental Data*, Wiley, New York, 1943, Chapter 6.
9. K. A. Brownlee, *Industrial Experimentation*, Chem. Publ. Co., New York, 1943, Chapter 2.
10. J. H. Perry (editor), *Chemical Engineer's Handbook*, 4th ed., McGraw-Hill, New York, 1963, pp. 1-39.
11. J. J. van Deemter, F. J. Zuiderweg, and A. Klinkenberg, *Chem. Eng. Sci.* **5**, 271 (1956).
12. A. Klinkenberg and F. Sjenitzer, *Chem. Eng. Sci.* **5**, 258 (1956).
13. W. L. Jones, *Anal. Chem.* **33**, 829 (1961).

Chapter 7

THE JONES-VAN DEEMTER EQUATION: COMMENTS

In the previous chapter we followed a derivation of the Jones-van Deemter equation. In this chapter we shall examine what the equation means. Examining the Jones-van Deemeter equation not only shows how to improve procedure but also illustrates some of the contradictions between theory and practice in gas chromatography.

Even considering the equation as merely a sum of three terms labeled only A, B, and C, is instructive. As we now know, the total spreading per unit length of column due to all independently acting, ideal causes is called the height equivalent to a theoretical plate, HETP, or, more simply, just h:

$$h = A + \frac{B}{u} + Cu \tag{7.1}$$

We wish to minimize h by minimizing each term separately.

The height h per theoretical plate varies with flow rate u as shown in Fig. 7.1 and as indicated by the equation. The first term is the A term which, according to the equation, does not change with flow rate. No matter how small the other terms get, h should remain equal to or larger than A. (We shall presently see that the true state of affairs is somewhat better than this.)

FIGURE 7.1 With increasing carrier gas velocity, the height h per theoretical plate first drops sharply to a minimum, then increases relatively slowly. Such a plot of h *versus* u expresses the van Deemter equation, h = A + B/u + Cu.

The B term increases sharply toward infinity as the flow rate u approaches zero. Obviously, low flow rates are bad, but how low? If the change of h with u is calculated by differentiation, we get

$$\frac{dh}{du} = -\frac{B}{u^2} + C \qquad (7.2)$$

At the minimum of the plot of h versus u, where h is constant for small changes in u, we can set dh/du = 0 and solve for u, finding that

$$u = u_{optimum} = \sqrt{B/C} \qquad (7.3)$$

Thus there is a flow rate u_{opt}, which is defined by B and C, and below which h increases sharply. With flow rates decreasing below u_{opt}, the column quickly becomes very inefficient.

Therefore the B term controls at flow rates less than u_{opt}, but diminishes steadily as u increases. It is important to use a flow rate that is high enough, but how high?

For any solute that is at all retained, the C term is not zero. However, as C becomes smaller, not only does u_{opt} usually increase but h also increases less for further given increases in u above u_{opt}, as we shall be seeing repeatedly. The coefficient C expresses the slope of the h versus u plot for $u > u_{opt}$. If the coefficient C can be reduced, then higher flow rates can be used. Separations then quicken, deteriorating little even with flow rates well in excess of the optimum flow rate, which itself is higher. Thus, learning how to minimize the resistance to mass transfer and applying this insight yield separations that are not only sharper but also much faster.

To summarize, we try to decrease A because it sets a floor under h. We learn about the B term primarily for academic interest: We avoid its effects by using fairly high flow rates. It is the C term that is the most important. With a vanishingly low C term and higher than optimum flow rates, the column yields separations about as efficiently as ever but several times faster.

7.1 THE TERMS OF THE JONES-VAN DEEMTER EQUATION

A. The A Term

The A coefficient was found to be

$$A = 2\lambda d_p \qquad (7.4)$$

where λ is a packing constant and d_p is the particle diameter. More will be said presently about the A term, but the implication of Eq. (7.4) is useful: Decrease the particle diameter.

The conclusion is good. Historically, columns have improved as particle diameter have both decreased (Fig. 7.2) and become more uniform (Fig. 7.3). A 100 to 120-mesh

Sec. 7.1 The Terms of the Jones-van Deemter Equation   99

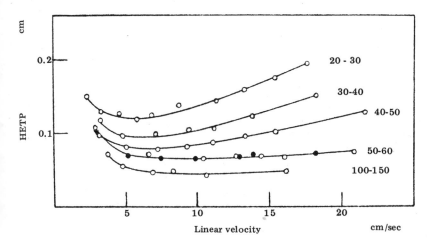

FIGURE 7.2 Over a wide range of particle sizes, the efficiency of packed columns uniformly improves--shows lower h--with decreasing inert support particle size. Solute: 0.4 µℓ acetone. Stationary phase: polyethylene glycol. Temperature: 47°C. Carrier gas: N_2. (Reproduced with permission of Academic Press. From J. Bohemen and J. H. Purnell, Ref. 1.)

(149 to 125-µm) packing, carefully sized, makes a much better column than the 35 to 80-mesh packings first used in gas chromatography. Still smaller particles can be used [2], but a practical minimum to particle size for gas chromatographic columns is about 100 to 120 or 120 to 140 mesh; such packing should be resized before it is used.

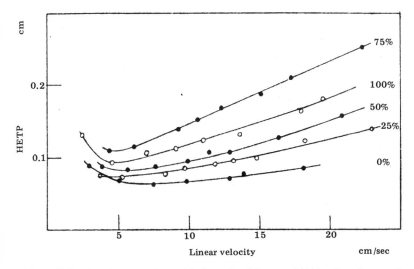

FIGURE 7.3 Inert support particles should vary little in size. The column efficiency here decreased and then passed through a minimum as 50 to 60 mesh Silocel was added to 30 to 40 (percentages of 30 to 40 shown). Solute: 0.5 µℓ benzene. Stationary phase: 20% polyethylene glycol, at 47°C. (Reproduced with permission of Academic Press. From J. Bohemen and J. H. Purnell, Ref. 1.)

B. The B Term for Molecular, or Axial, Diffusion

The B coefficient was derived as

$$B = 2\gamma D_G \quad (7.5)$$

where γ is a tortuosity constant and D_G is the gas diffusivity. Notice that the B term is directly proportional to the gas diffusivity.

The B term is minimized by using a packing of low tortuosity and a carrier gas of low diffusivity. But the important point is not with tortuosity or gas diffusivity, but with the carrier gas flow rate, which should be large, and certainly no smaller than u_{opt}. The carrier gas flow rate may be large if the C terms that express resistance to mass transfer are small and do not thus cause a prohibitively large loss in column efficiency at higher flow rates.

We shall see D_G appear in the denominator of a C term. This would pose a contradiction with the B term, except that the C terms are so much more important. The emphasis must be on reducing not the B term but each C term.

C. The C_{LP} Term, for Resistance to Mass Transfer in the Liquid Stationary Phase

The C_{LP} coefficient is usually the most important of the C terms, because the resistance to mass transfer in the liquid stationary phase is usually limiting. The C_{LP} coefficient is

$$C_{LP} = c_L \frac{k}{(1+k)^2} \frac{d_L^2}{D_L} \quad (7.6)$$

Here c_L is a constant, k is the partition ratio, d_L is the effective thickness of the stationary phase on the inert support, and D_L is the diffusivity of the stationary phase.

In this term-by-term, left-to-right examination of the Jones-van Deemter equation the stationary phase effective thickness d_L is the first variable we come to that is

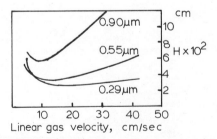

FIGURE 7.4 Reducing the thickness of the stationary phase effectively decreases the resistance to mass transfer in the liquid stationary phase. If that resistance is limiting--and it usually is--the reduction will not only increase column efficiency but will also permit the use of higher flow rates with little loss of the improved column efficiency. Solute: n-butane. Open tubular columns at 25°C, stationary phase thicknesses in micrometers shown on graph. (Data from Desty and Goldup, Ref. 3.)

Sec. 7.1   The Terms of the Jones-van Deemter Equation   101

exponentially effective. Moreover, it occurs in by far the most important term for gas-liquid chromatography, the one that expresses the resistance to mass transfer in the liquid stationary phase. As can be seen in Fig. 7.4, even the classical wall-coated open tubular column, in which the radial distance from the gaseous center of the column to the film on the wall is usually more than 100 times the distance through the liquid film on the wall [4], may be limited in performance by resistance to mass transfer, not in the gaseous but in the liquid phase: Gas diffusivity D_G is at least 100,000 times greater--and proportionately more effective--than liquid diffusivity D_L [5].

Packed columns are usually similarly limited by resistance to mass transfer in the liquid phase. Therefore the performance of packed columns can be strikingly improved within wide limits by decreasing the proportion of stationary phase in the column (see Fig. 7.5). This decreases the effective stationary phase thickness d_L and decreases, as the square of d_L, the resistance to liquid mass transfer term C_{LP}. The improvement is even more striking if the volume of stationary phase is held constant--by proportionately increasing the column length as the loading is decreased. For example, the column length may be doubled as the loading is halved: replace a 6-ft, 30% column with one 12 ft long having 15% stationary phase; or better yet, with one 24 ft long having 7.5% stationary phase.

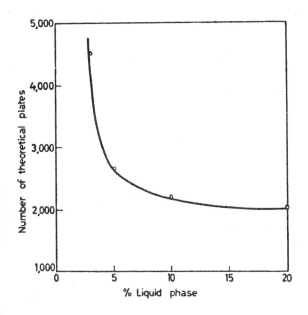

FIGURE 7.5 Packed column performance usually improves sharply as the percentage stationary phase is decreased, at least until support adsorptivity or sample size becomes limiting. Column: Hexanedione on firebrick, 1/8 in o.d. × 4.5 m. (From Amos and Hurrell, Ref. 6. Reprinted by permission of the authors and Butterworths Publications Ltd.)

Nevertheless there would seem to be a limit to the decrease of loading--some minimum percentage of stationary phase below which not only will column performance improve no further, but also no column will even function. History has shown, though, that what is at any given time considered to be the minimum usable stationary phase loading has decreased with the passage of time. Experience has revealed more and more of the deleterious effects that have tended to make given higher loadings seem necessary, and then has shown the ways of obviating these effects. The nominal minimum usable percentage of stationary phase has then decreased still further.

No given minimum--certainly no optimum--usable percentage of stationary phase can be realistically quoted. It is a function of technique, time, and field. As technique improves and time passes, the required percentage of stationary phase tends to decrease. But it varies most widely with field--a given gas chromatographic combination of tubing material, support, stationary phase, and, most important, sample. Thus, even in various current uses, let alone average practice as a function of time, the percentage of stationary phase varies over at least three orders of magnitude [7, 8] (see Fig. 7.6). For all these reasons, let us prefer to seek insight rather than some specific but essentially fortuitous minimum or optimum usable percentage of stationary phase.

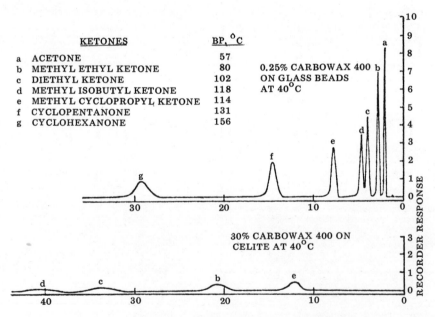

FIGURE 7.6 In the separation of these ketones, the number of theoretical plates, the time per analysis, and the sensitivity all improved sharply when the percentage stationary phase was decreased from 30 to 0.25%, a factor of 120. (Reprinted with permission from Hishta, Ref. 9. Copyright by the American Chemical Society.)

Sec. 7.1   The Terms of the Jones-van Deemter Equation

Now let us consider the liquid diffusivity D_L. Obviously the resistance to mass transfer in the liquid phase C_{LP} would be smaller with a stationary phase of higher rather than lower liquid diffusivity. Yet practicality dictates stability rather than column efficiency in a stationary phase: We want a given column to last a long time. Therefore a polymer of acceptably low vapor pressure [10] generally is used as the stationary phase. Before the study of Butler and Hawkes that was published in 1972 [11], this practical use of a viscous polymer seemed to entail a theoretically adverse, low diffusivity. But Butler and Hawkes found that with the methyl silicones, for example, "...better diffusivities apparently go with the longer chain and better thermal stability...the long-chain elastomer seems to be preferable [over a chemically similar light oil]...The use of the viscosity to predict diffusivity in polymers is completely erroneous." Because most stationary phases are going to be such polymers, this was an unexpected but most welcome finding.

We have just mentioned that gas diffusion is 100,000 to 1,000,000 times faster than liquid diffusion [12]. Negligibly little spreading occurs as a result of the molecules of a given solute diffusing away from each other while dissolved. No term directly proportional to liquid diffusivity and comparable to the molecular diffusion B term for the gas phase exists for the liquid phase in gas-liquid chromatography. Also, in gas chromatography the partition ratio, which expresses the ratio of solute weight in the liquid stationary phase to that in the gaseous mobile phase, generally is a number in the tens or hundreds. Thus the great majority of solute molecules exist in solution and are, compared to their diffusional and flowing motion in the carrier gas, motionless. If the carrier flow is stopped, the solute molecules do not undergo significant spreading, especially if the column temperature is lowered, exponentially increasing the partition ratio still more and decreasing the already negligible liquid diffusivity. Thus a chromatographic column can be used to preserve separations *in situ* as well as to obtain them.

We conclude our consideration of this term by examining the $k/(1 + k)^2$ function. The plot of this function *versus* k (Fig. 7.7) shows the value of the function rising rapidly from zero at k = 0 to a maximum of 1/4 when k = 1. Then it falls much less rapidly, eventually asymptotically approaching zero essentially as 1/k. At k = 2, the value of the function is 2/9--almost 1/4; at k = 3, 3/16. At k = 10, the value of the function has decreased to a little less than 1/10. Let us think of this in terms of the chromatogram.

The partition ratio k can be measured as t_L/t_G, that is, $(t_R - t_G)/t_G$. Generally the air peak (t_G) comes through quickly, in a matter of a few seconds. A delay usually follows of minutes or tens of minutes before solute peaks arrive, so that here k equals perhaps 30 to 300. So to have k = 1, there must be one unit of delay for the air peak and only one more for the solute peak. Now, compared to normal ones, that is a very fast chromatogram. However, here the $k/(1 + k)^2$ shows a maximum; here there is maximum resistance to mass transfer in the liquid phase. Also, this resistance decreases only slowly as peak retention times increase, the peaks moving away from the air peak.

How, then, with respect to the partition ratio k and the function $k/(1 + k)^2$, ca one get the best conditions for minimizing rather than maximizing C_{LP}? The $k/(1 + k)$ function exhibits its smallest values at the origin, where k = 0, and far out in the chromatogram, where k approaches infinity. In the first case, the solute peaks appea immediately adjacent to the air peak, well ahead of the peak position for which the partition ratio k equals unity, thus having spent less time in the liquid than in the gas phase. Although this case is unusual in practice, it is not without potential. For a given column, as the partition ratio k approaches zero, small absolute differences between the successive k's of two successive peaks can effectively yield a usably large ratio of k's, that is, a usably large relative retention α. But the probability that the relative retention α will become or remain usably large tends merely to decrease as the partition ratios of two difficultly separable peaks increase [13]. In sum, the first case--elution near the air peak--has good potential in theory but has had little or no application in practice.

In the second case, the function $k/(1 + k)^2$ approaches zero as k approaches infinity, where $t_L \gg t_G$. Notice that a large partition ratio k implies nothing about retention time magnitude, but only that the solute peak takes long to emerge relative to the air peak.

At this point, let us examine the partition ratio function $k/(1 + k)^2$ expressed as a function of the phase ratio β and the partition coefficient K. This approach ma yield more insight: The phase ratio β expresses the physical nature of the column--the gas/liquid volume ratio--without regard to the identity of the stationary phase. The partition coefficient K expresses the chemical nature of the column, that is, the solubility of the solute in the stationary phase, without regard to the proportion of stationary phase in the column.

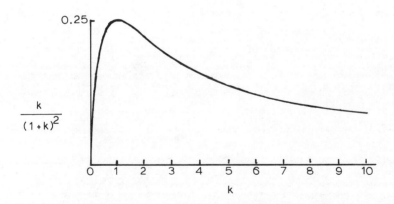

FIGURE 7.7 The partition ratio function in the C_{LP} term rises rapidly to a maximum at k = 1 (two air-peak times), then falls away more slowly. For most peaks in the usual gas chromatogram, k >> 10, and the value of the partition ratio function is very small, essentially 1/k.

Sec. 7.1 The Terms of the Jones-van Deemter Equation

The partition ratio equals the partition coefficient K divided by the phase ratio β:

$$k = \frac{K}{\beta} \tag{7.7}$$

We substitute this into the partition ratio function:

$$\frac{k}{(1+k)^2} = \frac{K/\beta}{[1+(K/\beta)]^2} = \frac{K/\beta}{[(\beta/\beta)+(K/\beta)]^2} = \frac{K/\beta}{(\beta+K)^2/\beta^2}$$

Therefore

$$\frac{k}{(1+k)^2} = \frac{K\beta}{(\beta+K)^2} \tag{7.8}$$

This reexpression shows that an exact trade-off exists between the physical and chemical parameters embodied in the partition ratio function. To explore this idea, we can hold either parameter constant. Specifying a given volume percentage of any stationary phase on any given support amounts to holding the phase ratio β constant for that column, regardless of any given solute-stationary phase solubility. On the other hand, specifying a given solute, stationary phase, and temperature amounts to holding the partition coefficient K constant for that column, regardless of any physical feature of the column such as length or weight percentage of stationary phase.

If either the partition coefficient K or the phase ratio β is held constant, then the reexpression as shown in Eq. (7.8) has the same form as the partition ratio function and thus the same characteristics: The locus of the function rises rapidly from zero to a maximum when the variable becomes unity, and slowly approaches zero again as the variable approaches infinity.

First we examine what can be done solely from the physical point of view, that is, with β variable and K constant.

The phase ratio β ranges in packed columns from about 5 at about 40 wt% stationary phase to about 100 at about 2 wt% [14]. At 30 to 40 wt%, packings just barely pour. Thus in packed columns the phase ratio cannot approach zero because packings corresponding to such low β values are physically unusable--they will not pour. And open tubular columns (described by a different but similar function) generally have a high ratio of gas volume to liquid volume, and thus never have a phase ratio that approaches zero. Thus from the physical point of view, considering only the phase ratio β as the variable, the partition ratio function $k/(1+k)^2$ never approaches zero for low β.

Therefore the function $k/(1+k)^2$, again considering only the phase ratio β as the variable, can become desirably small only as the phase ratio $\beta = V_M/V_L$ becomes large. Because in analytical gas chromatography we never wish to increase an internal cross-sectional area, we can increase β only by decreasing V_L, not by increasing V_M. We decrease V_L by using a thinner effective depth of stationary phase. We use a lower loading in a packed column or support-coated open tubular column, a thinner film in a wall-coated open tubular column.

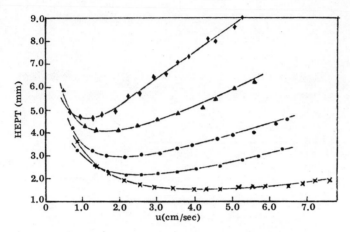

FIGURE 7.8 Increasing the phase ratio β by decreasing the loading of stationary phase improves column efficiency, increases the optimum flow rate, and permits flow rates in excess of the optimum to be used with little penalty. 2,3-Dimethylbutane on polyethylene glycol 400 at 25°C. Stationary phase loadings, w/w, top to bottom: 50%, 43%, 34%, 25%, and 18%. (Reprinted with permission from Duffield and Rogers, Ref. 15. Copyright by the American Chemical Society.)

Experimental evidence showing the advantages of using a thinner effective depth of stationary phase is clear and abundant. See, for example Fig. 7.4 for open tubular columns and Fig. 7.8 for packed columns. This conclusion accords with general experience: Loadings have decreased with time. High percentages were used during the early years of gas chromatography [16] but were increasingly abandoned in favor of lower percentages, usually not exceeding 5 wt%.

Thus we find that what can be done solely from the physical point of view is clear and effective and in accord with the historical development of gas-liquid chromatography: Decrease the proportion of stationary phase. As we have brought out earlier, specifying a minimum percentage of stationary phase for packed columns is difficult if not impossible. Specifying a maximum is much easier, if arbitrary: 5 wt% on an acid-washed, silylated, 100/120-mesh support. (See also Sec. 8.1D.)

Now, what can be done solely from the chemical point of view, that is, with the partition coefficient K variable and the phase ratio β constant? Let us remember also that in this instance we are seeking, not selectivity, that is, a larger relative retention $\alpha = K_j/K_i$, but only low resistance to mass transfer in the liquid stationary phase through some property of the partition coefficient K.

Unlike the phase ratio β for which a small value, let alone an approach to zero, is a physical impossibility, so that one is left with little choice but to increase it by decreasing the proportion of stationary phase, the partition coefficient can show almost any value. The partition coefficient K may easily range continuously from zero to infinity with any stationary phase and at any temperature. Over any stationary phase, some gases will exhibit essentially zero solubility (K = 0), and equally, some solutes will exhibit no usable vapor pressure (K → ∞). Moreover, each solute

Sec. 7.1 The Terms of the Jones-van Deemter Equation

separated on a given column exhibits a different partition coefficient and therefore a different resistance to mass transfer in the liquid stationary phase. It might seem, therefore, that increasing the partition coefficient would nicely decrease the C_{LP} term, as desired. Unfortunately, other effects become limiting. Increase in the partition coefficient alone may even lead to a greater height per theoretical plate, as shown by the results of Desty et al. shown in Fig. 7.9 (from which Fig. 7.4 was abstracted): As solute hydrocarbons increased in chain length and thus in partition coefficient, not only did the height per theoretical plate increase, the minimum in the h versus u curve narrow, and the slope of the higher velocity branch of the curve increase--all bad effects--but also the potential improvements from a thinner effective depth of stationary phase were canceled.

We can conclude: In gas-liquid chromatography not only the most striking but also the surest way to improve column performance is to increase the phase ratio β by decreasing the proportion of stationary phase. In successive consequence, this decreases the partition ratio function $k/(1 + k)^2$, the C_{LP} term for resistance to mass transfer in the liquid stationary phase, and the height per theoretical plate.

D. The C_G Term, for Resistance to Mass Transfer in the Gas Phase

Jones [17] showed the need for extending the van Deemter equation by three more terms. The first of these is the C_G term:

$$C_G = c_G \left(\frac{k}{1+k}\right)^2 \frac{d_G^2}{D_G} \tag{7.9}$$

The effective unit depth of the gas phase, as presented to an average solute molecule, is d_G. Thus d_G^2 is analogous to d_L^2, and is the second squared variable we come to in the Jones-van Deemter equation. The effective unit depth of the gas phase is best reduced by decreasing the particle size of the support. Most supports at this writing are coarser than 140 mesh, which corresponds to 100-μm diameter. Hawkes [18] has reviewed the nature of flow patterns in gas chromatography, and noted that even with large-pore (up to 15 μm in diameter) Type II supports such as Chromosorb W and G, most of the flow is inter- rather than intraparticle. As is pointed out at some length in the chapter on supports, the most efficient diatomaceous earth support is the much-smaller-pore (about 0.5 μm in diameter) Type I, which Hawkes concludes carries almost no internal flow. Thus, although the surface area of diatomaceous earths may actually increase with decreasing particle size [19], the gas depth is nevertheless effectively determined by the interparticle distance and thus primarily by particle size, which should be kept as uniform as possible and as small as is experimentally convenient--about 100 to 120 mesh at this writing.

The B term of the Jones-van Deemter equation is directly proportional to the gas diffusivity D_G. Here, however, the C_G term is inversely proportional to the gas diffusivity D_G. We have seen that good practice tends to lead not only to large flow velocities, thus decreasing any effective contribution from the axial molecular

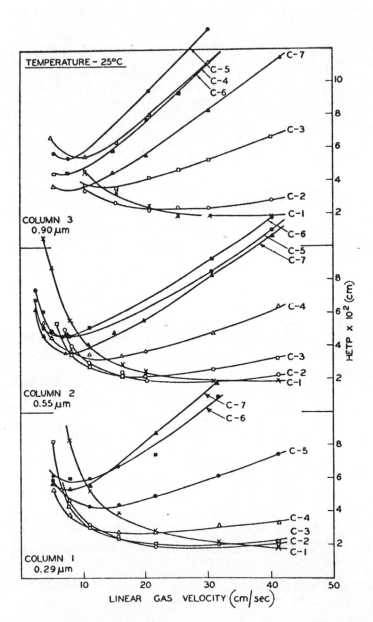

FIGURE 7.9 With any given column (β constant), longer solute chains lead to less volatility and thus to larger partition coefficients, but because other effects may be limiting, not necessarily to better results. Solutes: n-paraffins. (Reprinted from Desty and Goldup, Ref. 3, by permission of the authors and Butterworths Publications Ltd.)

Sec. 7.1 *The Terms of the Jones-van Deemter Equation*

diffusion B term, but also to low percentages of stationary phase on the support. Such low percentages increase the relative importance of the C_G and C_2 contributions to the theoretical plate height. Therefore it becomes more important to maximize rather than minimize the gas diffusivity D_G. The two gases of highest diffusivity are helium and hydrogen. These should be used in columns of high performance.

The partition ratio function was $k/(1 + k)^2$ for the C_{LP} term, but has become $[k/(1 + k)]^2$ for the C_G term. The value of this function increases from the origin, where $k = 0$, but asymptotically approaches unity. At $k = 1$, $[k/(1 + k)]^2 = 1/4$; at $k = 2$, 4/9; at 3, 9/16; at 10, 100/121; and by 100, very nearly unity. Thus the resistance to mass transfer in the gas phase is minimal only near the position of the air peak, at $k = 0$. As we have pointed out, peaks that would closely follow the air peak would lie in an area that is theoretically promising for fast chromatography but one that remains unexploited.

One may as well accept a value near unity for the C_G partition ratio function. Fortunately the C_G term is less effective than the C_{LP} term, so that having the value of the partition ratio function near unity in the C_G term is much less important than having it near zero--under the very same experimental conditions--in the C_{LP} term.

E. The Velocity Distribution C_2 Term

The C_2 term is brief:

$$C_2 = c_2 \frac{d_p^2}{D_G} \qquad (7.10)$$

Again, the parameter analogous to gas phase depth, d_p, the particle diameter, is squared, as were d_L and d_G in the C_{LP} and C_G terms, respectively. As in the C_G term, the "depth" contribution is minimized by minimizing the particle diameter.

The gas diffusivity D_G again appears in the denominator, as it did in the C_G term, thus implying that carrier gases of high diffusivity should be used.

However, much more is involved in "velocity distribution" than is suggested by the two parameters of the C_2 term. *Velocity distribution* means that any two typical solute molecules, at first aligned with each other by being in the same cross section of the column during the movement along the column, become mutually displaced if one of them temporarily gets into a faster stream. This could not occur if diffusion (or turbulence) were so effective that the two were always interchanging or interchangeable. What places these molecules out of diffusional communication with each other within the column?

Giddings [20] has described the interaction of column packing with the pattern of carrier gas flow, and Hawkes [18] has supplemented that description. Giddings discerns five malarrangements leading to flow inequality and to radial noninterchangeability of solute molecules--in short, to band spreading:

1. *Transchannel*: The center of a gas stream between two particles flows faster than the edges of the stream.

2. *Transparticle*: The gas within a support particle is essentially still, while outside the particle the gas flows on.
3. *Short-range interchannel*: Adjacent small groups of particles are experimentally observed as more tightly, then less tightly, packed; the gas moves faster through the irregularly spaced looser regions.
4. *Long-range interchannel*: The regions just described vary mutually, leading to different average gas velocities over groups of regions.
5. *Transcolumn*: The gas velocity at the column center can differ from that at the wall, and the velocity at one region near the wall can differ from that in another region also near the wall.

Of these mechanisms, the two interchannel inequalities account for most velocity distribution spreading. Transchannel spreading is negligible (1%); transparticle, small (about 10%), and dependent on particle structure. Transcolumn spreading due to bends in analytical columns is quite generally negligible [21-23], contrary to persistent myth, although it could become significant for columns an inch or more in internal diameter if they were bent--but they are not.

F. The Revised A Term: A and C_2 Coupling

The "classical" A term, for eddy diffusion,

$$A = 2\lambda d_p \tag{7.4}$$

is not a coefficient of the gas velocity u. The classical A contribution to the plate height is independent of the carrier gas velocity. However, there is enough difference between this simpler concept and reality to force the consideration of a more complicated and detailed description.

Revision of the simpler theory followed quickly enough [24, 25], and the whole Jones-van Deemter equation, including the A term, has been carefully and lucidly reviewed [26].

The independence from carrier gas velocity of the A contribution to plate height exists only at higher velocities. At lower velocities, the eddy-diffusion A term and the velocity-distribution (and velocity-dependent) C_2 term interact. Therefore reality is accurately describable only if the two terms are expressed together:

$$H_{A,C_2} = \frac{1}{(1/2\lambda d_p) + (D_G/c_2 d_p^2)u} \tag{7.11}$$

Thus the H of Eq. (7.11) is smaller than either H_A or H_{C_2}, approaching the lesser: At lower velocities, approaching but not equalling or exceeding H_{C_2}; and at higher velocities, approaching but not equalling or exceeding H_A. Under normal chromatographic conditions,

$$H_{A,C_2} = c_{A,C_2} \frac{d_p^2}{D_G} u \tag{7.12}$$

7.2 THE REDUCED PLATE-HEIGHT EQUATION

All these terms can be expressed more simply, more economically, and more generally if the salient parameters are made nondimensional [27]. The reduced plate height, for instance, becomes

$$h = \frac{H}{d_p^*} \tag{7.13}$$

Both numerator and denominator have the dimensions of distance, so the reduced plate height h is dimensionless. Similarly, the dimensionless, reduced velocity v becomes

$$v = \frac{d_p^* u}{D_G} \tag{7.14}$$

When similar condensations are applied to all variables, the resulting equation allows all forms of chromatography to be compared on the same fundamental basis and also permits the limiting performance of any chromatographic system to be predicted. Our purpose here is mainly to point out that such an approach and resulting equation, allowing such a comparison and prediction, exist [26, 27].

We do wish, however, to report that in the reduced plate height equation, one coefficient, $CD_m/(d_p^*)^2$, expresses by the inequality

$$C \leq \frac{(d_p^*)^2}{D_m} \tag{7.15}$$

a general requirement for all forms of chromatography: Equilibration must occur equally readily in the stationary and mobile phases. Thus the coefficient C, which expresses essentially the resistance to mass transfer by the stationary phase, must be equal to or less than the square of the particle diameter d_p^* divided by the mobile phase diffusivity D_m.

Because it does not mean for gas-liquid chromatography what it may seem to, we must pay special attention to the term *particle diameter*, and hence to its symbol d_p^* (changed from the usual d_p). Changes in the meaning of and the language for the particle diameter occur when the discussion centers not on gas-liquid chromatography but on chromatography generally—this is not clearly brought out in the original paper [27]. For chromatography generally, the "particle" is the discrete particle of the stationary phase. For liquid-solid and gas-solid chromatography, this particle is the one normally indicated by the term *solid particle*, and the particle diameter indicates the diameter of this solid particle.

However, the particle of gas-liquid chromatography in the language describing chromatography generally is the discrete unit of stationary phase existing on the solid support, rather than the inert support particle itself. This discrete unit exists, for most modern packings that have fairly low loadings, as a minute pool contained in a pore of the inert support. In the language of chromatography taken generally, then, the term *particle diameter* refers to the pore size of the inert support rather than to the diameter of the inert support particle.

In the general language of chromatography, the inequality of Eq. (7.15) states that the coefficient C, which has the dimension of time and expresses essentially the resistance to mass transfer by the stationary phase, must be equal to or less than the square of the particle diameter divided by the mobile phase diffusivity. If resistance to mass transfer in the stationary phase is limiting, then further reducing the particle size will improve matters little, given that, as we emphasize here, equilibration is considered to take place only or even primarily at the surface of the particle. The physically depthless stationary phase surface of gas-solid or liquid-solid chromatography determines the speeds of mass transfer that are characteristic of the stationary phase. If such a surface is already limiting by exhibiting a large and limiting time for equilibration, there is little use in reducing particle size to make the surface more readily accessible to the solute molecule.

However, the stationary phase surface of gas-liquid chromatography, if not completely functionless [28], in general is not limiting. Rather, as the van Deemter C_{LP} term brings out, the liquid resistance to mass transfer is proportional to the square of the effective thickness d_L of the stationary phase and inversely proportional to the comparatively minute liquid diffusivity of the liquid stationary phase. Using an inert support of small and uniform pore size increases the speed of equilibration in gas-liquid chromatography by decreasing the effective depth or thickness of the unit stationary phase pool. The uniform, slightly less than 1-μm pore size of the Type I inert support (Chromosorb P) is far better suited for gas-liquid chromatography than the nonuniform and much larger pore sizes of the Type II supports such as Chromosorb W and G, except at very low loadings. At low enough loadings, only the smallest pores of the Type II supports seem to be brought into use. The studies leading to these conclusions are amply cited in the chapter on supports. In sum, for the case of gas-liquid chromatography, it certainly does pay to reduce the particle diameter even though the stationary phase rate of equilibration is limiting: In this case, this reduction speeds stationary phase equilibration.

REFERENCES

1. J. Bohemen and J. H. Purnell, in *Gas Chromatography 1958* (D. H. Desty, editor), Academic, New York, 1958, pp. 6-22.
2. M. M. Myers and J. C. Giddings, *Anal. Chem. 38*, 294 (1966).
3. D. H. Desty and A. Goldup, in *Gas Chromatography 1960* (R. P. W. Scott. editor), Butterworths, Washington, D. C., 1960, pp. 162-183.
4. L. S. Ettre, *Open Tubular Columns in Gas Chromatography*, Plenum, New York, 1965, p. 25.
5. J. H. Purnell and C. P. Quinn, in *Gas Chromatography 1960* (R. P. W. Scott, editor) Butterworths, Washington, D.C., 1960, pp. 184-198.

References

6. R. Amos and R. A. Hurrell, in *Gas Chromatography 1962* (M. van Swaay, editor), Butterworths, Washington, D.C., 1962, pp. 162-177, Figure 2.
7. A. M. Filbert and M. L. Hair, *J. Chromatog. Sci.* 7, 72 (1969).
8. P. L. Davis, *J. Chromatog. Sci.* 8, 423 (1970).
9. C. Hishta, J. P. Messerly, and R. F. Reschke, *Anal. Chem.* 32, 1730 (1960).
10. N. Petsev and C. Dimitrov, *J. Chromatog.* 30, 332 (1967).
11. L. Butler and S. Hawkes, *J. Chromatog. Sci.* 10, 518 (1972).
12. R. C. Weast and S. M. Selby (editors), *Handbook of Chemistry and Physics*, 47th ed., Chemical Rubber Publishing Co., Cleveland, Ohio, 1966, p. F-43.
13. S. H. Langer and R. J. Sheehan, in *Progress in Gas Chromatography* (J. H. Purnell, editor), Interscience, New York, 1968, p. 296.
14. S. Dal Nogare and J. Chiu, *Anal. Chem.* 34, 890 (1962).
15. J. J. Duffield and L. B. Rogers, *Anal. Chem.* 32, 340 (1960).
16. J. J. van Deemter, F. J. Zuiderweg, and A. Klinkenberg, *Chem. Eng. Sci.* 5, 271 (1956).
17. W. L. Jones, *Anal. Chem.* 33, 829 (1961).
18. S. J. Hawkes, in *Recent Advances in Gas Chromatography* (I. I. Domsky and J. A. Perry, editors), Dekker, New York, 1971, pp. 13-48.
19. J. Tadmor, *Anal. Chem.* 36, 1565 (1964).
20. J. C. Giddings, *Dynamics of Chromatography*, Dekker, New York, 1956, pp. 42-47, 50-52, and 150-151.
21. J. C. Giddings, *J. Chromatog.* 3, 520 (1960).
22. J. C. Giddings, *J. Chromatog.* 16, 444 (1964).
23. J. C. Giddings, *Anal. Chem.* 37, 1580 (1965).
24. M. J. Beran, *J. Chem. Phys.* 27, 270 (1957).
25. J. C. Giddings, *Nature* 184, 357 (1959).
26. C. L. De Ligny, *J. Chromatog.* 49, 393 (1970).
27. J. C. Giddings, *J. Chromatog.* 13, 301 (1964).
28. R. L. Martin, *Anal. Chem.* 33, 347 (1961).

Chapter 8

SUPPORTS

In gas chromatography, the support holds the stationary phase in space, exposing it to the carrier gas and thus to the solute molecules. The support becomes more important and can become crucial as the separation becomes in any way more difficult. In many applications of gas chromatography, separations are very difficult indeed, so the person attempting such an application may save considerable time by first becoming familiar with the nature and historical development of supports.

8.1 DIATOMACEOUS EARTH

A. Origin and Structures

The support used by James and Martin in their first experiments [1, 2] with gas chromatography was a diatomaceous earth [3-5]. Diatomaceous earth remains today by far the most widely used type of support. "How strange," it has been remarked, "that the best support should be just a type of dirt, just something dug out of the ground." Is it so strange?

Diatomaceous earth, also known as diatomite or kieselguhr, is strip-mined from great deposits that stretch for miles. These deposits comprise the remains of thousands of species of single-celled plants that lived in ancient seas and lakes. See Fig. 8.1 [3]. Obviously, such minute plants were and are too extended and too small to have the water move through them. The nutrients and waste products of diatoms moved to and from them not in currents but by diffusion.

The skeletons of diatoms efficiently accommodate diffusion: Gas chromatography depends on diffusion. That the skeletal forms evolved for the support of life by diffusion should be highly efficient for achieving separations by diffusion thus seems reasonable enough, not strange.

These skeletons that now support the stationary phase once supported plant tissue. The chief difference is that where in life nutrients and waste products diffused through water, in gas chromatography solute molecules diffuse through the carrier gas

Sec. 8.1 Diatomaceous Earth   115

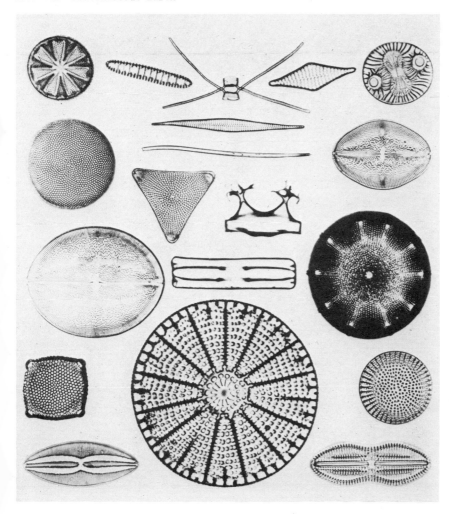

FIGURE 8.1 Typical diatom skeletons. (Reproduced from the *Journal of Gas Chromatography*, by permission of Preston Publications, Inc., and the author, Ref. 3. Photograph reproduced here supplied courtesy of the Johns-Manville Corporation.)

A second difference is that the polarity and concentration by volume of the water suspending the plants exceeded the polarity and concentration by volume of the nutrients and waste products. The polarity of the skeletal surface thus did not matter to the plants. But in gas chromatography, polar and polarizable solute molecules can interact with the skeletal surface. Therefore in gas chromatography the surface of the skeleton must be made inert to minimize any adsorption of polar and polarizable solutes. Let us now begin to describe diatomaceous earth-based supports.

The basic or "primary" structure of a diatom contains holes 1 to 5 µm in diameter. These holes may contain 50 to 100 subsidiary holes each--the "secondary" structure. See Fig. 8.2 [3].

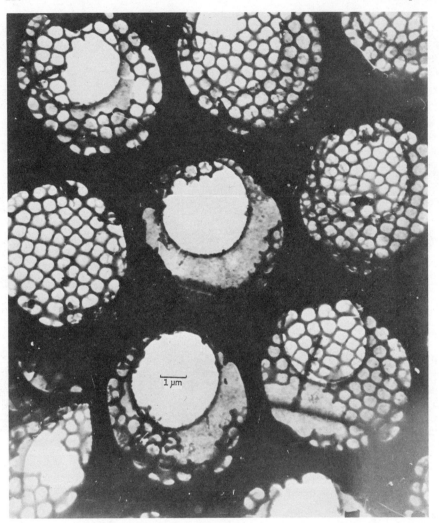

FIGURE 8.2 An electron micrograph of a typical diatom valve. From the 1-μm scale, it can be seen that the larger holes are about 2 μm in diameter. The smaller holes--the secondary structure--must be about 4/10 or 5/10 μm in diameter. (Reproduced from the *Journal of Gas Chromatography* with permission of Preston Publications, Inc., and the author, Ref. 3. Photograph reproduced here supplied courtesy of the Johns-Manville Corporation.)

Ottenstein [4] has referred to diatomaceous earth supports as Type I and II, a terminology we shall follow.

Type I (pink) supports are made by crushing, blending, and pressing the originally gray-white diatomaceous earth into brick, then calcining (heating) to over 900°C. Some of the amorphous silica becomes cristobalite. Mineral impurities react, and the brick turns pink. It is these bricks, largely used for high-temperature insulation, that are broken up to become the Type I supports.

Sec. 8.1 Diatomaceous Earth

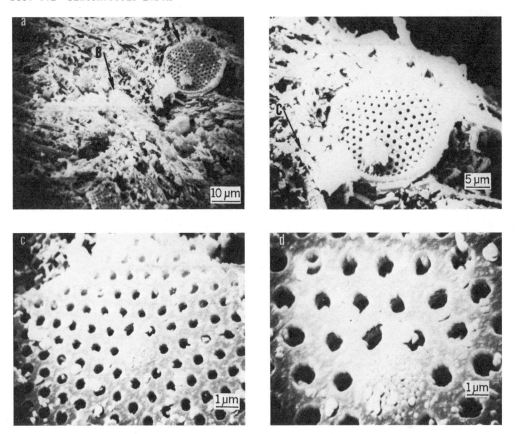

FIGURE 8.3 These scanning electron micrographs shown in Figs. 8.3 through 8.5 show that some diatom skeletons survive intact the processing into firebrick, from which Chromosorb P is made. These are views of a surface of Chromosorb P. (Reproduced from Bens and Drew, Ref. 6, with permission of Macmillan Journals, Ltd. Photograph for reproduction here supplied courtesy of E. M. Bens, Naval Weapons Research Center, China Lake, Calif.)

Type II (white) supports are made, not with the preliminary crushing and pressing into brick, but by mixing flux--such as sodium carbonate--with the diatomaceous earth and then calcining to over 900°C.

The resultant structures of the Type I and II supports differ radically from each other, as might be expected. The crushing and brick-pressing (Type I) operations may break down some of the structurally weaker parts of the original skeletons, but examination of the accompanying scanning electron micrographs reveals that the skeletons often remain more or less intact--see Figs. 8.3 to 8.5 [6] and Fig. 8.6 [3]. It is the secondary diatom structures that have been found especially optimum for gas-liquid chromatography, and these tend to survive the processing.

In contrast, the flux-calcining of the diatomite causes the diatomaceous earth skeletons to be stuck together by a glass. See Fig. 8.7 [3]. The resultant structure is essentially unrelated to that of the diatom.

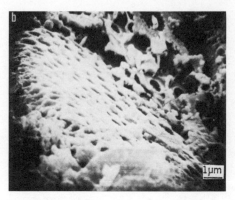
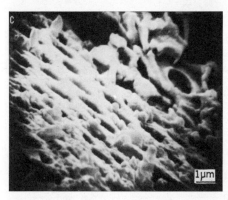

FIGURE 8.4 The particle indicated by arrow B of Fig. 8.3 is shown here. Both Figs. 8.3 and 8.4 show that the spacing between holes (primary structure) in the diatom skeleton surface is 0.7 to 1.1 μm. The diameters of the uniformly spaced holes are 0.5 to 0.8 μm. (Reproduced from Bens and Drew, Ref. 6, with permission of Macmillan Journals, Ltd. Photographs for reproduction here supplied courtesy of E. M. Bens, Naval Weapons Research Center, China Lake, Calif.)

It is as if we had a room full of folding chairs. The chairs would be made of a porous material, and would all be unfolded. In the case typical of the Type I operation, we crush the chairs together in a big press until they form an inextricably compressed mass. We then break this up. The particles emerging still have the original porous structure, but the space originally between the chairs has disappeared—has been pressed out, as it were. This would be a Type I particle.

In the case typical of the Type II operation, on the other hand, we would throw the still unfolded chairs in random order onto a big pile without breaking them. Then we would pour a thick, thermosetting glue over the pile and heat it until it hardened completely. In this process we would also be destroying much of the valuable fine structure of the porous material from which the chairs were made. Then we would break up the fused pile. This time the porous chair material would break before the glass-like fused glue. The new (Type II) particle now would have large internal voids that before the operation had been the spaces between the chairs.

Thus the Type I particles retain those secondary internal structures that facilitate diffusion and that are so useful in gas chromatography. The Type II flux-calcined particles have large internal structures that correspond to the *outsides* of the original diatoms. These outside spaces had nothing to do with either evolution or diffusion. Pores of the Type I are small, with a small size distribution; pores of

Sec. 8.1 Diatomaceous Earth   119

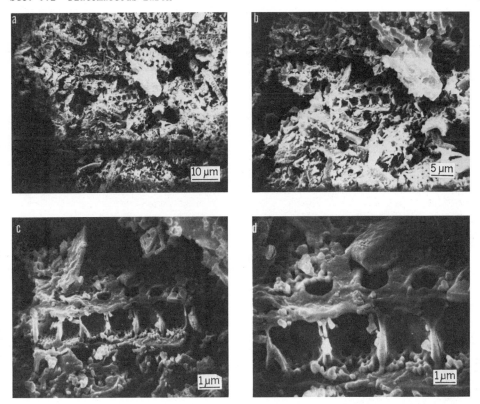

FIGURE 8.5 These scanning electron micrographs of the surface of a particle of Chromosorb P show how a number of broken fragments of diatoms may be arranged in the particle. The chambers shown have an apparent volume of about 1.5×10^{-8} $\mu \ell$; the lower holes (secondary structure) vary from 0.1 to 0.4 μm in diameter. (Reproduced from Bens and Drew, Ref. 6, with permission of Macmillan Journals, Ltd. Photographs for reproduction here supplied courtesy of E. M. Bens, Naval Weapons Research Center, China Lake, Calif.)

the Type II particles have a wide range of sizes with no particularly characteristic size, as is to be expected of a structure synthesized from a set of originally extended structures that were then altered and destroyed. The flux-calcined Type II structure consists mostly of large voids; the pressed Type I structure has considerable microstructure--see Fig. 8.8 [7].

About 1965, a new Type II support called Chromosorb G was introduced to combine by more severe fluxing the strength of the Type I support with the inertness of the first Type II, but with less of the largest void-space of the first Type II. The more melted-down, denser structure of Chromosorb G is shown in Fig. 8.9 [7].

The Type I support is generally known by the Johns-Manville Corporation trade name, Chromosorb P; the first Type II support is called Chromosorb W, and the more recent Type II support, as just mentioned, Chromosorb G.

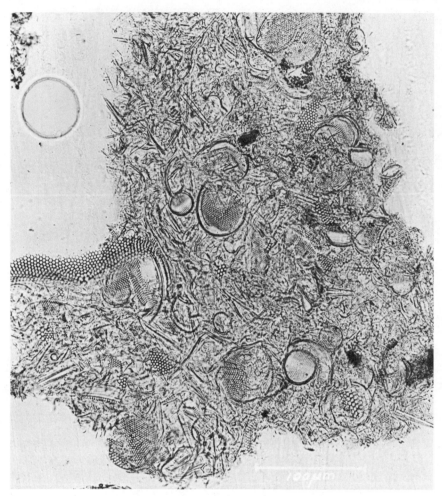

FIGURE 8.6 The structures visible in the scanning electron micrographs (Figs. 8.3 to 8.5) can be discerned in this thin-section micrograph of Chromosorb P. Little space is left between the fragments, but the spaces internal and natural to the fragments largely remain. (Reproduced from the *Journal of Gas Chromatography* with permission of Preston Publications, Inc., and the author, Ref. 3. Photograph for reproduction here supplied courtesy of the Johns-Manville Corporation.)

B. Pore Size and Pore Size Distribution

We can now examine some data and studies on pore size and pore size distribution in supports [8-12]. In a fine, early study, Baker et al. concluded, "The pore size appears to be the most important solid support physical property...a pore structure of fairly uniform pore diameter in the range of 1 micron or less, possibly much less, [seems] optimum" [8]. Later, in their study entitled, "Solid Support Characteristics and Plate Height in Gas Chromatography," Harper and Hammond concluded, "The ideal gas chromatographic solid support should have an open microstructure made up of large pores to minimize [the resistance to mass transfer in the gas phase] C_G. These large

Sec. 8.1 Diatomaceous Earth   121

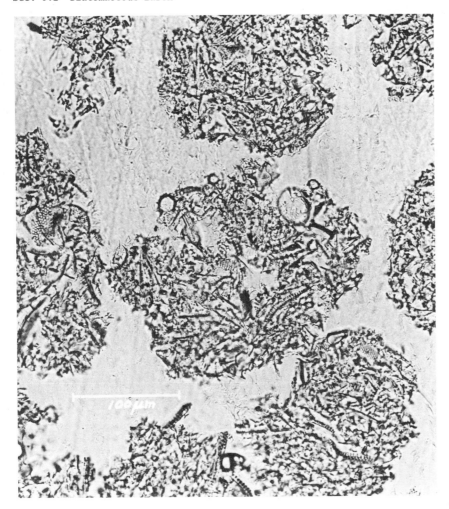

FIGURE 8.7 This thin section of Chromosorb W suggests the wide range of pore sizes and the friability of this flux-calcined material. (Reproduced from the *Journal of Gas Chromatography* with permission of Preston Publications, Inc., and the author, Ref. 3. Photograph for reproduction here supplied courtesy of the Johns-Manville Corporation.)

pores should be covered with short fine pores to reduce [the resistance to mass transfer in the liquid, stationary phase] C_L'' [9].

What do the measurements show? Figure 8.10 [10] shows the results of a pore size distribution study for a Type I support (Chromosorb P) and for three Type II supports (Chromosorb W and G and Gas Chrom S). Each curve shows the total pore volume for a given support as a function of increasing pore size of that support. The total pore volume is expressed in cubic centimeters with respect to the weight in grams of support used, namely, in cubic centimeters per gram.

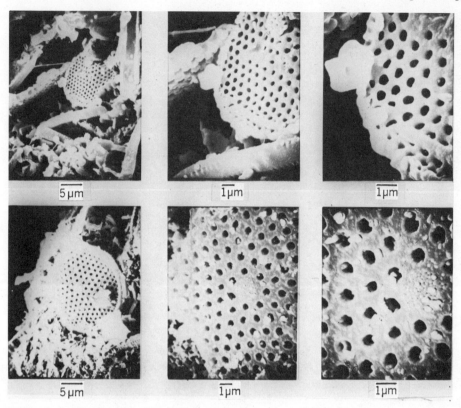

Figure 8.8 The flux-calcined Type II Chromosorb W shown in the top photographs is compared with the Type I Chromosorb P shown in the bottom photographs. Note the lack of fine surface detail, the washed-out appearance, and the fused-on appendages of the flux-calcined Chromosorb W, compared to the highly detailed surface and structure of of the Chromosorb P. (These scanning electron micrographs supplied courtesy of E. M. Bens of the Naval Weapons Center, China Lake, Calif.)

The pores of Chromosorb P are seen to be generally less than 1 μm in diameter; the curve is flat beyond the 1-μm pore diameter, showing that the total pore volume is made up only of pores less than 1 μm in diameter, and that larger pores are absent. The bulk of the pores, as indicated by the steepest part of the curve, have a diameter of 0.5 or 0.6 μm. This seems ideal, according to the passages just quoted. And Saha and Giddings, in the companion study entitled, "Comparative Column Efficiencies of Some Common Solid Supports in Gas Liquid Chromatography," concluded, "Of the four supports studied [P, W, G, and S], Chromosorb P is clearly the best in terms of column efficiency parameters...A large part of this is due to the very fine pore structure of Chromosorb P...It is somewhat surprising that Chromosorb P is not even more efficient..." [11].

On the other hand, the data presented in Fig. 8.10 [10] show that the white, Type II supports have, as the method of manufacture suggests, little or no fine pore structure. The curve for Chromosorb W wanders variously upwards and is still climbing at the upper 12-μm pore size limit of that study: With respect to pore size, Chromosorb W has a little of everything bad. Gas Chrom S is similar. Chromosorb G is

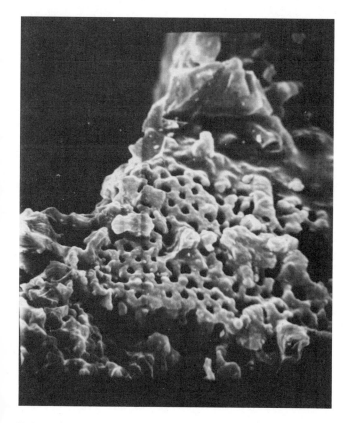

FIGURE 8.9 The melted-down look of Chromosorb G is well shown in this photograph of Chromosorb G-AW-DMCS. Magnification 2,400. (This scanning electron micrograph supplied courtesy of E. M. Bens, Naval Weapons Center, China Lake, Calif.)

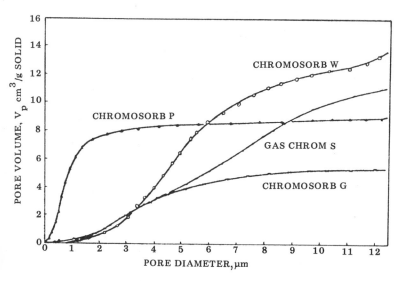

FIGURE 8.10 For gas-liquid chromatography, the diameters of the Chromosorb P pores are ideal, mostly about 0.5 μm, few over 2 μm. The diameters of the pores of the other materials, however, are too large and too widely distributed. (Reprinted with permission from Saha and Giddings, Ref. 10. Copyright by the American Chemical Society.)

somewhat less bad, if not having more of the more useful small pores, at least fewer of the less useful larger pores. (In Sec. 8.1D, we point out the selective use of these pores as a function of loading--a matter of profound importance in this area.)

Note that the ordinate of Fig. 8.10 is expressed as pore volume per weight of support. The packed densities of the supports are W, 0.24 g/cc; P, 0.47 g/cc; and G, 0.58 g/cc [4]. Thus, P is just about twice as dense as W, but about five-sixths as dense as G. If the curves of Fig. 8.10 were plotted more meaningfully for our point of view, with the ordinate showing total pore volume per packed volume of support, the curve for P would easily surmount all the others, and do it--as we have seen--with pores of optimum diameter for gas chromatography. Yet, as we shall show in Sec. 8.1D, this advantage is increasingly lost as lower loadings are used in columns that are consequently more efficient.

The surface area per volume [4] similarly shows the more natural origin of the Type I support. Chromosorb P presents more than 7 times greater than W (0.24 m^2/cc), and more than 6 times greater than G (0.29 m^2/cc). Yet that greater surface area per volume may be the very characteristic that causes the greater adsorptivity of Chromosorb P relative to Chromosorb W or G.

C. Adsorptivity and Deactivation

Adsorption has always been a principal source of trouble in gas-liquid chromatography. The peak of a polar or polarizable solute such as ethanol or benzene, respectively, is distorted in several ways by passage over an adsorptive support. First, the distorted peak is much broader because it now has a tail, more or less elongated and thick. Second, the distorted peak emerges later than it would in the absence of adsorption-- and moreover the shift in retention time is a function of sample size. Third, the distorted peak is far less high--the given area that might have gone toward increasing peak height now appears more and more in the tail. Fourth, if the amount of solute charged is extremely small, as it gets to be in the more demanding applications of gas chromatography, then the solute introduced may apparently simply vanish.

These troubles resulting from adsorption range from irritating to catastrophic. A broadened peak may ruin a separation. The sensitivity to detection and applicability to quantitation by peak height are decreased or canceled. The results become in every way more variable, less dependable, less usable. It is no wonder that the inertness of the Type II support outweighs its apparent disadvantages with respect to pore size and pore size distribution, when inertness is needed.

The fundamental answer to adsorptivity is not to overcome it, as many did in many ways, some of which we shall review presently, but to eliminate it. This more fundamental approach was first taken by J. Bohemen, S. H. Langer, R. H. Perrett, and J. H. Purnell, and reported in 1960 in "A Study of the Adsorptive Properties of Firebrick in Relation to its Use as a Solid Support in Gas-Liquid Chromatography" [13]. As they concluded in the last sentence of their introductory paragraph, "Since firebrick

Sec. 8.1 Diatomaceous Earth

possesses all the desirable attributes of a column support material, except that of adsorptive inertness, it is worthwhile to study means whereby its adsorptive capacity can be reduced to an acceptable level" [13].

Diatomaceous earth surfaces contain siloxane (Si-O-Si) and silanol (Si-O-H) groups [3, 4]. The silanol groups, which adsorb, are the troublemakers. Langer previously had studied a quantitative means of determining surface hydroxyl concentration. Bohemen et al. applied the technique here, replacing the H of the Si-O-H with CH_3. The surface became one of nonadsorptive methyl groups.

One reagent for this is hexamethyldisilazane (HMDS):

$$-\underset{|}{\overset{\overset{H}{|}}{O}}-Si-O-\underset{|}{\overset{\overset{H}{|}}{O}}-Si- \;+\; CH_3-\underset{\underset{CH_3}{|}}{\overset{\overset{CH_3}{|}}{Si}}-N-\underset{\underset{CH_3}{|}}{\overset{\overset{H}{|}}{Si}}-CH_3 \;\longrightarrow\; -\underset{|}{\overset{\overset{O}{|}}{Si}}\!\!-\!\!\underset{}{\overset{\overset{CH_3-\overset{\overset{CH_3}{|}}{Si}-CH_3}{|}}{O}}\!\!-\!\!\underset{|}{\overset{\overset{CH_3-\overset{\overset{CH_3}{|}}{Si}-CH_3}{|}}{O}}\!\!-\!Si- \;+\; NH_3$$

HMDS

(The reaction is quantitative, and does not require the adjacency of silanols that this representation suggests.)

The other reagent is dimethyldichlorosilane (DMCS):

$$-\underset{|}{\overset{\overset{H}{|}}{O}}-Si- \;+\; CH_3-\underset{\underset{Cl}{|}}{\overset{\overset{CH_3}{|}}{Si}}-Cl \;\longrightarrow\; -\underset{|}{\overset{\overset{O}{|}}{Si}}\!\!-\!\!\underset{}{\overset{\overset{CH_3-\overset{\overset{CH_3}{|}}{Si}-Cl}{|}}{O}}- \;+\; HCl \;\xrightarrow{CH_3OH\;wash}\; CH_3-\underset{\underset{Si}{|}}{\overset{\overset{CH_3}{|}}{Si}}-O-CH_3$$

DMCS

DMCS releases HCl rather than NH_3, also requires a methanol wash to replace the reactive Cl with the nonadsorptive OCH_3. (We shall see later that this apparent defect eventually became very useful.)

These reactions, called silylation, are the basis of modern deactivation. Supports to be so deactivated are called *silanized* and are ordered with the suffix -HMDS or -DMCS. But just silylation is not enough.

Silanized supports show a residual adsorption, not detectable without silylation, that apparently is caused by residual impurities [4]. Acid washing removes these mineral impurities and the further reduction in adsorption then becomes readily apparent. Supports to be acid-washed and then silanized are ordered with the suffix

-AW-HMDS or -AW-DMCS. The AW-DMCS supports are the less adsorptive and are therefore to be preferred [4, 14].

The final adsorptivity still not removed by acid washing and silanizing can be quite successfully countered by a trace of surface-active agent added to the stationary phase. This approach was included in the original paper on silylation [13].

As the proportion of stationary phase--the liquid loading--on the support decreases, those most demanding applications such as the separation and determination of steroids or pesticides that may require very low loadings tend to become impossible because minimum adsorptivity is also required. Conversely, when the most demanding applications in gas chromatography are not in question, higher liquid loadings can be tolerated. As the quantity of stationary phase on each particle increases, the effective adsorptivity of the resultant packing decreases. Eventually, with a high enough loading of a polar stationary phase, along with the concomitant relatively large sample and relatively low detector sensitivity, the effective adsorptivity becomes negligible. (All this certainly does not imply that any packing can be used for any application so long as enough stationary phase is used. Rather, more and more exacting work often tends to require less and less stationary phase, and therefore more and more care and insight.)

Adsorption also varies with the interaction of the solute with the stationary phase polarity and loading, and with the type of support, I or II. Adsorption is essentially no problem at any loading with solutes that are neither polar nor polarizable--normal paraffins, for example. Also, less stationary phase is required to cover adsorptive sites adequately if the stationary phase is polar. This is particularly true if the support is the less adsorptive Type II (W or G). For example, the polarizable solute benzene begins with decreasing loading to show tailing on a Type I support only at 0.25 wt% loading of the polar stationary phase polyglycol, but tailing appears at 8 wt% loading of the nonpolar stationary phase squalane. However, again with benzene, no tailing was seen on a Type II support at down to 0.25 wt% of the essentially nonpolar stationary phase squalane [15].

Thus the fundamental answer to adsorptivity is the Type II-AW-DMCS support, an answer made still more effective by a trace of surfactant in the stationary phase. But even all this does not always suffice. In such cases the stationary phase can be still further and more massively altered. Dimethylsulfoxide, for instance, is apparently completely adsorbed on, i.e., will not pass through, a column holding simply the vacuum grease Apiezon on a silanized support; yet it will pass readily, apparently without adsorption, if the same column also carries 8% KOH in the packing [16]. The situation is similar for amines [17, 18].

This more massive alteration of the stationary phase, although neither fundamental nor general, and necessarily always addressed only to the separation at hand, was the first approach used to combat adsorption and related troubles such as catalysis. It was devised and used, indeed, not only by A. J. P. Martin, the inventor of gas chromatography [2, 19], and A. T. James, his associate, but also on the very firs problem to which they applied gas chromatography [1]. They added several percent stearic acid to the silicone stationary phase to prevent in-column dimerization of the fatty acids they were trying to separate. (Phosphoric acid was later used as an appreciable portion of the stationary phase in a similar separation of the very troublesome fatty acids [20].)

Another effective but infrequently applied way of eliminating observable adsorption is to carry in a site deactivator with the carrier gas. This is especially useful if the detector is insensitive to the substance added to the carrier gas. Polar substances such as steam [21, 22], formic acid [23], ammonia [24], and carbon dioxide [25] have been used.

Still another approach to deactivation is to inject a deactivator such as water, or indeed the solute itself that is being adsorbed, once or repeatedly before the solute-containing sample is to be injected. This latter, always applicable, much-used technique is specific, easily applied, easily tested, and effective. A dramatic instance of its use occurred in work reported by Curtius et al. [26]. They "observed complete loss of microgram amounts of catecholamines after trace analysis of arsine and germane" in their gas chromatograph-mass spectrometer, used for mass fragmentography. "This phenomenon could be eliminated by the injection of several micrograms of dopamine-TFA [TFA: trifluoroacetyl] prior to the analysis. No memory peak was observed after these injections." The analysis in question involved samples resulting from brain biopsies yielding 1 mg of tissue and requiring 10-pg sensitivity for the dopamine-TFA.

D. The Preferred Diatomaceous Earth

The preceding sections indicate that Type I supports yield packings and columns that are more efficient over wide ranges of loading, but that adsorb unacceptably, despite deactivation. Type II supports, though less efficient at higher loadings, adsorb far less. Therefore, in practice, the Type II supports are used. Though true, this is not the whole story.

Two further factors strongly influence the choice of diatomaceous earth support type: retention reproducibility, and the variation of Type II efficiency with loading.

With respect to retention reproducibility, Evans and Smith [27] found that adsorptive supports "can cause large changes in retention by either the adsorption of solute molecules at the liquid-solid interface or by the adsorption of the polar functional groups of hydroxylic stationary phases." They suggested that for reproducible retention, supports should have the following properties:

(i) they should possess the minimum adsorptivity consistent with the maintenance of coherent films of liquid phase,
(ii) the surface should be free of fine pores, i.e., those with diameters less than 50 [angstroms],
(iii) they should have the **minimum surface area** consistent with the preparation of columns with good efficiency and resolution characteristics. In addition they should be homogeneous, have good handling properties, and should be batch invariant.

Of the supports thus far examined silanized Chromosorb G appears to fulfill the above requirements better than the remainder [27].

The other factor, the variation of Type II efficiency with loading, although not well known, is striking, and is favorable to gas chromatography in its overall effect.

Filbert and Hair [28] showed that the column efficiency of the highly efficient Type I Chromosorb P is relatively invariant with loadings of 5, 10, and 15% by weight (equivalent in stationary phase volume to 9.3, 17.9, and 26% by weight loadings on Chromosorb W). However, for Chromosorb W, a Type II support, column efficiency improved sharply for the 5% loading over the equal-performing 10 and 15% loadings. This agrees with the much earlier work shown in Fig. 11 [8]: The lower the loading, the smaller the pore size occupied. Thus lower-loaded supports correspond effectively to supports having pores of small size and small size distribution. This conclusion becomes more and more valid as loadings decrease.

Filbert and Hair did not investigate the behavior of Chromosorb G in this respect, nor does the manufacturer's literature indicate such variation, at least at this writing. That variation does exist, however. In fairly detailed testing, the results of which were not published previously, I found that the efficiency of Chromosorb G packings continues to increase with decrease of loading over the 5 to 0.5 wt% range investigated. At about 0.8% by weight, the column efficiencies of Chromosorbs P and G are about equal.

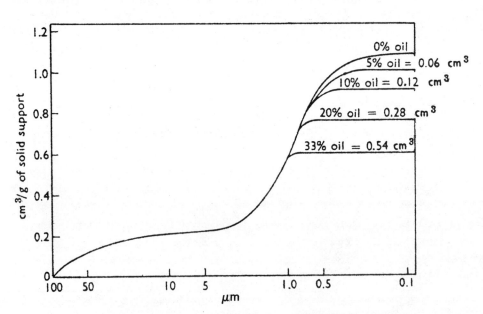

FIGURE 8.11 The pore size of Chromosorb P increases with the loading of the "oil," diethyl sebacate: "As the oil was added it selectively filled the finest available pores." This early study (1961) reflects the high loadings (5 to 33%) typical of those years. However, it shows by implication why flux-calcined supports can be efficient with low loadings. With loadings of 1% or less, these supports seem to have only favorable pore sizes within a favorably narrow size distribution. (From Baker, Lee, and Wall, Ref. 8, reproduced with permission from Academic Press.)

Sec. 8.1 Diatomaceous Earth

Thus at the low loadings that absolutely require the nonadsorptive Type II support, and with the recommended Chromosorb G in test, the column efficiency of Chromosorb G increases to quite acceptable levels--a most fortunate concurrence.

In summary, silanized Chromosorb G is the preferred diatomaceous earth. For good column efficiency, it should support no more than 1 to 2% by weight of stationary phase, preferably no more than 1%.

E. Equivalent Stationary Phase Loading

The different packed densities of Chromosorbs P (0.47 g/cc), W (0.24 g/cc), and G (0.58 g/cc) cause identical weight percent loadings of stationary phase to yield nonidentical amounts of stationary phase per given volume within the column. In consequence, the same weight percents of a given stationary phase in a given column envelope, but on different supports, give different retention times. Recognizing this problem, Durbin [29] derived an equation "from which...equivalent stationary phase loadings...could be calculated for any support at any phase loading:"

$$P_2 = 100\rho_1 P_1 / [\rho_2(100 - P_1) + \rho_1 P_1] \tag{8.1}$$

Let us derive this equation and then rearrange it. In the derivation we shall follow Durbin's approach.

The grams of stationary phase per unit column volume is given by the product $\rho P/(100 - P)$. This product is equal to the grams of inert support per unit volume-- the support density ρ--times the grams of stationary phase per gram of support, $P/(100 - P)$. (In every 100 gms of packing, P grams come from the stationary phase and thus $(100 - P)$ from the support.)

We let the grams of stationary phase per unit column volume with support 1 equal that with support 2:

$$\rho_1 P_1 / (100 - P_1) = \rho_2 P_2 / (100 - P_2)$$

For convenience, we invert both sides, then multiply by ρ_2:

$$(100 - P_1)/\rho_1 P_1 = (100 - P_2)/\rho_2 P_2$$

$$\rho_2(100 - P_1)/\rho_1 P_1 = (100 - P_2)/P_2 = (100/P_2) - 1$$

Now we solve for $1/P_2$ and then invert again, to get the (Durbin) Eq. (8.1).

$$100/P_2 = [\rho_2(100 - P_1)/\rho_1 P_1] + 1 = [\rho_2(100 - P_1) + \rho_1 P_1]/\rho_1 P_1$$

$$1/P_2 = [\rho_2(100 - P_1) + \rho_1 P_1]/100 \rho_1 P_1$$

$$P_2 = 100\rho_1 P_1 / [\rho_2(100 - P_1) + \rho_1 P_1] \tag{8.1}$$

Now let us rearrange this equation, partly to simplify it but primarily to express it in terms of the ratio ρ_2/ρ_1 of densities.

$$P_2 = 100\rho_1 P_1 / [100\rho_2 - \rho_2 P_1 + \rho_1 P_1] = 100 P_1 / [100(\rho_2/\rho_1) - (\rho_2/\rho_1)P_1 + P_1]$$

$$P_2 = 100 P_1 / [100(\rho_2/\rho_1) + (1 - \{\rho_2/\rho_1\})P_1] = P_1 / [(\rho_2/\rho_1) - (\{\rho_2/\rho_1\} - 1)(P_1/100)]$$

This last result is much clearer if we spread it out vertically:

$$P_2 = \frac{P_1}{\left(\frac{\rho_2}{\rho_1}\right) - \left(\frac{\rho_2}{\rho_1} - 1\right)\left(\frac{P_1}{100}\right)} \tag{8.2}$$

If we omit the second term in the denominator, we have a simple calculation:

$$P_2 = (\rho_1/\rho_2)P_1 \tag{8.3}$$

Durbin used the equation he derived to compile a table of equivalent stationary phase loadings for Chromosorbs P, W, and G. Similar numbers are presented in Table 8.1, which also shows the effect of method of calculation.

TABLE 8.1 Equivalent Stationary Phase Loadings for Three Chromosorbs and by Two Methods of Calculation

Chromosorb P	Chromosorb W		Chromosorb G	
wt%	Eq. (8.3) wt%	Eq. (8.2) wt%	Eq. (8.3) wt%	Eq. (8.2) wt%
0.1	0.2	0.2	0.1	0.1
0.2	0.4	0.4	0.2	0.2
0.5	1.0	1.0	0.4	0.4
1.0	2.0	1.9	0.8	0.8
2.0	3.9	3.8	1.6	1.6
5.0	9.8	9.3	4.1	4.1
10.0	19.6	17.9	8.1	8.3
15.0	29.4	25.7	12.2	12.5
20.0	39.2	32.9	16.2	16.8
25.0	49.0	39.5	20.3	21.3

Table 8.1 shows that, up to 5 wt% on Chromosorb P, equivalent stationary phase loadings as calculated approximately by Eq. (8.3) are indistinguishable from those calculated rigorously by Eq. (8.1) or its restatement, Eq. (8.2). Thus dropping the second term in the denominator of Eq. (8.2) simplifies the calculation. It has little effect on the number calculated for the equivalent loading, and no real effect in practice. As will be brought out more fully in Chap. 10, packing preparation is not as precise as the weight percent numbers imply. For instance, although a 6 wt% packing probably can be distinguished from a 5 wt% packing, a 9.8 may or may not be

distinguishable from a 9.3. A 3.9 cannot be made different from a 3.8 wt% packing, except perhaps statistically.

One final point: As has been emphasized in Chapters 5 and 7 and again earlier in this chapter, weight percent loadings above 2 to 5 wt% are death to column efficiency. They are particularly uncalled for in the naturally inert Chromosorbs W and G, which can yield acceptable column efficiencies at low loadings of 2 wt% and less.

8.2 OTHER SUPPORTS

A. Teflon

The Johns-Manville Teflon support Chromosorb T was another attempt to solve the problem of adsorption of polar solutes, and it did; but it proved difficult to pack. Today it is not in wide use. The Teflon support particle is an aggregate of smaller particles averaging 0.5 μm in diameter. The surface area is about 8 m^2/g and the packed density is about 0.49 g/cc, so that the surface area per volume is about 4 m^2/cc, or somewhat more than twice that of Chromosorb P. The Teflon support can be used only to about 250°C.

B. Glass Beads

Whether or not they will be, textured glass beads should be one of the supports of the future.

Glass beads have been known and used in gas chromatography since 1960. Their properties, initially striking enough [30-32], were continually improved [33-39]. The qualities desirable in a gas chromatographic support are by now fairly well defined. Glass beads, a human invention, can be altered with these qualities in mind, and so surpass diatomaceous earth.

The original appeal of glass beads lay in their low adsorptivity. It permitted extremely low stationary phase loadings--down to 0.05%. Such low loadings permitted the gas chromatographing of otherwise unprocessible high boilers at temperatures as much as 250°C below their boiling points [31]. This property of glass beads was unprecedented in 1960 and is still remarkable.

(Why should very low loadings permit "the gas chromatographing of otherwise unprocessible high boilers"? Remember that $V_N = KV_L$: The net retention volume of a given solute is directly proportional to the volume of stationary phase in the column. If the loading is 0.05% rather than 5%, for instance, this means that only 1% as much stationary phase is present. Or, 1% of the length of a 6-ft column is about 3/4 in.-- not a very long column, yet the retention times of the high boilers on low-loaded columns are normal. To go down a normal length of a 5% column, the retention times of a high boiler would be 100 times normal and the peak would be indistinguishable from the baseline. Thus very low loadings become necessary and useful if the support is adequately inert.)

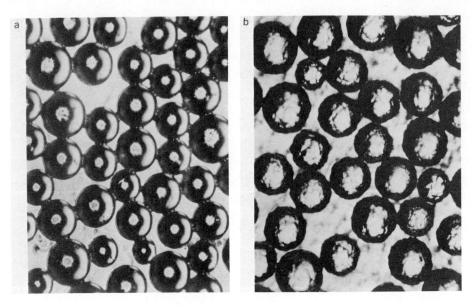

FIGURE 8.12 The stationary phase is distributed uniformly over roughened glass bead surfaces and the roughened beads also become easier to pack. 60 to 80-mesh beads, 0.1% DC 550 silicone oil. Magnification, 133. Left: Unetched beads. Right: Etched beads. (Reprinted with permission of Marcel Dekker, Inc., and the author, Ref. 43. Photograph reproduced here supplied courtesy of A. M. Filbert.)

The original Corning Code 0201 glass beads [40] were DMCS-treated, had a surface area by weight of 0.01 to 0.1 m^2/g, and a packed density of 1.5 g/cc. Thus the surface area by volume was, similarly, 0.01 to 0.1 m^2/cc, and these glass beads could take loadings from below 0.01 up to 1%. (By comparison, the packed density of W is about 0.24 g/cc; of P, about 0.47 g/cc. A 0.01% loading on glass beads corresponds to about 0.1% on W. As usual in speaking of packings, when we say percent, we mean weight percent.) The Code 0201 glass bead surfaces were described as *uniformly textured*.

The surface texture of glass beads can be controlled, as can the chemical composition. Ordinary glass beads have smooth surfaces, so that the stationary phase forms little pools at the bead contacts. These pools increase the resistance to mass transfer in the liquid phase and make the column relatively inefficient. However, roughening the surface of the glass beads [41, 42] causes the stationary phase to spread evenly over the beads--see Fig. 8.12 [43]. The efficiency of columns packed with roughened glass beads can be very good indeed, as we shall see.

However, although surface-roughened glass beads are efficient, the leaching process that is used to roughen the glass bead surface will also expose Ca^{++} ions if they are present within the bead. Such ions are highly adsorptive and act as acids. Therefore, at least for chromatography, a glass bead to be roughened by leaching must be free of Ca^{++}; a sodium silicate glass bead is suitable [38].

Sec. 8.2 Other Supports

In 1969, the Corning GLC-110 glass bead became commercially available [44]. The surface of this sodium silicate glass bead was rougher than that of its predecessor, was nonadsorptive because it was free of Ca^{++} ions, and was silanized as well. These beads made possible not only the gas chromatographic elution of free steroids [44] and endrin [39] without appreciable tailing, but also the attainment of over 1,000 theoretical plates per foot--an excellent column efficiency. Such a combination of support inertness and column efficiency is difficult to achieve with a column packed with a diatomaceous earth support.

Thus the glass bead support by 1969 had clearly surpassed the diatomaceous earth support with regard to technical characteristics. However, Corning stopped producing the beads. The beads now lack effective promotion, seem increasingly overlooked and neglected. Perhaps someone will pick up the fallen torch.

C. Porous Polymer Beads

In 1966, O. L. Hollis discovered and reported [45, 46] a most welcome and unexpected answer to the long-standing and exasperating problem of gas chromatographing very polar substances such as water and glycols. The answer: Porous polymer beads. No other means had ever been found before, nor has any been found since, that would produce similar results. For instance, water comes through a porous polymer bead column to yield a symmetrical peak between the peaks for ethane and propane. See. Fig. 8.13 [45]. The porous polymer beads are also useful, though not so strikingly, in separating the lowest hydrocarbons and also some fixed gases.

These polyaromatic porous polymer beads were quickly made commercially available, and are now obtainable in several forms, such as Porapak from Waters Associates and Chromosorbs 101 through 110 from the Johns-Manville Corporation. Both are obtainable through suppliers of gas chromatographic equipment.

Each type of porous polymer bead tends to have special applicability to what would otherwise be a very sticky problem. Chromosorb 103, for instance, easily resolves amines; Porapak Q, alcohols. But that no one type is generally applicable is more or less obvious from the commercial availability of a large number of them. In general, if a given type of porous polymer bead is at all applicable to a problem, it is likely to be very clearly the packing of choice. Obtaining full descriptions and properties of these materials from the manufacturers is therefore quite worthwhile.

Although porous polymer beads can be used as a support for a conventional stationary phase, they are normally used bare. Two aspects of their behavior indicate that they do not function either as a solid--that is, by adsorption--or as an inert support. First, increasing stationary phase loading decreases rather than increases retention times. Thus the solutes must be partitioned more effectively by the beads themselves than by the added stationary phase liquid. Second, separations effected by the beads are no function of either total micropore volume or average pore size [47]. Apparently, then, the solid beads actually dissolve the solutes, so that a sort of gas-liquid-solid chromatography is taking place.

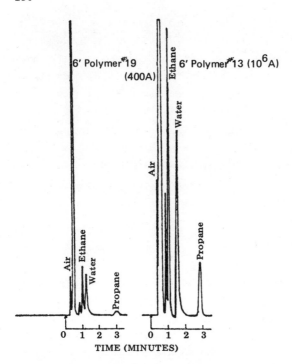

FIGURE 8.13 With some porous polymer beads, water is eluted quickly and without tailing, between ethane and propane. (Reprinted with permission from Hollis, Ref. 45. Copyright by the American Chemical Society.)

Porous polymer beads were not invented just for gas chromatography. Rather, some 25 types were presented to Dr. Hollis to see if the beads just might be useful in gas chromatography. The first four he tried--out of the order in which they were presented to him--showed the unusual and useful behavior we have been describing. The rest, including No. 1, were useless. Dr. Hollis has remarked that if he had tried the beads in their numerical order, he would have found the first few types useless, probably would not have persisted, and thus would not have discovered them. Whether anyone would ever have then discovered the utility of porous polymer beads for gas chromatography can only remain an interesting conjecture.

8.3 MOLECULAR SIEVES

This book concerns gas-liquid chromatography, not gas-solid. Also, it is intended more to afford insight than to be comprehensive, which it is not. Nevertheless, the need to separate certain fixed gases such as the components of air arises from time to time. The only easy way to do this conveniently, even now, is to use molecular sieves, materials interesting enough in their own right. Therefore we will now have a small discursion into this part of gas-solid chromatography.

Molecular sieves [48-51], commercially available from the Linde Company and from suppliers for gas chromatography, are a type of silicate with a peculiar structure. In this, an inner room is connected to the outside through a tunnel. The tunnel-room opening is partially blocked by a cation, so that only molecules smaller than a certain size can squeeze by. The sieves are named according to the apparent diameter in angstroms of those molecules that can just get through the partially blocked opening: 4A, 5A, 10X, and 13A.

The sieves are extremely effective drying agents. If a molecular sieve column is continually exposed to traces of moisture, it will become ineffective as a means of separating fixed gases. However, such a molecular sieve column can be easily regenerated by heating at 350°C for 1 hr under a flow of dry gas.

With molecular sieves, particularly the 5A sieve, which is the most popular, the nitrogen of air is easily separated from the oxygen, at room temperature. A 6-ft 5A column at room temperature will separate and yield, in that order, a noble gas such as neon, oxygen, nitrogen, methane, and, a little later, carbon monoxide. If a 20-ft column is used, helium, neon, and hydrogen can be separated and eluted in that order, preceding oxygen.

If the 6-ft 5A column, which at room temperature will not separate argon from oxygen, is cooled to —72°C, then argon will elute well before oxygen. If the column is steadily heated from room temperature, neon, argon(or oxygen), nitrogen, methane, carbon monoxide, nitrous oxide, and carbon dioxide will elute in that order, clearly separated. Or the normal paraffins can be eluted in order of increasing chain length from such a 5A column, again steadily heated from room temperature. (The branched paraffins, or other more bulky hydrocarbons, would not be retained at all but would pass right through the column. Thus the column can be used to subtract normal paraffins from a mixture or from an eluting sequence of separated hydrocarbons.)

Bombaugh [52] demonstrated that molecular sieves produce very much sharper peaks if they are ground before use into a 200-mesh flour. The flour is then loaded onto a diatomaceous earth. Bombaugh used a 28 wt% loading of 5A on Chromosorb P. The improvement in column efficiency is striking.

Hundreds, perhaps thousands, of 5A molecular sieve column lengths, operating temperatures or temperature programs, and complementary gas-solid columns such as silica gel or charcoal, or gas-liquid columns, have been reported over the years for various special separations involving fixed gases or normal paraffins. Some variation of almost any such problem is therefore very likely by now to be found in *Gas Chromatography Abstracts* under Sec. 4.3, "Active Solids" [53].

8.4 GAS-LIQUID *VERSUS* GAS-SOLID CHROMATOGRAPHY

Why is gas-solid chromatography (GSC) not more popular? Before we leave the subject of GSC, let us answer this. We will consider two important papers and also the history of GSC since 1965.

GSC was originally emphasized much more than now. The emphasis reached a peak in 1964 when Giddings [54] examined GSC on theoretical grounds and found it promising indeed. It offered unusually high column efficiency to the skilled practitioner and retention indices higher than those available with gas-liquid chromatography.

It has been generally characteristic of gas chromatography that its techniques, to be popular, must be fairly easily usable. Achieving the promised high efficiency of GSC was not at all easy, and remains essentially undemonstrated. What about the promised high selectivity?

The high selectivity of GSC seems obvious to the experimenter who finds himself standing and waiting for substances to be eluted from the GSC column. Benzene, for instance, will just vanish into a 6-in. column of activated charcoal. The partition coefficients are very high, and one tends to feel that the relative retentions must be also. One should not.

L. R. Snyder et al. [55] examined GSC versus GLC with respect to general utility and, especially, selectivity. They concluded in favor of GLC. GSC offers little if any advantage. This conclusion was perhaps the more impressive because Dr. Snyder is an authority on adsorption chromatography [56].

The practice of GSC since 1965 has borne out these conclusions. GSC remains generally limited to the separation of fixed gases, done mostly with molecular sieves. Silica gel and charcoal are used only sparingly, and proportionately less with advancing time. GSC has become, and will probably remain, of minor importance.

8.5 THE BONDED STATIONARY PHASE

The bonded stationary phase is a direct outgrowth of the silylated support [13]. Depending on the approach, the reactant with the siliceous support Si-O-H surface group can be either the stationary phase itself or an intermediate to which the stationary phase can then be attached. In either case, the result should be a molecularly thin coating bonded to an inert support of desirable physical configuration. In short, an ideal packing.

The bonded stationary phase seems so clearly ideal that it is worthwhile to consider the target characteristics it should display. An ideal packing implies an ultimate method for holding an ideal stationary phase on an ideal support. Ideal stationary phases would be thermally stable, show low resistance to mass transfer, and be obtainable in various polarities. An ideal support would be inert and would hold the stationary phase in a physically efficient fashion. The ideal packing would combine those characteristics.

For gas-liquid chromatography, ideal packings would combine, in the order of decreasing importance, thermal stability, inertness, and column efficiency. Various polarities, replicability, and purchasability should also characterize such packings.

Sec. 8.5 The Bonded Stationary Phase

A chemically bound stationary phase does not bleed, by definition. Zero bleed, however, does not ease the requirement for great thermal stability. It is not the zero-bleed stationary phase but the cumulative residue from ever-so-slightly contaminated carrier gases and pyrolyzed or bleeding septa, as well as from the uneluted higher boiling components of previously injected samples, that disturbs the baseline during later temperature programming. Unless a column can be cleaned easily and effectively by heating it to 300 or 350°C, that column can exploit its zero bleed quality only with long conditioning at lower temperatures with extremely pure carrier gas. Such a column is too much trouble. To be fully useful, a chemically bound stationary phase must withstand extended heating at 300 to 350°C.

A column holding a zero-bleed, bound stationary phase should be usable at full detector sensitivity for the separation of trace components. If traces are not to be adsorbed, the packing must be inert. Once zero bleed is postulated, inertness becomes a requirement.

Nominally molecularly thin, a bound stationary phase on an efficient support should show minimal resistance to mass transfer, and therefore yield at least as efficient a column as any conventional stationary phase on any conventional support. A target of 1,000 theoretical plates per foot should be readily achievable.

With these criteria in mind, we can survey the chemically bonded stationary phase. Grushka has edited a helpful survey, *Bonded Stationary Phases in Chromatography*, published in 1974 [57].

In 1969, Halasz reported having chemically bonded silanol groups to long-chain compounds [58]. He called them *brushes*. These are commercially available as Durapak. The temperature limits of the three varieties available are as low as 135°C and no higher than 200°C--clearly too restrictive. Also, the materials are subject to hydrolysis. Otherwise, they do exhibit zero bleed, adsorptive inertness, and good column efficiency.

Abel [59] tried polymerizing hexamethyldecyltrichlorosilane on a diatomaceous earth surface and obtained thermal stability but no chemical bonding to the support; and little use in gas chromatographic separations. But these experiments stimulated Aue, Hastings, and co-workers to perform repeated and, this time, successful experimentation [60, 61]. Industry has shown active interest [62, 63].

The packings pioneered by Aue and Hastings require an elaborate but commercially reproducible treatment to produce silanols and then react them with silicones *via* the remaining active -Cl from DMCS or trimethylchlorosilane (TMCS). The ultimate polymeric coatings may be either thermally stable or gas chromatographically useful, or both. They do not cause polar solutes, either alcohols or amines, to tail (Fig. 8.14) [60]. They can differ in polarity. In favorable cases, they exhibit as low a height per theoretical plate as can be achieved comparably with SE-30.

(Remember that, as pointed out in the chapter on stationary phases, the high viscosity of a polymeric stationary phase does not imply an impaired chromatographic efficiency [64].)

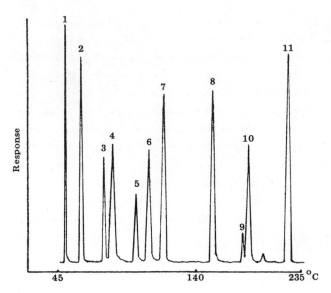

FIGURE 8.14 Some substances shown in this gas chromatogram normally show great tailing, yet do not with this chemically bound silicone. The peaks: acetone, 1; octene-1, 2; n-hexyl alcohol, 3; di-n-butylamine, 4; phenol, 5; tri-n-butylamine, 6; nitrobenzene, 7; dodecyl alcohol, 8; n-octadecene, 9; tri-n-butyl phosphate, 10; and 1-chloro-octadecene, 11. The stationary phase is 16.3 wt% $[C_6H_5(CH_3)SiO]_n$ on 60/80 Chromosorb G. (Reproduced from Aue and Hastings, Ref. 60, in the *Journal of Chromatography*, with permission of the Elsevier Scientific Publishing Company.)

Aue and Kapila [65] see the bonding of their phases as primarily physical, yet stronger than chemical. They quickly found in their initial work that the main criterion for successful "bonding" was resistance to exhaustive extraction. This resistance could well be attained if the long polymer chains could be allowed by prolonged heat treatment to settle in an intimate, molecular fashion onto the supporting surface. Once there, the polymer not only would adhere stably but also could react, becoming from both, almost completely insoluble. Further, the now thoroughly deactivated support surface would be intimately and completely covered by an "ultra-thin (1-20 angstrom) polymer film": "free cholesterol, for instance, can be gas-chromatographed in nanogram amounts." Such supports would be particularly well suited, they suggest, as "highly deactivated supports for other liquid phases."

All this is not to suggest that chemically support-bonded stationary phases do not exist, but rather that the chemical bonding of bonded stationary phases need not be the primary goal. More to the point may well be the elaborate and exhaustive surface preparation of Aue and his co-workers that precedes the support coating by polymer. Certainly, the ultimate products of their approach tend to display quite fully the target characteristics of the ideal packing. They remain to be made widely usable.

References

RECENT DEVELOPMENTS

Nestrick and Stehl [66] have developed a type of bonded stationary phase packing that has made possible a number of recently reported trace determinations [67-75]. This packing apparently exhibits the target characteristics cited in Sec. 8.5: "significantly more efficient separations...and...shorter retention times...compared to conventional coated packings" [68]; "high efficiency, high solvent capacity...minimum liquid phase bleed" [74] "permitting temperature-programmed separations using an [electron capture] detector" [67]. The Dow Chemical Company has licensed the manufacture and sale of these packings to hnu systems, inc., 30 Ossipee Road, Newton, MA.

REFERENCES

1. A. T. James and A. J. P. Martin, *Biochem. J. 50*, 679 (1952).
2. A. T. James, in *Gas Chromatography* (H. J. Noebels, R. F. Wall, and N. Brenner, editors), Academic, New York, 1961, pp. 247-254.
3. D. M. Ottenstein, *J. Gas Chromatog. 1*, 11 (1963), and also Figures 1 to 4.
4. D. M. Ottenstein, in *Advances in Chromatography*, Vol. 3 (J. C. Giddings and R. A. Keller, editors), Dekker, New York, 1966, pp. 137-196.
5. J. Janak and R. Staszewski, *J. Gas Chromatog. 2*, 47 (1964).
6. E. M. Bens and C. M. Drew, *Nature 216*, 1046 (1967), and also Figures 1, 2, and 4.
7. E. M. Bens, Personal communication, 1978.
8. W. J. Baker, E. H. Lee, and R. F. Wall, in *Gas Chromatography* (H. J. Noebels, R. F. Wall, and N. Brenner, editors), Academic, New York, 1961, pp. 21-32, and also Figure 1.
9. J. M. Harper and E. C. Hammond, *Anal. Chem. 37*, 490 (1965).
10. N. C. Saha and J. C. Giddings, *Anal. Chem. 37*, 822 (1965), and also Figure 4.
11. N. C. Saha and J. C. Giddings, *Anal. Chem. 37*, 830 (1965).
12. J. Viska, F. Kiss, M. Pollak, and O. Pospichal, *J. Chromatog. 51*, 103 (1970).
13. J. Bohemen, S. H. Langer, R. H. Perrett, and J. H. Purnell, *J. Chem. Soc. 1960*, 2444.
14. D. M Ottenstein, Pittsburgh Conf. on Anal. Chem. and Appl. Spectroscopy, March, 1966.
15. R. G. Scholz and W. W. Brandt in *Gas Chromatography* (N. Brenner, J. A. Callen, and M. D. Weiss, editors), Academic, New York, 1962, pp. 7-26.
16. R. Z. Muggli, Private communication, 1970.
17. E. D. Smith and R. D. Radford, *Anal. Chem. 33*, 1160 (1961).
18. Y. L. Sze, M. L. Borke, and D. M. Ottenstein, *Anal. Chem. 35*, 240 (1963).
19. A. J. P. Martin in *Gas Chromatography* (V. J. Coates, H. J. Noebels, and I. S. Fagerson, editors), Academic, New York, 1958, pp. 237-247.
20. L. D. Metcalfe, *Nature 188*, 142 (1960).
21. H. S. Knight, *Anal. Chem. 30*, 2030 (1958).
22. A. Davis, A. Roaldi, and L. E. Tufts, *J. Gas Chromatog. 2*, 306 (1964).
23. R. G. Ackman and R. D. Burgher, *Anal. Chem. 35*, 647 (1963).

24. H. A. Saroff, A. Karmen, and J. W. Healy, *J. Chromatog. 9*, 122 (1962).
25. A. Karmen, I. McCaffrey, and R. L. Bowman, *Nature 193*, 575 (1962).
26. H. C. Curtius, M. Wolfensberger, B. Steinmann, U. Redweik, and J. Siegfried, *J. Chromatog. 99*, 529 (1974).
27. M. B. Evans and J. F. Smith, *J. Chromatog. 30*, 325 (1967).
28. A. M. Filbert and M. L. Hair, *J. Chromatog. Sci. 7*, 72 (1969).
29. D. E. Durbin, *Anal. Chem. 45*, 818 (1973).
30. C. Hishta, J. P. Messerly, and R. F. Reschke, *Anal. Chem. 32*, 1730 (1960).
31. C. Hishta, J. P. Messerly, R. F. Reschke, D. H. Frederick, and W. D. Cooke, *Anal. Chem. 32*, 880 (1960).
32. D. H. Frederick, B. T. Miranda, and W. D. Cooke, *Anal. Chem. 34*, 1521 (1962).
33. S. J. Hawkes, C. P. Russell, and J. C. Giddings, *Anal. Chem. 37*, 1523 (1965).
34. R. W. Ohline and R. Jojola, *Anal. Chem. 36*, 1681 (1964).
35. I. Halasz and C. Horvath, *Anal. Chem. 36*, 2226 (1964).
36. C. Hishta and J. Bomstein, *J. Gas Chromatog. 5*, 395 (1967).
37. A. M. Filbert and M. L. Hair, *J. Gas Chromatog. 6*, 150 (1968).
38. A. M. Filbert and M. L. Hair, *J. Gas Chromatog. 6*, 218 (1968).
39. H. L. Macdonell and D. L. Eaton, *Anal. Chem. 40*, 1453 (1968).
40. Bulletin GC-2, Laboratory Products Dept., Corning Glass Works, Corning, N. Y. 14830 (May 2, 1967).
41. J. C. Giddings, *Anal. Chem. 34*, 458 (1962).
42. J. C. Giddings, *Anal. Chem. 35*, 439 (1963).
43. A. M. Filbert in *Recent Advances in Gas Chromatography* (I. M. Domsky and J. A. Perry, editors), Dekker, New York, 1971, pp. 49-98, and also Figure 2.
44. Bulletin CPG-5-11/69, Chromatography Products, Corning Glass Works, Corning, N. Y. 14830.
45. O. L. Hollis, *Anal. Chem. 38*, 309 (1966), and also Figure 2.
46. O. L. Hollis and W. V. Hayes, *J. Gas Chromatog. 4*, 235 (1966).
47. J. F. Johnson and E. M. Barrall, *J. Chromatog. 31*, 547 (1967).
48. R. M. Barrer, *Quart. Rev. 3*, 293 (1949).
49. R. M. Barrer, *Trans. Faraday Soc. 54*, 1074 (1958).
50. R. M. Barrer, *Zeolites and Clay Minerals as Sorbents and Molecular Sieves*, Academic, New York, 1978.
51. C. K. Hersh, *Molecular Sieves*, Reinhold, New York, 1961.
52. K. J. Bombaugh, *Nature 197*, 1102 (1963).
53. C. E. H. Knapman, editor, *Gas Chromatography Abstracts*, Applied Science Publishers, Essex, England.
54. J. C. Giddings, *Anal. Chem. 36*, 1170 (1964).
55. L. R. Snyder and E. R. Fett, *J. Chromatog. 18*, 461 (1965).
56. L. R. Snyder, *Principles of Adsorption Chromatography*, Dekker, New York, 1968.
57. E. Grushka (editor), *Bonded Stationary Phases in Chromatography*, Ann Arbor Sci. Publ., Ann Arbor, Mich., 1974.
58. I. Halasz and I. Sebastian, *Angew. Chem. (Int. Ed.) 8*, 453 (1969).
59. E. W. Abel, F. H. Pollard, P. C. Uden, and G. Nickless, *J. Chromatog. 22*, 23 (1966).

References

60. W. A. Aue and C. R. Hastings, *J.Chromatog.* **42**, 319 (1969).
61. C. R. Hastings, W. A. Aue, and F. N. Larsen, *J. Chromatog.* **60**, 329 (1971).
62. A. M. Filbert and D. L. Eaton, Joint Conference of the American Chemical Society and the Canadian Institute of Chemistry, Toronto, Canada (May 1970).
63. J. J. Kirkland and J. J. DeStefano, *J. Chromatog. Sci.* **8**, 309 (1970).
64. L. Butler and S. Hawkes, *J. Chromatog. Sci.* **10**, 518 (1972).
65. W. A. Aue and S. Kapila, in *Bonded Stationary Phases in Chromatography* (E. Grushka, editor), Ann Arbor Sci. Pub., Ann Arbor, Mich., 1974, pp. 13-26.
66. T. J. Nestrick and R. H. Stehl, US 4 199 330 (April 22, 1980).
67. L. L. Lamparski and T. J. Nestrick, *J. Chromatog.* **156**, 143 (1978).
68. P. W. Langvardt, T. J. Nestrick, E. A. Hermann, and W. H. Braun, *J. Chromatog.* **153**, 443 (1978).
69. C. D. Pfeiffer, T. J. Nestrick, and C. W. Kocher, *Anal. Chem.* **50**, 800 (1978).
70. L. L. Lamparski, T. J. Nestrick, and R. H. Stehl, *Anal. Chem.* **51**, 1453 (1979).
71. M. L. Langhorst and T. J. Nestrick, *Anal. Chem.* **51**, 2018 (1979).
72. T. J. Nestrick, L. L. Lamparski, and R. H. Stehl, *Anal. Chem.* **51**, 2273 (1979).
73. L. L. Lamparski, R. H. Stehl, and R. L. Johnson, *Environ. Sci. Tech.* **14**, 196 (1980).
74. P. W. Langvardt and R. G. Melcher, *Anal. Chem.* **52**, 669 (1980).
75. L. L. Lamparski, M. L. Langhorst, T. J. Nestrick, and S. Cutié, *J. Assoc. Off. Anal. Chem.* **63**, 27 (1980).

Chapter 9

DETECTORS

Dozens of types of gas chromatographic detectors have come and gone. Three have survived and become standard: The hot-wire thermal conductivity detector (TCD), the flame ionization detector (FID) and its thermionic variations, and the electron capture detector (ECD). These three are described in this chapter after the requirements that such widely used detectors must meet are set forth.

A fourth type of detector is also surely here to stay, but it is not so much a detector as a technique in its own right: The mass spectrometer. The general nature and use of the gas chromatograph-mass spectrometer are treated in considerable detail in the chapter on qualitative analysis (Chap. 13).

Gudzinowicz [1] has reviewed gas chromatographic detectors fairly comprehensively but with a resultant limitation in depth. Westlake and Gunther [2] have not only reviewed a number of selective detectors for gas chromatographic pesticide residue evaluation but also tried to get some of them to work--no easy task. In 1971, Hartmann [3] reviewed "only 23" types of gas chromatographic (GC) detectors, six of them in some detail but without a discussion of theory. Published in 1974, David's book *Gas Chromatographic Detectors* contains detailed reviews of eight detectors [4].

9.1 THE GENERAL REQUIREMENTS FOR A WIDELY USEFUL DETECTOR

To be generally useful for gas chromatography, a detector must have adequate sensitivity for most substances (unless it is selective), stability, linear dynamic range, and usable temperature range, as well as adequately small effective internal volume and short response time. It must be reasonably easy to use and rugged, capable of tolerating operator mistakes. Each of these requirements becomes more stringent as the quality of the GC separation increases. Failure to meet any one of them will prevent the detector from ever becoming very useful, that is, really widely adopted. Let us take up each requirement in more detail.

Sec. 9.1 *The General Requirements for a Widely Useful Detector*   143

A. Adequate Sensitivity

Before we discuss what is meant by adequate sensitivity, we must first distinguish concentration from mass sensitivity, and the sensitivity S from the detectability D.

A detector that responds to the presence of a substance in the carrier gas is mass sensitive. The FID, for instance, which responds to substances that burn, is mass sensitive, whereas a detector that does not necessarily respond to the presence of a substance in the carrier gas but rather to a difference or change in the concentration of that substance is concentration-sensitive. The TCD, for instance, gives no response unless the composition of the gas flowing through it changes, so it is a concentration-sensitive detector.

The sensitivity S expresses the ratio of detector response to quantity detected, whereas the detectability D expresses the smallest quantity that will cause a detector response equal to twice the noise level. For examples, we consider the TCD for S and the FID for D. The concentration-sensitive TCD responds in volts to a change in concentration. A good TCD will show an S of 50 µV/ppm, where ppm indicates one part of nitrogen in 10^6 parts of helium, *versus* pure helium. The mass-sensitive FID can detect a certain minimum combustible mass that is presented to it within a unit time: The FID D is 10^{-12} g/sec.

Despite these definitions, we shall generally use the terms *sensitive* and *sensitivity* more loosely. When we use the exact meanings just defined, we shall say so and also use the symbols. Now, back to adequate sensitivity.

What "adequate" sensitivity is does not at first seem to make very good sense, because the basic sensitivities of the three most common detector types differ mutually by about 10^7. Yet they are all widely used, so the question must be examined more closely.

"Adequate" sensitivity must mean "generally satisfactory for the task at hand." The sensitivity of the FID is extreme and generally unsurpassed for any substance that will burn, so it is therefore suitable for the most demanding separations and analyses of most organic substances. The sensitivity of the TCD is far lower but completely general, as we shall be discussing presently, and adequate for most GC tasks. (A tribute to the improving level of GC technology is the ongoing replacement of the TCD by the less wide-ranging but more sensitive and considerably more expensive FID.) The sensitivity of the ECD is even higher than that of the FID for certain substances such as pesticides, but in such trace detection, probably no detector will ever have a satisfactory sensitivity/selectivity/ease-of-use combination.

B. Adequate Stability

Adequate stability means that the detector will produce a stable and narrow baseline when operated at its highest sensitivity. Thus the stability requirement for a detector with extreme, FID-ECD-type sensitivity is far more demanding than that for the TCD. (For many years, the FID sensitivity was known to exist--here we refer to the

FID detectability D of 10^{-12} g/sec--and was so indicated on instrument control panels, but in practice it was unusable because of low stability. Turning the detector sensitivity to higher levels produced nothing usable, only a high and intolerable noise level or, under temperature programming, a spectacularly swooping baseline. Only recently has improved instrument design of at least some models stopped imposing a stability limit on sensitivity in the use of the FID and the ECD, as it had previously with the TCD.)

C. Adequate Linear Dynamic Range

Suppose that a detector produces a certain noise level when operated at highest sensitivity. Now, for a concentration-sensitive detector, a certain minimal increase in the concentration of the substance being detected will produce a signal equal to this noise level. Twice this concentration increase will produce twice the signal increase, and 3 times the concentration, 3 times the signal. (Because both the signal increase and the noise level are signal levels, the ratio of the two is a pure number. Also, the minimum noise level should refer to the detector type, not to a particular instrument, make, or model.)

Thus we define the *linear dynamic range*: For each minimal unit increase in concentration (referred to the noise level), there is a corresponding minimal unit increase in signal. The number that describes the linear dynamic range is the number of these corresponding minimal units.

(We spoke of a *concentration*-sensitive detector. If *mass* were substituted for concentration, the definition of linear dynamic range would then apply to a mass-sensitive detector. The TCD is concentration-sensitive; the FID, mass. The ECD is sometimes one, sometimes the other, depending on the manner of its use.)

A *linear* plot of the concentration on the abscissa *versus* the resulting signal on the ordinate will be a straight line passing through the origin and extending throughout the linear dynamic range. Or, expressed numerically rather than graphically (because the graphic expression would quickly become physically unwieldy, even impossible), the ratio of the concentration to the resultant signal is invariant over the linear dynamic range.

Ultimately, a further unit increase in concentration will not produce a corresponding further unit increase in signal. The concentration-signal plot, now no longer a straight line, has passed beyond the linear dynamic range. (In practice, linearity is *defined* to cease when the deviation from linearity exceeds a certain percentage, say 5% [3].)

In practice, the linear dynamic range is a function of several characteristics of the particular gas chromatograph that supplies and houses the detector. It is also a function of the sensitivity and stability of the type and model of the detector itself. Nevertheless, the number cited for the linear dynamic range of a given type of detector should express a limit set by the detector type, not the limit set by the instrument.

Sec. 9.1 The General Requirements for a Widely Useful Detector   145

The linear dynamic range indicates a usable concentration range: That concentration range over which the detector can be used for quantitative analysis without diminution of sensitivity or redetermination of the effective response factor. (The determination and use of response factors are described in Chap. 14, Quantitative Analysis.) The wider the linear dynamic range, the more widely usable the detector. The linear dynamic ranges of the modern TCD, FID, and ECD, although mutually very different, are rarely limiting to the practical quantitative analyses to which they are individually applied. (However, the studies described at the end of Chap. 14 show that if highest precision is to be obtained, response factors must be determined specifically for the sample and instrument at hand.)

D. Adequate Usable Temperature Range

The detector must retain adequate sensitivity throughout its temperature range. The hot wire rather than the thermistor has become the predominant TCD active element because its sensitivity has always been better at temperatures over approximately 100°C and has also been greatly improved for lower temperatures.

The detector must tolerate a temperature so high that any material reaching the detector from the chromatographic column will not condense in the detector. As a current rule of thumb, the detector should remain usable and undamaged at 400°C.

The detector should be housed and used so that its baseline is not affected by changes in column temperature.

Finally, the detector should not require an excessive time to attain stability at some given temperature. A detector should reasonably be expected to produce a stable baseline at full sensitivity within 1 hr from the time a change to a new and higher detector temperature is instituted.

E. Adequately Small Effective Internal Volume

Due to less than optimal design, certain parts of the gas chromatograph may have internal volumes that do not take part in actually separating the sample components, and inside diameters that are considerably larger than the generally minimal inside diameters of the connecting lines. Such larger volumes broaden peaks, limit resolution.

With detectors such as the FID or the ECD, another gas is often added to the column effluent. This is especially important with the FID, the usual detector used with the 0.01-in., minute inner diameter, open tubular columns. The effective detector volume V_e then becomes $VF/(F + F_a)$, where V is the effective volume without the added gas and F and F_a are the flow rates of the column effluent and added gas, respectively.

F. Adequately Short Response Time

The detector with its associated electronics must track the peaks precisely if the resultant signal is to be usable for quantitative analysis. In precise terms, *response time* means the time required for the detector output to reach 0.632 of its final value if the substance being detected changes abruptly and discontinuously from one concentration to a given new concentration. (For a discussion of this, see any text on differential equations, or Ref. 5.) Within three response times, the output will be within 5% of its final value; in five response times, within 1%; and in ten response times, within 0.1%.

The number n of theoretical plates is

$$n = 16 \left(\frac{t_R}{w_b}\right)^2$$

where t_R is the total (unadjusted) retention time and w_b is the time for the peak of base-width 4σ to pass through the detector. If the peak is to be tracked precisely, the response time of the detector and the associated electronics should not be greater than $w_b/20$ [5]. For instance, if after a 200-sec delay a peak showing 10^4 theoretical plates appears, then w_b equals 8 sec and the detector system must show a response time equal to or less than $w_b/20$ or 0.4 sec--a requirement easily met. On the other hand, accurately recording a peak of 50,000 theoretical plates 60 sec after injection is *not* easily done.

G. Easy to Use, and Foolproof

Because gas chromatography is so widely and generally easily applicable, many wish to use it without what they consider unnecessary care or understanding. One result of this attitude is that any component that really does require care or understanding is not likely to become widely used in gas chromatography. In mute confirmation, modern thermal conductivity detectors often protect themselves: If the detector is turned on when it should not have been, so that it might burn out, it turns itself off.

9.2 THE HOT-WIRE THERMAL CONDUCTIVITY DETECTOR (TCD)

The already well-developed technology of thermal conductivity detectors progressed still further under the impact of gas chromatography. The TCD became the first widely used GC detector, whereupon GC immediately began pressing TCD limits in sensitivity, response time, and stability. Because it has become so much more sensitive and more quickly responsive than it was, and yet correspondingly more stable, the TCD is still widely used in new gas chromatographs. It also remains the most rugged and least expensive of the most-used detectors. A good review of the thermal conductivity detector as used in GC was presented by Lawson and Miller [6].

The TCD depends on the transfer of heat from its hot sensing element through the column effluent gas, which it is monitoring, to the TCD cell wall surrounding the element. Thus the TCD tends also to respond to many operational variables to which it ideally should be insensitive. These variables include room and column temperature, carrier gas flow rate, and ambient pressure.

Much of the improvement in TCD design for GC has been directed at increasing TCD stability in order to make its full modern sensitivity more usable. Making the TCD insensitive to room and column temperature and thus more stable in these regards starts by making the TCD cell massive, i.e., of high heat capacity. Then this massive cell is thermally isolated by surrounding it with a layer of still air. Thick, thermally conductive walls usually come next. These walls are held at the desired temperature of the detector by a heater responding to a temperature sensor also located on these walls. Sometimes the air between the conductive walls and the TCD cell proper is filled with relatively large metal balls. These decrease the time required for the detector to come to a given temperature without unduly increasing the thermal coupling of the cell to the walls. The whole assembly is further thermally isolated by a thick layer of insulation.

The tubes bearing the carrier and reference gases to and from the detector are often made to take a circuitous path through the still air surrounding the TCD cell. This is done to bring the gas to be sensed to thermal equilibrium with the cell. Theoretically, it is unnecessary. In practice, it is not a bad idea because the equations that express the theory do not describe well the phenomenal thermal sensitivity of these detectors. (I have put my finger on a tube bearing gas at ambient temperature to an equilibrated TCD. The finger was placed on the tube at a point a foot away from the cell. According to theory, the gas should have returned to ambient temperature long before it reached the cell. In fact, however, the cell showed a strong response, rapidly causing the recorder pen to go off the scale.)

The hot wires of the TCD are usually welded to much coarser wire supports. These supports are based on a hermetic seal through which they pass, as shown in Fig. 9.1. The heated wire, usually a filament, is ordinarily held in a chamber of minimal volume. Normally, gas exchange at the sensing element is brought about directly by the full flow of the carrier gas rather than indirectly by diffusion to and from this flow. This minimizes response time but increases sensitivity to changes in flow and the possibility of mechanical damage to the filament. To minimize such flow changes, especially when the column temperature is being changed as in temperature programming, flow controllers can be used.

More rugged construction of the TCD filaments and filament mounts not only has allowed the hot, coiled, extended filaments to be exposed to the full force of the column effluent flowing through a chamber of minimal volume but has also greatly increased the current that can be passed through and tolerated by the filament. Because the sensitivity of the TCD is roughly proportional to the cube of the filament current [6], increasing the nominal maximum current from about 130 mamp in 1956 to about 1 amp at present has increased the available sensitivity by a factor of about 500. (Note, however: The higher the filament current, the shorter its life. The filament current should always be held to that minimum that will yield adequate sensitivity for the task at hand.)

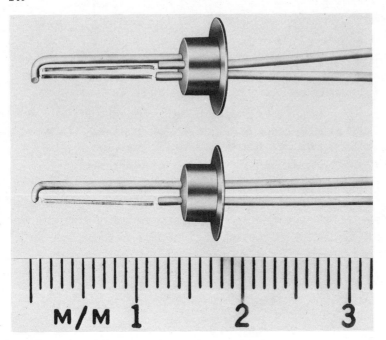

FIGURE 9.1 The sensing elements of the thermal conductivity detector are filaments supported by coarse, hermetically sealed wires. The scale shown is divided into millimeters; each filament is about one centimeter long. (Photo courtesy of the Gow-Mac Instrument Company, Bound Brook, New Jersey.)

The TCD compares the thermal conductivity of a mixture of the solute gas or vapor in the carrier gas with the thermal conductivity of the carrier gas itself. This is a comparison in time rather than in space: the reference elements need only be stable during the making of the gas chromatogram, not necessarily be bathed in the carrier gas. Before we examine the electrical nature of the TCD, let us consider the thermal conductivities of the gases and gas mixtures to which it responds, and its sensitivity.

Hydrogen and helium conduct heat far more efficiently than other gases. In consequence, dilution of either of these gases by another gas immediately and sharply decreases the effective thermal conductivity. Qualitatively, this was put very well in the abstract of a study by Cowling, Gray, and Wright [7]: "[There are] two principal effects operating in the transport of heat or momentum through gaseous mixtures. The first (and larger) effect is that molecules of one species impede transport of heat or momentum by other species. The second effect is a transfer of the transport of heat (or momentum) from one species to another."

When the one species is the very efficient helium and the other is some far less efficient solute vapor, the picture painted by Cowling, Gray, and Wright may perhaps be likened to a group of men (the helium) moving a pile of fairly heavy rocks by hand. Dilution of the helium by a small proportion of solute vapor is like substituting two

Sec. 9.2 The Hot-Wire Thermal Conductivity Detector (TCD)

or three children for two or three of the men. The group becomes less efficient. The children not only get in the way (the more important effect) but also cannot move the rocks nearly so well.

Theory [8-10] suggests that the TCD response is directly and linearly proportional to the difference between the thermal conductivity of the mixture and that of the reference, which in this case is the carrier gas. We list in Table 9.1 some representative thermal conductivities [11] and also the differences between these and the thermal conductivity of the carrier gas. The carrier gas is taken in one case as helium, in the other as nitrogen.

TABLE 9.1 Thermal Conductivities and Thermal Conductivity Differences of Some Gases

Gas	k_{TC}[a]	Δk_{TC}[a] Carrier gas	
		Helium	Nitrogen
Hydrogen	45.8	−8.9	−39.4
Helium	36.9		−30.5
Methane	8.6	28.3	−2.2
Nitrogen	6.4	30.5	
Oxygen	6.6	30.3	−0.2
Ethane	5.5	31.4	0.9
Propane	4.5	32.4	1.9
Ethanol	3.7	33.2	2.7

[a] cal/[(sec)(cm^2)(°C/cm)] × 10^{-5}, at 100°F. Source: Data from Ref. 11.

This table shows why helium and hydrogen are used so generally as carrier gases when the TCD is the detector. The TCD responses are proportional to the figures in the second and third columns. The sensitivity with helium as carrier, with respect to that with nitrogen as carrier, is higher by the ratios of the figures in the second column to those in the third. Also, the figures in the second column are quite similar to each other, so that the response with helium as carrier is seen to be pretty much the same for all solutes. However, with nitrogen as carrier, the much lower response not only changes drastically from one solute to another in magnitude but also in sign, so that the peaks do not necessarily all lie on the same side of the baseline.

With helium as carrier, a quantitative analysis that is good enough for most purposes can be obtained simply by taking the peak areas as linearly proportional to weight [12]. Dividing a given peak area by the sum of all the peak areas and multiplying by 100 gives the area percent for a given component. In this case, it also gives the weight percent for that component to within perhaps 5% of the amount involved. This is very convenient. (Care and calibration greatly improve precision.)

The early literature of gas chromatography is replete with elaborate theoretical studies of TCD response. However, this effort has just about vanished. For one thing, the attempts were not successful, as Lawson and Miller pointed out [6]. (For this reason, various impressive but really not very instructive equations will not be cited here.) For another thing, the FID brought a shift in reliance and emphasis. Obtaining a good theoretical description of the TCD is no longer so important.

I am also reluctant to cite sensitivity formulas for the TCD. One frequently cited formula gives TCD sensitivity values ranging over almost two orders of magnitude rather than one given value, as such a formula might legitimately be expected to do. This same formula does not show the sensitivity as a function of a number of important experimental variables which determine the sensitivity, for instance the filament current, with which the TCD sensitivity varies as the cube.

Among the variables affecting TCD sensitivity is one not generally appreciated or deliberately used for this purpose: The flow rate. At least two experimental studies have shown that the TCD peak area is inversely proportional to the flow rate [13, 14]. This suggests that when a TCD is the detector, the flow rate should be restricted to no more than the optimum van Deemter flow rate, especially when columns of very small diameter or open tubular columns are used. Such a minimum flow rate would maximize the response available from the TCD and in consequence minimize the sample necessary for adequate sensitivity. With the TCD, overloading--charging too much sample to the column--can be taken for granted. With other parameters such as the TCD internal volume and diameter being kept in good order, overloading becomes the most important determinant of the resolution achieved. Therefore minimizing necessary overloading would improve the achievable resolution quite sharply. Especially with the TCD, then, retention times with a given column should be controlled primarily by column temperature rather than flow rate, which should be set only slightly if at all higher than the van Deemter optimum flow rate.

The ultimate TCD sensitivity seems to be about one part per million in helium. This is a very rough estimate, taken from reported ultimate sensitivities ranging from 0.2 to perhaps 20 ppm [6]. For a solute with a molecular weight of about 20, this is a sensitivity of about 10^{-9} g solute per milliliter of carrier. (Hartmann [3], writing in 1971, agrees, quoting 2×10^{-9} g/ml.) For such a solute presented in a peak lasting 10 sec, this would be an ultimate sensitivity of about 10^{-8} g/sec.

An ultimate sensitivity of 10^{-8} g/sec for the TCD compares with 10^{-12} g/sec for the FID and perhaps 10^{-15} g/sec for optimal cases for the ECD. Thus the TCD, operated at ultimate sensitivity, has a sensitivity range overlapping for three orders of magnitude the 10^7 linear dynamic range of the FID; this range extends from 10^{-12} to 10^{-5} g/sec. It would seem that the effective applicability of the modern TCD is not being fully demonstrated or utilized in present GC practice.

Hartmann gives the linear dynamic range of the TCD as 10^5 [3].

Let us now examine some electrical and physical aspects of the TCD, starting with the concept of the voltage divider and moving from there to the Wheatstone bridge and to filament behavior.

Sec. 9.2 *The Hot-Wire Thermal Conductivity Detector (TCD)*   151

In electrical diagrams, a connector is assumed to have no resistance. Therefore a voltage at one point on a connector exists unchanged at all points on that connector. So if we connect a battery of voltage e_i (the voltage in) to a resistor, that voltage e_i exists between the terminals of the battery and also, because the connectors are ideal, equally at the ends of the two resistors.

Let the resistor be a combination of the two resistors, resistances r_1 and r_2, connected in series and to the battery:

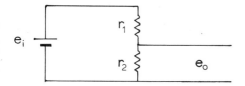

A proportional voltage appears across each resistor. The voltage appearing across resistor r_2 is $e_i[r_2/(r_1 + r_2)]$. This arrangement of a voltage source and resistors is called a voltage divider: The voltage in, e_i, is conveniently divided to produce the voltage "out," e_o.

The battery can equally well be connected to two more resistors of resistances r_3 and r_4:

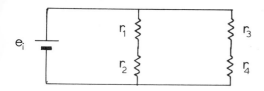

By the same argument, the voltage appearing across r_4 is $e_i[r_4/(r_3 + r_4)]$.

Let a meter be connected to the r_1r_2 and r_3r_4 junctions:

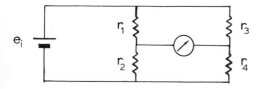

If the voltages at both ends of the meter are equal, no current will flow through the meter. But the voltages are $e_i[r_2/(r_1 + r_2)]$ and $e_i[r_4/(r_3 + r_4)]$, so no current will flow through the meter if the ratio $r_2/(r_1 + r_2)$ is equal to the ratio $r_4/(r_3 + r_4)$.

In this case, we can equate the ratios:

$$\frac{r_2}{r_1 + r_2} = \frac{r_4}{r_3 + r_4}$$

We invert both sides, divide by r_2 and r_4, and subtract unity from each side:

$$\frac{r_1 + r_2}{r_2} = \frac{r_3 + r_4}{r_4}$$

$$\frac{r_1}{r_2} + 1 = \frac{r_3}{r_4} + 1$$

$$\frac{r_1}{r_2} = \frac{r_3}{r_4}$$

The third diagram, just shown, is the basic diagram of the Wheatstone bridge, a device very widely used for the measurement of resistance and, in realted forms of the bridge, impedance. In it, only the ratios of the resistances are measured, not the resistances themselves. However, if three of the resistances are known, the fourth is readily determined by means of the last equation cited. In other words, and in the actual operation, once the meter is observed to be carrying no current, then the equality of the ratios holds: $r_1/r_2 = r_3/r_4$. The bridge is then said to be *at balance*

In the hot wire TCD used in gas chromatography, the resistors having the resistances r_1, r_2, r_3, and r_4 are all heated wire filaments that monitor gas streams, and they are nominally equal to each other. Each carries such a current--100 to 1,000 milliamperes out of the voltage source into the bridge--that it becomes very hot. The resistance of a metal increases linearly with temperature. The temperature coefficient of resistance for tungsten, for instance, is 0.0045 °C^{-1}. The usual GC filament has a resistance of about 20 Ω when cool, about 400 Ω when heated and in use. When a peak is eluted from the gas chromatographic column and passes through the detector, thus surrounding the filament with gas of poorer thermal conductivity, the filament gets hotter and displays a higher resistance. The Wheatstone bridge therefore exhibits a voltage of unbalance. The recorded course of this voltage becomes the peak on the recorder chart.

When the gas chromatograph has not been in use, it will be full of air. Of course, when the filament is hot it is chemically more reactive and will tend to oxidize to destruction relatively quickly (practically at the throw of a switch). Therefore on the one hand the TCD should not be turned on until all the air in the gas chromatograph has been flushed out with helium; on the other hand, manufacturers continuously attempt to find and use filament materials or surfaces that when hot will not react with air so quickly. One choice has been, for instance, tungsten-rhenium

Theoretically, the response of the TCD is directly and linearly proportional to the temperature difference between the hot wire and the cell wall. Experimentally, this effect has not always been observed. One way to increase the temperature difference is to make the wire hotter by increasing the current through it, but although this increases sensitivity as the cube of the current, it shortens the life of the filament, and repairs can be very inconvenient. Another way is to cool the cell wall, but this is neither very effective--it leads to an increase by a factor of only 6 when the cell is cooled to $-65°C$ [15]--nor convenient. Nor is the approach general; it would make the detector a condenser for any condensable gas, so that most GC effluents to be detected would not even reach the filaments. In practice, then, the detector should be kept 15 to 20°C above the maximum temperature the column may reach during the analysis; this is true for any detector.

In the use of the Wheatstone bridge in the TCD, the opposing resistors are as a rule used together to double the sensitivity: r_1 and r_4 are immersed in the effluent gas from one column, and r_2 and r_3 in the other. If the two columns are identical, then the same carrier gas composition and flow should emerge from each column, even though that composition and flow rate may have been changed by change in the column oven temperature. Because such changes should affect the $r_1 r_4$ and $r_2 r_3$ combinations equally, the bridge should remain in balance and the baseline remain straight.

The smooth baseline resulting from this two-column, two-opposed-detectors idea has been viewed by some as analogous to a double-beam spectrophotometer in which one beam can be used to cancel the solvent spectrum. The analogy does not hold. The two-column, two-detector idea does *not* imply that two injections should be made simultaneously, one of solvent into one column and the other of the solution into the other column, with the hope of canceling the solvent gas chromatogram. Neither the columns nor the column carrier gas flow rates precisely replicate each other, so that the solvent retention times and peak shapes do not either.

Earlier it was remarked that the TCD compares gas compositions in time rather than in space, that the temporarily unused portions of the Wheatstone bridge need only be stable, not necessarily bathed in the same carrier gas, nor at the same temperature, nor even near each other. Numerous specific applications can be and have been made of this feature. For instance, if a small concentration of hydrogen is to be determined in air, one column carrier can be helium and the other, nitrogen. The air components will be measured with good sensitivity with reference to helium; the hydrogen, with reference to nitrogen. The only requirement is the avoidance of simultaneity in peak arrival. Helium is the carrier gas in one column, nitrogen in the other. Duplicates are injected into both columns at about the same time. With separate retention times for each column, the detector for one column can always serve as the reference for the other column.

The TCD is much more amenable to this approach than the dual-flame FID. The TCD is much less expensive, responds to all gases whether they burn or not, and exists as a bridge that has just one power supply and one output. The really elaborate possible ramifications of this concept do not seem to have been used. For instance, each of the four filaments could sense a different gas stream, yet be part of the same bridge.

In closing this description of the TCD, let us finish the description of the Wheatstone bridge. The last previous diagram showed the four filament resistors arranged as two voltage dividers connected by a meter, with a battery as the power supply. The Wheatstone bridge is rarely shown this way in the literature. Usually it is presented in a diamond-shaped configuration, electrically identical to what we have just seen, but showing the resistors of resistances r_1 and r_3 joined, and also those of resistances r_2 and r_4:

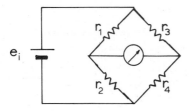

Next: The filaments are supposed to be identical in construction, mounting, and environment. In practice, such identities never hold. Thus the bridge as shown above would always show some voltage difference, that is, it would never be in balance. Therefore some means for achieving balance must be adopted. A resistor called a potentiometer is installed so that its resistance can be distributed by the operator as necessary to cause the bridge meter to show no current:

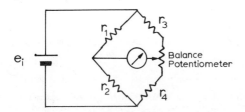

Two such balance controls, a coarse and a fine, are usually used. The resistance of the coarse balance potentiometer is made perhaps 10 times that of the fine. We also wish to be able to switch the bridge off and on, and to control and measure the bridge current. For these purposes we install a switch between the power supply and the bridge, a variable current-controlling resistor called a rheostat, and a current meter called an ammeter. A diagram of this is shown on p. 154.

Finally, in sketches, the leads that carry the signal from the bridge are customarily brought outside the diamond-shaped diagram of the bridge. In practice, they are connected across an elaborate voltage divider called the attenuator. Here the bridge is the source of voltage into the attenuator, and the operator selects the voltage to be sent to and measured by the recorder.

Sec. 9.2 The Hot-Wire Thermal Conductivity Detector (TCD)

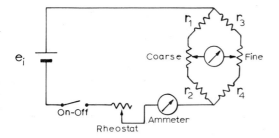

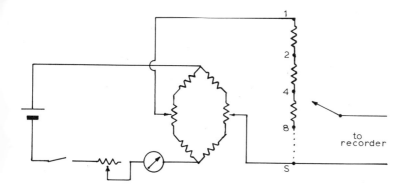

The voltage divider that is the attenuator is set up, in almost all gas chromatographs, to divide exactly the signal from the bridge in powers of two. Precision wire-wound resistors of minimum temperature coefficient of resistance are used and mounted directly on a selector switch of good quality. The resistance from "1" to "2" is made equal to the sum of the resistances of all the other resistors:

$$r_{1,2} = r_{2,4} + r_{4,8} + r_{8,16} + \cdots + r_{last}$$

This follows throughout. The resistance of a given resistor is always equal to the sum of the resistances of all the remaining resistors:

$$r_{x,2x} = \sum_{i=2x}^{n} r_{i,2i}$$

Because of this, the resistances of the few and small resistors that correspond to high attenuations become limited by the increasing relative importance of contact resistance and lead resistance. With too wide-ranging an attenuator, the maximum attenuations become imprecise because the resistances involved are too small relative to the remaining and finite contact and connection resistances. Therefore the range of a voltage-dividing attenuator for a TCD usually does not exceed 1024, i.e., 2^{10}.

9.3 THE FLAME IONIZATION DETECTOR (FID)

The flame ionization detector (FID) was introduced by McWilliam and Dewar [16], and its properties were investigated definitively by Desty, Geach, and Goldup [17] and Sternberg, Gallaway, and Jones [18]. As we shall see, the FID is just about ideal for gas chromatography. But it was the extraordinarily thorough and well-presented work of Desty et al. that made this clear. McWilliam [19] later reported a further study of the FID with great attention to the working details of his experimentation--a most useful contribution.

The list of characteristics of the FID reads like one for an ideal GC detector: Nearly complete insensitivity to fluctuations in operating variables, extreme sensitivity to the substances to be detected (10^{-12} g/sec), extremely wide linear dynamic range (10^7), vanishingly low noise level (10^{-12} amp), and vanishingly small effective internal volume (perhaps 10^{-2} µℓ, at most). The FID is easy to build and easy to use. The worst misuse does not damage it, but only necessitates cleaning the electrodes. Even this small trouble can be avoided by proper design. Let us look more closely at this paragon.

The FID consists of a jet of hydrogen burning in a sheath of air.

The FID is a mass-sensitive detector. That is, the FID responds to the presence of burnable material in the flame. A zero signal indicates an absence of such material in the flame, as would be the case for a zero signal in a concentration-sensitive detector such as the TCD.

The FID is so sensitive that the hydrogen, air, and carrier gas all must be ultrapure with respect to organic materials, and the tubes carrying these gases must be completely free of contamination.

The flow of a carrier gas emerging from the column is joined to that of the hydrogen (decreasing still further the already-minute effective internal volume of the FID). The hydrogen-carrier gas mixture then emerges from the jet into the hydrogen flame. Figure 9.2 shows the arrangement used by Desty, Goldup, and Whyman [20].

The FID has two electrodes. As a rule, the metal tube or "jet" from which the hydrogen-carrier gas mixture emerges and on which the flame is based is one electrode. This metal tube is usually 0.01-in. in inner diameter. It is also relatively thick-walled, so as to carry heat away and thus hold down background current from thermal emission [19]. (Having the jet as one electrode produces relative response factors which do not vary with changes in the design of the upper electrode. Whereas, if the jet is not one electrode, the relative response factors do reflect the design of the electrodes [21].)

Thus the jet is the lower electrode in the FID. The upper electrode is held a centimeter or so above the jet and is usually polarized positive with respect to the jet.

The shape of the upper electrode is important to the linear dynamic range: It must not be merely a vertical rod, but must have some more encompassing shape. It ma

Sec. 9.3 The Flame Ionization Detector (FID)

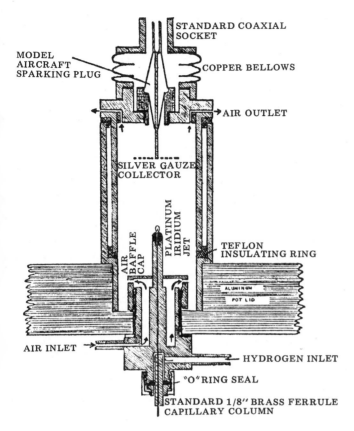

FIGURE 9.2 This diagram of an early flame ionization detector shows especially that each peak "from the column is mixed at the end of the capillary [column] with a faster stream of hydrogen emerging from around the top of the sleeve fixed at its end. Thus all dead spaces are avoided." (Reproduced by permission of the Institute of Petroleum from Desty, Goldup, and Whyman, Ref. 20.)

be a spiral of wire, or a hat-shaped piece of wire mesh, or perhaps a wide horizontal ring or cylinder above the flame and coaxial with it. If the upper electrode is merely a vertical rod, then higher concentrations of the positively charged molecular fragments produced in the flame produce a positive space charge around the rod, repelling ions further off. Such ions are then swept away and lost. If the upper electrode has an extended form, this loss is prevented [22, 23].

The voltage applied between the jet and the electrode is not critical. A plateau called the "proportional" region [24] is reached with only a few tens of volts. Perhaps 250 volts are needed to collect all the ions produced [19], if the jet is one electrode and the upper electrode is cylindrical (see Fig. 9.3). (Also note in Fig. 9.3 that no plateau is reached if the electrodes are parallel plates--plates parallel to each other and to the vertical axis of the flame [19].)

If 250 volts are adequate, well over 350 volts can be tolerated, given these electrode spacings and smooth electrode surfaces, before additional and unwanted

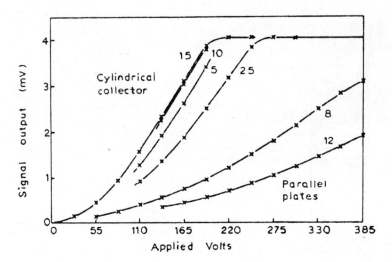

FIGURE 9.3 In a flame ionization detector with an upper cylindrical electrode that is coaxial with the flame jet as lower electrode, a plateau is attained and 250 volts are adequate for ion collection. No plateau occurs with parallel-plate electrodes. Numbers on the upper curves represent the cylinder length in millimeters; on the two lower curves, the distance in millimeters between the parallel plates. (From McWilliam, Ref. 19, in the *Journal of Chromatography*, reproduced with permission of the Elsevier Scientific Publishing Company.)

ionization is produced by electron impact leading to avalanche [22]. In other words, with jet-and-cylinder electrodes just about any voltage between 250 and 350 volts gives pretty much the same results. Therefore fluctuation in the voltage source is relatively unimportant, and the voltage source need not be designed for high stability. Usually about 300 volts are applied between the electrodes, with the jet negative

The sheath of air within which the hydrogen burns generally emerges into the flame chamber through an annulus of sintered metal centered below the jet. An excess over the stoichiometric requisite is used. The FID response rises to a plateau as the air flow rises from zero. Thus we find not critical still another operational parameter of the FID. Usually the ratio of air flow rate to hydrogen flow rate ranges between 10 and 15. The large flow of air also helps to keep the jet cool and the background current minimal [19].

The hydrogen flow rate is the most nearly critical operational parameter. The response can be maximized as a function of it (Fig. 9.4). These response maxima are also a function of the molecular structure of the substance being burned (Fig. 9.5) [19]. Thus it is impossible to optimize the flame rigorously for a sample comprising different types of molecular structures. Fortunately the FID usually does not require such rigorous optimization.

In good designs, the hydrogen flow rate-flame response peak is quite broad. For a given FID assembly and carrier flow, the response should be maximized as a function of hydrogen flow rate. For this maximizing, pick a hydrocarbon with a retention time of less than 5 min but more than 2 or 3 min. Inject replicate quantities--small

Sec. 9.3 The Flame Ionization Detector (FID)

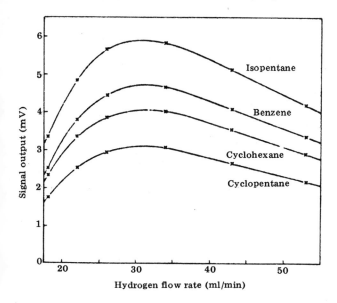

FIGURE 9.4 The response of the flame ionization detector should be maximized as a function of the hydrogen flow rate, which is the most nearly critical parameter of this detector. (From McWilliam, Ref. 19, in the *Journal of Chromatography*, reproduced with permission of the Elsevier Scientific Publishing Company.)

enough so that the peak is recorded at about 10^{-5} of full sensitivity--of the hydrocarbon as the hydrogen flow rate is varied, one flow rate per injection, to find the flow rate that produces the maximum response.

(Important note: When hydrogen and oxygen burn together, water is formed. This water will condense inside the flame chamber unless the whole chamber is kept hot enough. The condensate can and will run down, short out the electrodes and thus ruin the run, or, eventually, through corrosion, the instrument. The flame itself will heat the chamber adequately if the chamber is insulated, or an auxiliary heater can be installed around the chamber walls. But massive cold walls will lead to trouble and failure.)

The current measured across the H_2-air flame is about 10^{-14} amp [17]. Thus the amplifier used with the flame should also be characterized by an internal noise level of about 10^{-14} amp. To use a much quieter amplifier is to throw away money, because the expensive margin of decreased internal noise is not usable. To use a less expensive amplifier than one with only 10^{-14} amp of internal noise is to throw away some of the sensitivity and linear dynamic range available from the FID.

For all its sensitivity, the hydrogen flame is strikingly inefficient: About 10^{-12} g/sec of a hydrocarbon can be detected with respect to the noise level of 10^{-14} amp. For n-heptane, for instance, there are 100 g/mole, or about 6×10^{21} molecules/g. Thus the 2×10^{-12} g/sec of n-heptane that is detectable [17] is about 12×10^9 molecules/sec, and these produce 10^{-14} amp, which is 10^{-14} coulomb/sec. One coulomb is

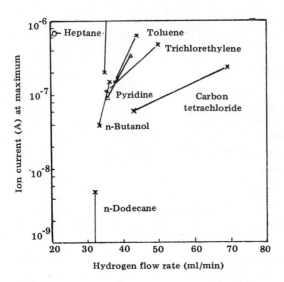

FIGURE 9.5 The flow rates of the hydrogen flow rate response maxima for the flame ionization detector vary with the molecular structure of the compound being burned, particularly if the compound is chlorinated. (From McWilliam, Ref. 19, in the *Journal of Chromatography*, reproduced with the permission of Elsevier Scientific Publishing Company.)

6×10^{18} electrons, so the number of molecules producing one singly ionized fragment--one electron--is $(12 \times 10^9)/(6 \times 10^4)$, or 10^5: One ion per 100,000 [17]. This leaves a great deal of room for improvement. We shall come back to this during the discussion of the thermionic detector, which is a variation of the FID.

Although the proportion of ions produced is small, it is effective. But how ionization could occur at all, given only the chemical energy of the hydrogen flame, was for quite a time, quite a puzzle. However, a satisfactory answer seems to have been agreed on [18, 25, 26]. In the first place, the hydrogen-air flame theoretically produces no ions at all (about 10^{-14} amp of standing current is found with the flame either burning or out [17]):

$$H + O_2 \rightleftharpoons OH + O \qquad (9.1)$$

$$O + H_2 \rightleftharpoons OH + H \qquad (9.2)$$

$$OH + H_2 \rightleftharpoons H_2O + H \qquad (9.3)$$

The free radicals from these reactions accumulate and then recombine:

$$H + H + M \longrightarrow H_2 + M \qquad (9.4)$$

$$OH + H + M \longrightarrow H_2O + M \qquad (9.5)$$

Reactions (9.4) and (9.5) generate a lot of heat, speeding up reactions (9.1), (9.2), and (9.3), "so that, emerging from the reaction zone (about 0.1 millimeter thick at atmospheric pressure and traversed in about 50 microseconds) are gases in which there are excess free radicals... These excess concentrations persist several cm. downstream, a distance corresponding with a few milliseconds in time" [25].

Sec. 9.3 The Flame Ionization Detector (FID)

Notice there are no ions produced in reactions (9.1) through (9.5).

But ions *are* produced when organic substances burn. The mechanism accounting for this was proposed by Calcote [27]:

$$CH + O \longrightarrow CHO^+ + e^- \tag{9.6}$$
$$CHO^+ + H_2O \longrightarrow H_3O^+ + CO \tag{9.7}$$
$$H_3O^+ + e^- \longrightarrow H_2O + H \tag{9.8}$$

"The heats of formation of CH, O, and CHO^+ show reaction [9.6] to be thermoneutral... Reaction [9.7]...is...exothermic to the extent of 34 kcal., which again makes it a most probable process" [25]. "...the process of chemi-ionization involves the reaction of a small hydrocarbon fragment containing one carbon atom that is *not* double-bonded to an oxygen atom. It is presumably the strong exothermicity of the formation of this first CO double bond which liberates a large part of the necessary ionization energy" [26].

Thus the production of ions by the burning of organic substances has a theoretical base. Experimentally, the response of the FID to materials that burn is not uniform. The response can be viewed as depending primarily on how many saturated carbon atoms are being oxidized. The number of these has been called the *effective carbon number* N_c; N_c for n-heptane is 7.00 [18]. For hydrocarbons generally, N_c more or less equals the number of carbons [18]:

Class	Compound	N_c
n-Paraffins	n-Pentane	4.94
	n-Octane	7.91
Branched paraffins	2,2-Dimethylbutane	6.13
	2-Methylbutane	4.69
Olefins	2-Methylpentene	5.76
Naphthenes	Cyclohexane	6.04
Aromatics	Benzene	5.95
	Cumene	9.20

However, the response drops off as halogens or oxygen appear in the molecule [18]:

Class	Compound	N_c
Halogenated	Chloroform	0.68
	Carbon tetrachloride	0.48
	3-Chloropropene	2.90
	Tetrachloroethylene	2.25
Alcohols	Ethyl alcohol	1.70
	n-Butyl alcohol	3.56
Ethers	Diethyl ether	3.00
Ketones	Acetone	2.06
Esters	Ethyl acetate	2.49

Thus, for quantitative analysis, particularly of nonhydrocarbons, response factors must be measured.

A number of gases that do not burn cannot be detected directly by the FID under normal circumstances: H_2O, the noble gases, CO_2, the oxides of nitrogen, H_2S, SO_2, COS, CO, NH_3, and HCOOH. The FID is also very nearly insensitive to CS_2. All these gases can, however, be detected indirectly as negative peaks [28]. Such peaks occur when the column-separated inert components are carried through the FID "in an ionizing carrier gas stream" [28], and thus cause a decrease in ionization.

Before we leave the topic of FID response, consider flame overloading: Because the FID sensitivity is 10^{-12} g/sec and its linear dynamic range is 10^7, its linear response without special reoptimizing extends only to 10^{-5} g/sec. For a peak lasting 10 sec, this means that less than 10^{-4} g must be injected: Approximately 0.1 µℓ is the largest acceptable sample quantity for a 10-sec-wide peak. Suppose, for example, we decide for experimental ease to operate the FID at 1/1000 of its attainable sensitivity. This leaves us with 10^4 remaining parts of the 10^7 parts of the linear dynamic range. If we do inject 0.1 µℓ, then the concentration of the solute should be no larger than 0.01%--about one part in 10^4--if the full remaining linear dynamic range is to be available for the quantitative analysis of that solute.

A. The Thermionic Detector

In 1964, Giuffrida reported her discovery of the thermionic effect and her detector [29, 30]. This detector was and is particularly important because any FID can be readily converted into one (although not so readily converted back into an ordinary FID); and because it responds selectively to phosphorus-containing compounds. Because many pesticides contain phosphorus and because it is such a simple and usable device, the thermionic detector was at once a most welcome contribution. It has gained a lasting acceptance [1, 2].

The original device (Fig. 9.6) was a normal FID with the upper electrode shaped like a ring and polarized negative (rather than positive as in the normal FID). Fusing a sodium salt onto this ring made the detector respond 600 times higher than normal to a burnable compound containing one phosphorus atom. In use, then, the electrometer output should be attenuated by perhaps 500; all the other peaks would sink into the background, but the peak for the phosphorus-containing insecticide would remain--isolated, large, and usable (Fig. 9.7). Less-striking selectivity for halogen containing compounds was also observed--a factor of about 20.

Karmen [31] increased the selectivity markedly by placing a second flame detector directly above the first but separated from it by a salt-covered wire grid. The upper flame detector then received no organic ions, except from incomplete combustion, but only an increase in ion population when phosphorus or halogen atoms were carried into the lower flame. In this arrangement, the polarity from the first jet, upwards, must be + −, + − [32]. Modern commercial versions of this detector claim selectivity ratios for phosphorus of 10^5 to 10^6.

Sec. 9.3 *The Flame Ionization Detector (FID)*

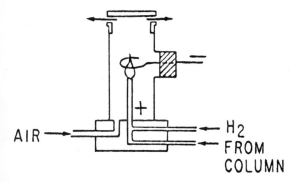

Figure 9.6 The phosphorus-selective thermionic detector "consists of a conventional flame detector except that a coat of sodium salt, such as sodium sulfate, has been fused onto the electrode." (Reprinted with permission from Giuffrida, Ref. 29. Copyright by The Association of Official Analytical Chemists, Inc.)

In the discussion of the FID, its low ionization efficiency of 10^{-5} was brought out. Aue has pointed out that the ionization efficiency for phosphorus in the thermionic detector can be made unity, namely, one ion per phosphorus atom [33]. (This implies a selectivity of 10^5 in such a single-stage detector.) He found further that the thermionic detector can be made selective toward nitrogen by a factor of 2 or 3×10^2 if the electrode position and the hydrogen flow rate are closely controlled and optimized [33].

Following Aue's lead, others (e.g., Refs. 34-37) developed further the selectivity, specificity, and mode availability of the thermionic detector. Commonly in these improvements, electrical heating is substituted for the flame in the heating of the alkali salt activator, and the alkali salt itself is selected to be relatively nonvolatile. In an approach reported by Burgett et al.[34], "the normal flame collector is replaced with a collector that has [an alkali salt activator-coated] ceramic cylinder suspended in [its] center. This cylinder...is electrically heated...to a dull red...by a step-down transformer... The flame is not ignited, but rather a low-temperature plasma is generated...for ion formation and dissociation of the organic compounds... A —240 [volt] potential is applied to the collector for species collection." Although the "exact nature of the response" mechanism for this detector "is not completely known...it is readily apparent that excellent response is achieved for a wide variety of chemical structures [including even] barbiturates and pesticides which contain vicinal carbonyl groups and no HCN bonding." Its selectivity with respect to carbon is not quite 10^5 for phosphorus, a little over 10^4 for nitrogen.

In the approach described by Kolb et al. [35, 36], the detector consists of two electrically independent parts. The activator is a glass bead that contains rubidium silicate and that has a constant potential of —180 volts with respect to the collector electrode. Primarily by changing the mutual polarities of the jet, the bead, and the collector, the detector mode can be changed easily and quickly. It can operate as an

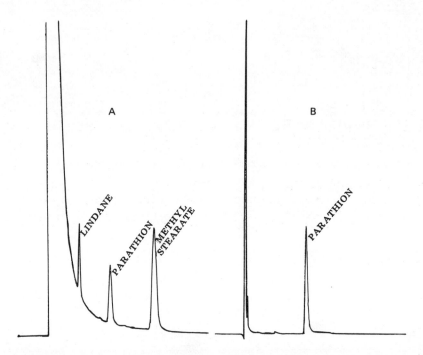

FIGURE 9.7 The phosphorus-selectivity of the thermionic detector is demonstrated by these gas chromatograms of the same sample. The electronic sensitivity for the B chromatograms (thermionic) has been decreased from that for the A chromatograms (conventional) by a factor of 300: "(A) An analysis of 1 μg lindane, 1 μg parathion and 2 μg methyl stearate by the hydrogen flame ionization detector. Electrometer range, 10^{-9} AFS. (B) The same sample analyzed by the sodium thermionic detector at 3×10^{-7} AFS." (Reprinted with permission from Guiffrida, Ref. 29. Copyright by The Association of Official Analytical Chemists.)

ordinary FID; as an FID with added sensitivity to phosphorus compounds ("by heating the bead electrically to dull-red"); as a detector sensitive only to phosphorus compounds ("by eliminating all other signals, which can be achieved simply by earthing the jet"); or, finally, "as a nitrogen detector, with additional sensitivity for phosphorus"(by changing "the flame conditions such that organic nitrogen compounds are pyrolyzed...rather than being oxidized in the normal way'). The detector has normal FID sensitivity, can show roughly 100 times that for phosphorus and 10 times that for N, and can show a selectivity with respect to carbon of 10^6 for phosphorus and 2.5×10^4 for nitrogen. Linearity in the selective modes is 10^5.

9.4 THE ELECTRON CAPTURE DETECTOR (ECD)

J. E. Lovelock [38-40] introduced the electron capture detector in 1960, helped improve its design in 1965 [41], and published a two-part review of the theory and practice of electron capture detection in 1974 [42] and 1978 [43].

Sec. 9.4 The Electron Capture Detector (ECD)

In a paper notable for valuable experimental detail on pesticide residue analysis, Burke cited the ECD as "probably the most widely used detector for residue analysis," its greatest advantage being its "extremely high sensitivity..." [44]. The comments of Fenimore and Davis in 1970 seem justified: "...electron capture detection remains the method of choice when high orders of sensitivity are required...interest in electron capture detection continues to grow..." [45]. Let us examine the ECD.

In the ECD, a cavity is lined with an ionizing source, usually radioactive. Two electrodes are present, more or less within the cavity. The column effluent--the carrier gas is preferably 90% argon-10% methane [46]--flows through the cavity and becomes ionized, producing positive ions and free electrons. Within about 0.1 sec after appearing, these electrons come to possess only thermal energy ("thermal" electrons) [47]. These thermal electrons do not recombine with either the positive ions or the carrier gas molecules, but stay free. They exist at the (cylindrical) source surface within an annular region about 2 mm deep at room temperature and atmospheric pressure, and perhaps 4 mm deep at 400°C and atmospheric pressure [41, 47].

Periodically but very briefly, a voltage is imposed between the electrodes, whereupon the electrons are electrically swept to the center of the cavity (if the positive electrode is a centered probe), collected, and measured. The time average of the electrons so collected constitutes the standing current I_b. If only the 90% argon-10% methane carrier gas is flowing through the ECD, the standing current is about 1×10^{-8} amp [41, 45], depending on the source in use, with an associated noise of about 5×10^{-12} amp [45].

When they are not being collected [46, 47], which is most of the time, the electrons in the ECD become readily attached to (i.e., captured by) any molecules that have a strong electron affinity. With this, the molecules become negatively charged molecular ions. These negative ions quickly recombine with the positive ions present. The electrons that become captured thus become unavailable for electrical collection. Detection of a component by the ECD comes from a decrease in its standing current. This decrease is the ECD signal.

Two defects of the original ECD were its parallel-plate design and its tritium source. The parallel-plate ECD was not dynamically self-cleaning. This defect was aggravated by the low temperature limit of the tritium source--about 200°C. When a component of both low volatility and high electron affinity got into such an ECD, trouble had really arrived. And, of course, the ECD is customarily applied to the determination of just such components.

These defects were largely corrected by a more temperature-resistant source and a generally improved design for the detector [45](Fig. 9.8). The new source, Ni-63, and the redesigned ECD can easily withstand 400°C. Moreover, the new design, called the concentric tube design, not only is self-cleaning but also can be easily disassembled for still more thorough cleaning. However, very few columns are used routinely at temperatures above 300°C, so the probability of contamination of the Ni-63 ECD by condensation of either sample components or stationary phase is much lower.

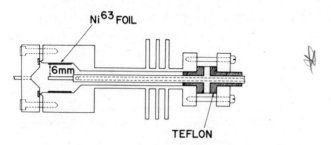

FIGURE 9.8 Electron capture detection was greatly facilitated by the Ni^{63} radioactive source that permits high (400°C) temperatures in the detecting region, and the concentric tube design that permits self-cleaning. (Reproduced by permission of Preston Publications, Inc., from Fenimore and Davis, Ref. 45, in the *Journal of Chromatographic Science*.)

The sensitivity of the ECD is extreme for compounds with high electron affinity. (We shall consider the variability of electron affinity presently.) Burke [44] in 1965 listed the ECD sensitivity for chlorinated compounds as 0.1 to 1 ng (*i.e.*, 10^{-10} to 10^{-9} g). For the improved ECD, Simmonds et al. [41] in 1965 stated that "the routine analysis of pesticides in the 10-to-100 picogram range [10^{-11} to 10^{-10} gram] would be a relatively modest objective," and illustrated this with a figure (Fig. 9.9) showing distinct and sizeable peaks for picogram quantities of lindane (A), aldrin (B), and dieldrin (C). This extreme sensitivity is a function of the way the ECD is operated--linearity is also of interest here--and of the electron affinity of the substances being detected and determined.

Earlier, the periodic but brief sampling of the electron population in the gas within the ECD was mentioned. The reason for such sampling was well stated by Lovelock and Gregory in one of the early papers [40] on the ECD:

> The most serious objection to the use of a simple DC ion chamber for electron attachment measurements is the variation of electron energy with applied potential for it is well known that the affinities of different molecules for electrons varies greatly with electron energy. The simplest method of maintaining a constant electron energy at atmospheric pressures is to conduct the reaction between free electrons and vapor molecules under zero field conditions, that is with no applied potential to the chamber.

Indeed, Wentworth, Chen, and Lovelock later ascertained that the attachment or capture of the electrons by the molecule occurs for the most part only under zero field [47]. Thus, the reasons for ECD pulse-sampling are fundamental.

Pulse sampling can be done by periodically driving one electrode (the gas inlet) negative with respect to the other electrode. This other electrode is a probe that by its presence forms a 6-mm annulus within the Ni-63 source lining the concentric-tube ECD. The probe is monitored by an electrometer through a simple smoothing circuit. Using this approach and this equipment, Simmonds et al. [41] investigated the effect of varying the pulse width, height, and period. They found that the number of

Sec. 9.4 The Electron Capture Detector (ECD)

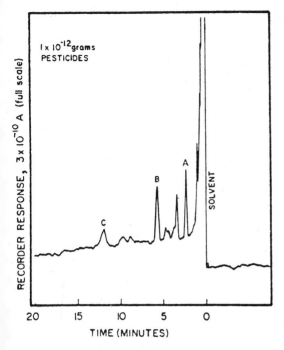

FIGURE 9.9 Publication in 1965 showed electron capture detection of picograms of lindane (A), aldrin (B), and dieldrin (C). (Reprinted with permission from Simmonds et al., Ref. 41. Copyright by the American Chemical Society.)

electrons rises quickly to a plateau as the pulse width is increased from zero to about 1/4 μsec *if* the argon-methane carrier contains not less than 10% methane (see Fig. 9.10); therefore they suggest a pulse width of 1.0 μsec, well onto the plateau (Fig. 9.11).

Similarly, the number of electrons collected rises to a plateau as the interelectrode potential rises from zero to about 8 volts; therefore they suggest a pulse amplitude of about 30 volts, again well onto the plateau, as shown in Fig. 9.12.

Each pulse reduces the electron concentration to zero. After a pulse, the electron concentration does not rebound discontinuously to its former level, but instead increases relatively slowly. As shown in Fig. 9.13, this increase continues linearly for about 1,000 μsec, finally reaching a plateau at about 1,500 μsec.

The more frequently the electron concentration is sampled and thus wiped out before the concentration increase has been completed, the fewer electrons are present to be collected. Conversely, the longer the interval between pulses, so long as the electron concentration is still increasing during that interval, the more electrons are present to be collected at the time of the pulse.

The sensitivity of the electron measurement is related, for a given source, not to the total number of collected electrons averaged over all time (*i.e.*, the standing current) but to the number collected per pulse. This number is higher, and thus the

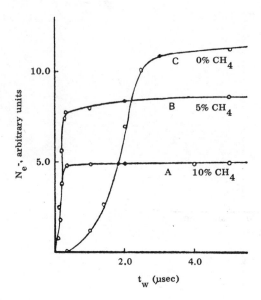

FIGURE 9.10 In the electron capture detector, the number of electrons captured rises with increasing pulse width quickly to a plateau if the argon carrier gas contains at least 10 vol% methane. (Reprinted with permission from Wentworth, Chen, and Lovelock Ref. 47. Copyright by the American Chemical Society.)

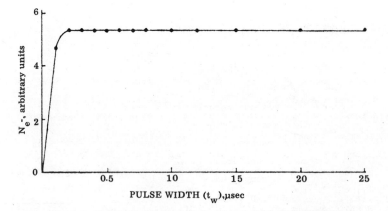

FIGURE 9.11 A pulse width of 1.0 μsec is adequate for sampling the electron population in electron capture detection. The carrier gas is argon plus 10 vol% methane. The pulse amplitude is 30 volts. (Reprinted with permission from Simmonds et al., Ref. 41. Copyright by the American Chemical Society.)

Sec. 9.4 The Electron Capture Detector (ECD)

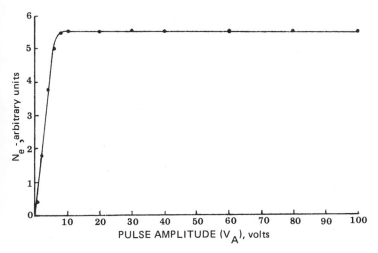

FIGURE 9.12 A pulse amplitude of 30 volts suitably samples the electron population in electron capture detection. Pulse width: one μsec. Carrier gas: argon with 10 vol% methane. (Reprinted with permission from Simmonds et al., Ref. 41. Copyright by the American Chemical Society.)

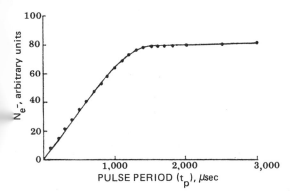

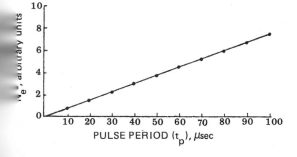

FIGURE 9.13 In pulse-sampled electron capture, the electron concentration increases linearly with the interval between pulses, up to an interval of about 1,200 μsec. (Reprinted with permission from Simmonds et al., Ref. 41. Copyright by the American Chemical Society.)

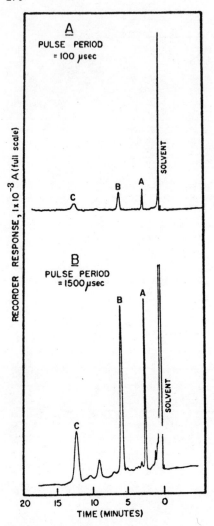

FIGURE 9.14 Up to perhaps 1,500 μsec pulse period, electron capture sensitivity is linearly proportional to the interval between pulses, as shown in these otherwise replicate determinations of 10 pg quantities of lindane (A), aldrin (B), and dieldrin (C). Detector temperature: 300°C. Pulse width: 1 μsec. Pulse amplitude: 30 volt (Reprinted with permission from Simmonds et al. Ref. 41. Copyright by the American Chemical Society.)

sensitivity is greater, as the interval between pulses increases. Figure 9.14 shows the striking increase in sensitivity gained by lengthening the pulse interval from 1 to 1,500 μsec.

It is consonant with this reasoning that the pulse interval required for a given sensitivity should increase as the source activity decreases. With 90% argon-10% methane, the pulse interval required for maximum sensitivity increases from about 50 μsec with a tritium source (tritium sources vary appreciably in radioactivity from

Sec. 9.4 The Electron Capture Detector (ECD)

source to another [48]) to 1,000 μsec with a weaker-than-tritium, 30-mCi, Ni-63 source [45](both intervals much longer than the 150 μsec more typical in the mid-70's). (1 Ci = 3.7 disintegrations/sec [10].)

For ECD pulse operation of the constant pulse width, amplitude, and interval, *i.e.*, frequency, that we have been discussing, the following expression for ECD response was derived [47] and verified [45, 48, 49]:

$$\frac{I_b - I_e}{I_b} = K^* E^*$$

where I_b is the standing current, I_e is the current observed while an electron-capturing species AB is in the ECD at concentration E^*, and K^* is the coefficient of electron capture of species AB. The expression has been found to describe both modes of electron capture: The nondissociative—$AB + e^- \rightarrow AB^-$ (for which it was derived)—and the dissociative—$AB + e^- \rightarrow A + B^-$. If the ECD signal is processed by an analog converter according to the expression just cited [48] and with adequately long pulse periods—up to 2,000 μsec—to allow the attainment of electron equilibrium and thus maximum electron concentration for the source and carrier gas at hand, then the expression is shown to describe accurately the ECD response for the capture of up to 98% of the electrons present [45]. Given the use of such a linearizing converter, and under these rather different conditions (pulse periods of 100 to 200 μsec are typical [45]), the ECD can have a "linear" dynamic range of 10^5 [45].

Wentworth [49] examined the dependence of ECD response and linear dynamic range on pulse interval and on the reactions following electron capture. For example, for pulse intervals down to 20 μsec the ECD response is described accurately by the function just cited if the ionized capturing species AB^- disappears either more quickly by electronic dissociation ($AB^- \rightarrow AB + e^-$) than by neutralization ($AB^- + \oplus \rightarrow$ neutral product) or more quickly by neutralization than by molecular dissociation ($AB^- \rightarrow A + B^-$). How nearly general this mechanism is *versus* other mechanisms that do not conform to the function was not established.

Principally because electron capture mechanisms can be strongly temperature-sensitive, ECD response can vary by a factor of as much as 5,000 over a 200°C range [50] (see Fig. 9.15). With a dissociative mode of electron capture ($AB + e^- \rightarrow A + B^-$), the capture coefficients may increase with increasing temperature (Fig. 9.16) or may remain constant [47]. On the other hand, the temperature dependence of the electron capture coefficient K^* for the nondissociative mode ($AB + e^- \rightarrow AB^-$) at higher temperatures is described by the expression

$$K^* = Z^* T^{-3/2} e^{-E/(k_B T)}$$

where Z^* is a temperature-independent factor, E is the electron affinity, k_B is the Boltzmann constant, and T is the ECD temperature [47, 51]. Thus in this case—the nondissociative mode at higher temperatures—the electron capture coefficient will

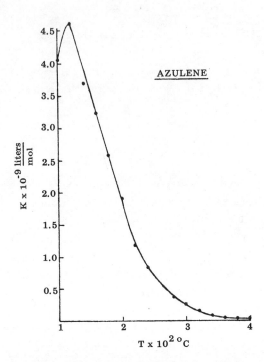

FIGURE 9.15 Electron capture sensitivity for azulene decreases by a factor of 20 a the detector temperature rises from 100 to 400°C--"behavior...characteristic...[fo: compounds which capture electrons by a nondissociative mechanism." Carrier: argon 10 vol% methane. Pulse width: 1 μsec. Pulse amplitude: 30 volts. (Reprinted w. permission from Simmonds et al., Ref. 41. Copyright by the American Chemical Soci

decrease by 20-fold from 100 to 400°C [41]. This temperature dependence thus may] strong, can vary from one molecular species to another, and can vary within a give molecular species from one temperature region to another [47, 52, 53].

Certainly two conclusions follow: First, extremely precise thermostatting of ECD is necessary for reproducibility [46, 47, 54] (±0.1°C at 100°C for 1% error, fo instance [50, 55]). (The need for such thermostatting has nothing to do with stab lizing the activity of the radioactive source, which is not affected by either tem ature or pressure.) Second, being able to vary the ECD temperature is necessary f maximizing the ECD sensitivity to a given compound. Bromobenzene, for instance, i far more detectable at 250 than at 83°C, whereas the opposite holds for benzaldehy [50].

Whatever ECD temperature is used, it should be some 25°C higher than the maxi temperature the column will reach, so as to prevent any type of condensation withi the detector. The advantage of Ni-63 over tritium in tolerating high temperatures and thus in keeping the detector clean, has been pointed out earlier. This advan is tremendously important in practical work.

Sec. 9.4 The Electron Capture Detector (ECD) 173

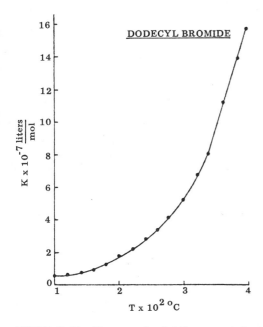

FIGURE 9.16 "Conversely [with respect to the electron capture temperature sensitivity for azulene, Fig. 9.15], the electron capture sensitivity for dodecyl bromide, which dissociates after electron attachment, shows a proportional increase." This shows a 20-fold increase as the detector temperature increases from 100 to 400°C. Experimental conditions of Fig. 9.15. (Reprinted with permission from Simmonds et al., Ref. 41. Copyright by the American Chemical Society.)

A few paragraphs back, during the discussion on pulse interval, the linear increase of electron concentration for about 1,000 μsec after a pulse was described. Thus, "the longer the interval between pulses..., the more electrons are present to be collected at the time of the pulse...and thus the sensitivity is greater." Thus varying the pulse period is a means of varying the ECD sensitivity conveniently, continuously, and linearly. Alternatively, in analytical determinations, the ECD sensitivity can be varied by varying the pulse period to indicate linearly the concentration of an electron-capturing component as it is carried through the detector. The instrumentation of this approach became commercially available in 1971 [55].

In this method [55], the interval between pulses--the pulse frequency--is electronically servoregulated. The standing current resulting from the ECD pulses is compared to a current generated within the instrument and held constant for a given setting (thus this approach is called *constant-current pulse-modulated*). Any difference between the standing and reference currents is amplified. The resulting voltage activates a voltage-to-frequency converter. The resultant frequency is applied to the ECD as pulses. In the work cited [55], the pulses were 1 μsec in width and 60 volts in amplitude.

The detection limits with this mode are similar to those observed with earlier approaches--fixed frequency pulse sampling and direct current. Its linear dynamic range is estimated to be about 4×10^4, a considerable improvement. Both pulse methods--fixed frequency and pulse-modulated--give better stabilities than the direct current method. In short, the pulse-modulated method is another way of using pulse sampling. It is based on the same physical principles as the earlier mode, and therefore shows similar characteristics.

In ECD, argon-methane is used as the carrier rather than nitrogen because electrons diffuse some 10 times faster in argon-methane [56]. In a cell such as that shown in Fig. 9.8 in which the electrons move essentially radially during collection, using nitrogen would require an inconveniently close electrode spacing. However, if the radioactive foil that is the electron source is made cylindrical and is placed downstream from a collector electrode that is both cylindrical and coaxial with the foil, then the end of the collector can be conveniently brought adequately near the foil [57]. With this design, nitrogen can be used as carrier. The electrons can be collected by 50-volt, 0.6-μsec pulses during which the electrons move upstream. Also with this design, the electrons that lead to nonlinearity (the long-range radiation from the foil) tend not to be collected. The detector shows an improved linear dynamic range, 7×10^5.

It is the extremely wide variation of electron affinity that underwrites ECD selectivity. However, opinions vary on the utility of ECD selectivity. Simmonds et al. [41] speak of the ECD as "highly selective," whereas Westlake and Gunther [2] say, somewhat glumly, "For pesticide residue analysis, the electron-capture detector is the next-to-last choice, primarily because of its nonspecificity. Data obtained by use of this detector can be considered valid only if they are confirmed by at least one other method. This confirmation may be achieved by the use of a different column, having a different elution pattern, or preferably by a different detector or detectors, or most realistically by a completely different method."

What the analyst would like is complete specificity, but electron affinity is a natural quality in which molecules differ in degree, and through a range that includes negative affinity. The entire range of difference is enormous. Zlatkis and Lovelock [58] presented this difference very clearly, and the following treatment rests heavily on that presentation.

ECD response to benzene, to aliphatic hydrocarbons, and to most simple oxygen- and nitrogen-containing derivatives of these is either zero or too low to measure. Alcohols, phenols, amines, aldehydes, and thioethers show absorption coefficients less than 0.05 (counting the chlorobenzene absorption coefficient as unity). An even lower response, too low to measure, comes from "aliphatic and alicyclic-saturated, ethenoid ethinoid, and conjugated hydrocarbons; benzene and alkyl benzenes; styrene; diphenyl aliphatic and aromatic ethers; and fatty acid esters"[58].

Electrons do attach to halogen and nitro substituents, so that ECD response is observed for molecules bearing these. ECD response increases with halogen weight: F < Cl < Br < I. Multiple substitution on the same or nearby carbon greatly enhances the effect; ECD response to CCl_4 is 10^4 that to $CHCl_3$. Also, the weaker the halogen carbon bond, the stronger the ECD response, e.g., the absorption coefficients of vinyl chloride, allyl chloride, and benzyl chloride are 0.2, 55, and 110, respectively.

Certain conjugated systems, primarily oxygenated, also produce an ECD response ranging from 500 to 1,000 (again with chlorobenzene response taken as unity). In such systems, a given group is found at each end of a conjugated bridge. For instance, in diethyl fumarate, the group is R-O-CO and the conjugated bridge is -CH=CH. In acetylene dicarboxylate the group is the same but the bridge is just one double bond, -C=C. In quinone, the group is CO= and the bridge is the quinoid ring.

If the system is conjugated but not symmetrical, or symmetrical but not oxygenated, ECD response exists but is much lower. The response to cinnamaldehyde is 310; to benzaldehyde, 48. The response to the symmetrical but not oxygenated stilbene (phenyl-CH=CH-phenyl) is 4; to the analogous azobenzene, 9.

Very importantly, the electron capture detector shows, as we have seen, generally negligible response to almost all materials but exceptionally high response to certain substances that are very important to life and tend to exist in the environment. The ECD response to the halogenated pesticides such as DDT and endrin ranges from 300 to 3,000; to carcinogens, to 10,000; and to certain fungistatic compounds, also to 10,000.

The heightened ECD response to halogenated materials [59] has caused numerous workers to make derivatives specifically for the ECD. For example, trifluoroacetylation of indole aromatic acids was successful: ECD sensitivity to these derivatives is 1,000 times greater than FID sensitivity to the corresponding methyl esters [59]. Chloroacetylation of phenols [60] has been similarly used for improving sensitivity and selectivity at the same time.

In a completely different field, the rapid ultratrace determination of metals has been developed. Determination of beryllium by complexing and solvent extraction with trifluoroacetylacetone, separation by gas chromatography, and detection by ECD constitutes a technique of sensitivity surpassing neutron activation or atomic absorption. The sensitivity extends down to 4×10^{-13} g in the case of beryllium [61].

There is no doubt that the "interest in electron capture detection" [39] will continue in both physical and analytical chemistry.

REFERENCES

1. B. J. Gudzinowicz, in *The Practice of Gas Chromatography* (L. S. Ettre and A. Zlatkis, editors), Interscience, New York, 1967, pp. 240-333.

2. W. E. Westlake and F. A. Gunther, in *Residue Reviews*, Vol. 18 (F. A. Gunther, editor), Springer-Verlag, New York, 1967, pp. 176-217.

3. C. H. Hartmann, *Anal. Chem. 43*, 113A (1971).

4. D. J. David, *Gas Chromatographic Detectors*, Wiley-Interscience, New York, 1974.

5. E. B. Pearson, *Technology of Instrumentation*, Van Nostrand, New York, 1957.

6. A. E. Lawson and J. M. Miller, *J. Gas Chromatog. 4*, 273 (1966).

7. T. R. S. Cowling, P. Gray, and P. G. Wright, *Proc. Roy. Soc.* **276A**, 69 (1963).
8. S. Dal Nogare and R. S. Juvet, *Gas-Liquid Chromatography*, Interscience, New York, 1962, p. 194.
9. L. J. Schmauch and R. A. Dinerstein, *Anal. Chem.* **32**, 343 (1960).
10. C. Bokhoven and A. Dijstra, *Nature (Lond.)* **186**, 743 (1960).
11. R. C. Weast and S. M. Shelby, *Handbook of Chemistry and Physics*, 47th ed., Chem. Rubber Pub. Co., Cleveland, Ohio, 1966.
12. D. M. Rosie and R. L. Grob, *Anal. Chem.* **29**, 1263 (1957).
13. G. Guiochon, *J. Chromatog.* **14**, 378 (1964).
14. C. C. Neto, J. T. Koffer, and J. W. DeAlencar, *J. Chromatog.* **15**, 301 (1964).
15. R. L. Hoffman and C. D. Evans, *J. Gas Chromatog.* **4**, 198 (1966).
16. I. G. McWilliam and R. A. Dewar, in *Gas Chromatography 1958* (D. H. Desty, editor), Academic, New York, 1958, pp. 142-152.
17. D. H. Desty, G. J. Geach, and A. Goldup, in *Gas Chromatography 1960* (R. P. W. Scott, editor), Butterworths, Washington, D. C., 1960, pp. 46-64.
18. J. C. Sternberg, W. S. Gallaway, and D. T. L. Jones, in *Gas Chromatography* (N. Brenner, J. E. Callen, and M. D. Weiss, editors), Academic, New York, 1962, pp. 231-269.
19. I. G. McWilliam, *J. Chromatog.* **51**, 391 (1970).
20. D. H. Desty, A. Goldup, and B. H. F. Whyman, *J. Inst. Petrol.* **45**, 287 (1959).
21. J. Novak, P. Bocek, L. Keprt, and J. Janak, *J. Chromatog.* **51**, 385 (1970).
22. I. G. McWilliam, *J. Chromatog.* **6**, 110 (1961).
23. R. A. Dewar, *J. Chromatog.* **6**, 312 (1961).
24. S. C. Brown, in *Handbook of Physics* (E. U. Condon and H. Odishaw, editors), McGraw-Hill, New York, 1958, Chapter 10.
25. T. M. Sugden, in *Ionization in High Temperature Gases* (K. Shuler and J. Fenn, editors), Academic, New York, 1963, p. 153.
26. A. van Tiggelen, in *Ionization in High Temperature Gases* (K. Shuler and J. Fenn, editors), Academic, New York, 1963, p. 180.
27. H. F. Calcote, *Combust. Flame* **1**, 385 (1957).
28. W. C. Askew, *Anal. Chem.* **44**, 633 (1972).
29. L. Giuffrida, *J. Assoc. Off. Anal. Chem.* **47**, 293 (1964).
30. L. Giuffrida, in *Recent Advances in Gas Chromatography* (I. I. Domsky and J. A. Perry, editors), Dekker, New York, 1971, pp. 125-136.
31. A. Karmen, *J. Gas Chromatog.* **3**, 336 (1965).
32. V. Svojanovsky, M. Krejci, K. Tesarik, and J. Janak, *Chromatog. Rev.* **8**, 91 (1966).
33. W. A. Aue, C. W. Gehrke, R. C. Tindle, D. Stalling, and C. D. Ruyle, *J. Gas Chromatog.* **5**, 381 (1967).
34. C. A. Burgett, D. H. Smith, and H. B. Bente, *J. Chromatog.* **134**, 57 (1977).
35. B. Kolb and J. Bischoff, *J. Chromatog. Sci.* **12**, 625 (1974).
36. B. Kolb, M. Auer, and P. Pospisil, *J. Chromatog.* **134**, 65 (1977)
37. R. A. Hoodless, M. Sargent, and R. D. Treble, *J. Chromatog.* **136**, 199 (1977).
38. J. E. Lovelock and S. R. Lipsky, *J. Am. Chem. Soc.* **82**, 431 (1960).
39. J. E. Lovelock, *Anal. Chem.* **33**, 163 (1961).
40. J. E. Lovelock and N. L. Gregory, in *Gas Chromatography* (N. Brenner, J. E. Callen, and M. D. Weiss, editors), Academic, New York, 1962, pp. 219-229.

References

41. P. G. Simmonds, D. C. Fenimore, B. C. Pettit, J. E. Lovelock, and A. Zlatkis, *Anal. Chem. 39*, 1428 (1965).
42. J. E. Lovelock, *J. Chromatog. 99*, 3 (1974).
43. J. E. Lovelock and A. J. Watson, *J. Chromatog. 158*, 123 (1978).
44. J. A. Burke, *J. Assoc. Off. Agr. Chem. 48*, 1037 (1965).
45. D. C. Fenimore and C. M. Davis, *J. Chromatog. Sci. 8*, 519 (1970).
46. W. E. Wentworth and E. Chen, *J. Phys. Chem. 71*, 1929 (1967).
47. W. E. Wentworth, E. Chen, and J. E. Lovelock, *J. Phys. Chem. 70*, 445 (1966).
48. D. C. Fenimore, A. Zlatkis, and W. E. Wentworth, *Anal. Chem. 40*, 1594 (1968).
49. W. E. Wentworth, in *Recent Advances in Gas Chromatography* (I. I. Domsky and J. A. Perry, editors), Dekker, New York, 1971, pp. 105-124.
50. W. E. Wentworth and E. Chen, *J. Gas Chromatog. 5*, 170 (1967).
51. W. E. Wentworth and R. S. Becker, *J. Amer. Chem. Soc. 84*, 4263 (1962).
52. G. Briegleb, *Angew. Chem. Int. Ed. 3*, 617 (1964).
53. G. Briegleb, *Angew. Chem. 76*, 326 (1964).
54. W. E. Wentworth, R. S. Becker, and R. Tung, *J. Phys. Chem. 71*, 1652 (1967).
55. R. J. Maggs, P. L. Joynes, A. J. Davies, and J. E. Lovelock, *Anal. Chem. 43*, 1966 (1971).
56. T. E. Bortner, G. S. Hurst, and W. G. Stone, *Rev. Sci. Instr. 28*, 103 (1957).
57. P. L. Patterson, *J. Chromatog. 134*, 25 (1977).
58. A. Zlatkis and J. E. Lovelock, *Clin. Chem. 11*, 259 (1965).
59. J. L. Brook, R. H. Biggs, P. A. St. John, and D. S. Anthony, *Anal. Biochem. 18*, 453 (1967).
60. R. J. Argauer, *Anal. Chem. 40*, 122 (1968).
61. W. D. Ross and R. E. Sievers, *Talanta 15*, 87 (1968).

Chapter 10

THE COLUMN

Within the column, the stationary phase is presented to the passing solutes. The efficiency of this presentation varies radically with column design.

Columns can vary and have varied in every conceivable parameter, for instance, length (from a few centimeters to hundreds of meters [1]), internal diameter (from the few inches or even feet of a packed preparative giant [2] down to the 34.5 μm open center of a research capillary [3]), enevelope material (glass, metals, plastics), and stationary phase support (the column wall itself, diatomaceous earths, steel spirals [4] and gauze [5], leached Tide [6]--to cite just a very few--and beads-- smooth beads [7], rough beads [8], and beads with porous layers [9]).

Two main types of columns have emerged. We limit this chapter to these two. Most gas chromatographic separations are done with the familiar and easy-to-use conventional columns that are packed with stationary phases supported on diatomaceous earths. Meanwhile the open tubular column continues to gain adherents, as it has been doing for over two decades. But why so slowly?

Why does not the *best* design for gas chromatographic columns supersede other, poorer designs? To get at the answer, consider another question: What is the gas chromatographic column supposed to offer? We quickly realize that that depends on what the column is going to be used for--general use, for instance.

"General" use implies that the sample to be injected may vary in kind--that the nature of the next sample is not really predictable. To be used with such samples, a column must be easily usable: (1) inert, so that components do not smear or vanish (2) reasonably efficient--yielding at least 250 effective plates per foot, let us say (3) obtainable with the necessary selectivity; (4) tolerant of mistakes; and (5) if necessary, expendable. Conventional packed columns meet all these requirements. Conventional packed columns are the type best suited for general use.

In another common situation, high efficiency is not required, but the column must above all be inert. Such a column is used for the determination of highly labile or adsorbable components, often present only as traces. Here, in general, the choice is a packed column in the form of a packed glass tube into which the sample is injected

However, if the mixture to be analyzed contains a great many components, as almost every natural mixture does, then a high number of theoretical plates is required or the components will not become separated. Given the requisite high number of theoretical plates, the column must then show inertness and a certain requisite selectivity. Primarily because of its high permeability and thus its characteristically low pressure drop in practice, the open tubular column, especially the glass or the fused silica open tubular column, most readily meets these requirements.

The column design that is "best" varies with the use of the column and the state of the art, that is, technical criteria. The use specifies a certain mix among the qualities of efficiency, selectivity, inertness, ease of use, and general applicability. Technical criteria for column performance, considered later in this chapter, merely complement these qualities. Use, more than technical criteria, determines the general acceptability of a given column design.

We now consider the characteristics, preparation, use, and technical evaluation of conventional packed columns and of open tubular columns.

10.1 THE CONVENTIONAL PACKED COLUMN

A. History and Characteristics

As we shall see soon, the credentials of the conventional packed column are impeccable. Yet it could hardly be simpler to make.

The stationary phase is deposited onto the support (we consider the methods of deposition in more detail, presently) and the impregnated support, almost always a free-flowing powder, is poured with gentle tapping into an empty tube until the tube will take no more. The ends of the tube are usually plugged with glass wool to hold the powder in, but no more tightly than that. The column is then ready to be installed.

Such conventional packed columns--generally 1 to 2 m long, about 4 mm in internal diameter, packed with 3 to 4 parts by weight of a diatomaceous earth supporting 1 part by weight of a stationary phase--were used:

In 1951 by James and Martin [10, 11] to demonstrate gas-liquid chromatography some 9 years after its conception by Martin [12]
In 1956 to demonstrate the van Deemter equation [13]
In 1958 to attain 1,000 theoretical plates per foot [14]
In 1960 to demonstrate support deactivation by silylation [15]
In 1961 to demonstrate solute adsorption at the gas-liquid interface [16]
In 1961 to demonstrate linear temperature programming [17] and
In 1962 to extend, by silylation, the applicability of gas chromatography to the analysis of sugars [18] and, implicitly, many normally nonvolatile substances (this was a tremendous extension).

The conventional packed column served to introduce, improve, and extend gas chromatography. It remains the primary form of the gas chromatographic column [19, 20].

Although the initial emphasis in the development of conventional packed columns was on selectivity (*i.e.*, the chemical characteristics of the stationary phase), the more lasting emphasis was on column efficiency (*i.e.*, the physical means leading to the most efficient use of the stationary phase). Therefore the following questions became of increasing though not overriding importance: How efficient can a conventional packed column become? What physical factors define the efficiency of a conventional packed column?

The comments (Chap. 7) on the terms of the van Deemter equation apply mainly to the packing of the conventional packed column. From these, we saw that that packing becomes more efficient as the support particle size, the support particle size distribution, and the support loading decrease. The lower levels to which these variables are taken, however, are set primarily by operating convenience or commercial competition rather than by theory or performance maximization.

The pressure difference Δp necessary to produce a desired flow of carrier gas varies inversely as the square of the particle diameter d_p [21, 22]. Now, using a large pressure difference across a column does not impair column performance [23], although it does make more critical the attainment of optimal gas velocity [22]. Pressures up to 500 psig have been used with good success for conventional packed columns [24] (and, indeed, pressures up to 200 psig for open tubular columns [3], which are generally thought never to require more than a very few psi). Nevertheless high inlet pressures are inconvenient. Rarely does the modern gas chromatographer elect to use a column inlet pressure much over 30 psig. As a result, the usual support particle diameter is held to not less than about 150 μm, *i.e.*, either 60-80 mesh corresponding to a 250-170 μm range in diameter, or 80-100 mesh, corresponding to a 170-149 μm range in diameter. (The mesh designation gives the number of openings per inch in a screen of standard wire diameter for that mesh. The mesh range indicates size range of particles that pass the coarser screen but not the finer [25].)

Column efficiency increases as the range of support particle sizes decreases [1]. Therefore, particle size distribution has steadily decreased. At this writing, 10-mesh cuts have become commercially available [26].

Column efficiency increases as loading decreases because of the resulting improvement of mass transfer into and out of the stationary phase. There would be no further improvement once the resistance to mass transfer within the mobile and the stationary phase became balanced, but loading decrease is generally halted short of reaching that balance. Decreasing column loading decreases the amount of sample that may be injected into a column of a given length; makes necessary, for a given retention time difference, a longer column; and increases the probability of adsorption by the increasingly bared support. Most users avoid these effects and instead choose the convenience of loadings that are somewhat higher than optimal.

Sec. 10.1 *The Conventional Packed Column*   181

Finally, we mention a 1958 paper of R. P. W. Scott [27]. Scott advocated several changes in the conventional packed column. Possibly because the numerous, interacting parts of his evidence were presented without evaluation of relative weight, not all the changes were adopted. In agreement with J. Bohemen and J. H. Purnell [14], he recommended "small particle size, narrow grades of particle size, and low quantities of liquid phase," and these came to be. He further recommended a reduction in column diameter from the 1/4-in. o.d., then prevalent, to 1/8-in. (*i.e.*, from about 4 mm to about 2 mm i.d.). Probably as a result, such narrower columns superseded the earlier, wider columns. (The true relationship between column diameter and column efficiency, if there is such a relationship, remains unsettled [28, 29].)

To summarize the discussion thus far: The inception of and the most important developments in gas chromatography depended on the conventional packed column. Most gas chromatographic separations have been and are still being done on conventional packed columns.

The modern conventional packed column is usually 2 to 3 m in length and about 2 mm in internal diameter. It contains minute, uniformly sized, acid-washed, silylated particles of diatomaceous earth. These particles support a very small proportion of their weight of stationary phase, carefully applied and well distributed. (As brought out on p. 129 in Sec. 8.1, we specifically recommend 90-100 mesh Chromosorb G-AW-DMCS holding no more than 1% by weight of stationary phase.)

B. Preparation of Packings

The methods of preparation of packings for the conventional packed column may or may not require a volatile solvent for the stationary phase.

Simple Addition

The method of simple addition does not require a volatile solvent. Simple addition involves merely pouring the neat stationary phase onto the support in the desired proportion. The mixture is then gently tumbled, as in a closed-mouth jar. After a few minutes of such tumbling, the packing becomes apparently homogeneous, looks like the original support, and is ready to pack.

A packing prepared by simple addition shows improvement in column efficiency until the stationary phase stops redistributing itself. This period of improvement may last several weeks at ambient temperatures, or can be shortened by column conditioning at higher temperatures. In any event, if the stationary phase has an adequately low surface tension and flows easily, and if the initial efficiency of the packing is not particularly critical, simple addition has much to recommend it. It is by far the simplest, easiest, and safest method for preparing a packing.

Slurry

If for whatever reason the stationary phase must first be dissolved, then the support may be impregnated either by slurry or filtration. With slurry impregnation, a known weight of the support is initially mixed with 3 to 4 times its volume of a solution that contains a known weight of the stationary phase. The solution solvent is one that is volatile, dissolves the stationary phase well, is recommended by the manufacturer of the stationary phase, and is listed with the stationary phase in vendors' catalogues.

Once this initially highly fluid slurry has been formed, most of the volatile solvent is then gently removed from it. The resulting wet cake is finally dried. (Methods of solvent removal and of drying are described presently.) In any event, with the slurry technique the whole of the stationary phase weighed out becomes deposited onto the known weight of the support. With both the simple addition and the slurry techniques, the resultant weight percent of stationary phase is unambiguous. It is not unambiguous with filtration.

Filtration

In filtration, a known but excess volume of solution containing a known but excess weight of stationary phase is slurried with a known weight of support. The support is assumed not to extract the stationary phase from the solution. The wet, nominally impregnated support is then separated from the excess solution by vacuum filtration. The filtrate volume is measured. From these measurements, the nominal weight percent of stationary phase in the packing is deduced.

To use the filtration method, first we need to know what to do. Then we need to know what has been done. Thus, before the filtration, we need to know how much stationary phase should be used for the corresponding amounts of solvent and support. However, these figures do not yield the result. Therefore, after the filtration, we still need to determine either the nominal or the actual weight percent of stationary phase in the resultant packing. We now consider these points.

Assume that we wish to make a weight W_P of packing that will contain a weight fraction x of stationary phase of weight W_L. (The weight percent of stationary phase will be 100 x.) We wish to use a weight W_S of the support. We want to know what total weight $W_{T,L}$ of stationary phase to put into the total volume V_T of the solution to be used for the impregnation.

The solution volume V_S retained on the support is some fraction n of the total solution volume V_T and some volume-per-weight multiple m of the weight W_S of the support. The retained volume V_S is determined by difference: It is measured as the total volume V_T minus the filtrate volume V_F.

But the filtrate volume cannot be precisely known. An indeterminate quantity of the volatile filtrate solvent is lost to vacuum. Also, the multiple m varies from solvent to solvent and from run to run. These experimental difficulties render unreliable any predictions from the equation we are about to derive. Nevertheless, we need *some* guide to the relative quantities of solvent, support, and stationary phase to use. Therefore, we proceed.

We have that

$$W_P = W_S + W_L \tag{10.1}$$

Sec. 10.1 The Conventional Packed Column

$$xW_P = W_L \tag{10.2}$$

$$V_S = mW_S \tag{10.3}$$

and $V_S = nV_T$ (10.4)

Because

$$W_S = W_P - W_L$$

and because by Eq. (10.2)

$$W_S = W_P - xW_P$$

$$W_S = W_P(1 - x) \tag{10.5}$$

However, also by Eq. (10.2)

$$W_P = \frac{W_L}{x}$$

so $W_S = \frac{1-x}{x} W_L$ (10.6)

We can now equate Eqs. (10.3) and (10.4):

$$nV_T = mW_S \tag{10.7}$$

From this and from Eq. (10.6), we have

$$nV_T = m \frac{1-x}{x} W_L \tag{10.8}$$

Because the stationary phase is nominally not extracted from the solution by the support, the proportion of stationary phase in the volume of solution retained on the bed is nominally the same as that in the original solution. Thus,

$$\frac{W_L}{V_S} = \frac{W_{L,T}}{V_T} \tag{10.9}$$

and $W_L = \frac{V_S}{V_T} W_{L,T}$

$$W_L = nW_{L,T} \tag{10.10}$$

From Eqs. (10.8) and (10.10), we have

$$nV_T = m \frac{1-x}{x} nW_{L,T}$$

or, $W_{L,T} = \frac{x}{1-x} \frac{1}{m} V_T$ (10.11)

Eq. (10.11) tells us what relative quantities to use.

To illustrate the use of Eq. (10.11) and the filtration technique itself, suppose we wish to make a packing that contains 3 wt% of a silicone (thus, $x = 0.03$; $1 - x = 0.97$--we take our numbers from reference [30]). We either know from experience or determine by dry run (corresponding to the operations given below) that 50 g of the support retains 100 ml of the particular solvent we use. Therefore, in this case the (solvent volume/support weight) ratio m is 100 ml/50 g, or 2.0 ml/g. So, at the direction of Eq. (10.11), for 250 ml (V_T) of solvent we weigh out the following amount $W_{L,T}$ of stationary phase:

$$W_{L,T} = \frac{0.03}{0.97} \frac{1}{2.0 \text{ ml/g}} \, 250 \text{ ml} = 3.87 \text{ g}$$

We dissolve this amount of the stationary phase in the 250 ml of solvent. To this solution (as to the solvent in the dry run), we add the support, meanwhile stirring gently the resultant slurry. After the support settles, we stir the mixture again and, while the support is suspended, pour the mixture onto a Buchner funnel mounted over a vacuum flask. We then apply vacuum, forcing the solution through the support and the Buchner frit, until dripping stops. We measure the volume V_F of filtrate in order to determine by difference the volume of solution retained on the support and thus the stationary phase loading nominally attained.

The proportion of solution volume retained on the support should also be the proportion of stationary phase retained. The stationary phase weight should therefore be

$$W_L = \frac{V_T - V_F}{V_T} W_{L,T}$$

Exactly, the weight fraction x of the stationary phase on the packing would be $W_L/(W_S + W_L)$, but the approximation W_L/W_S is quite close enough in view of the numerous sources of error. Thus we have

$$x = \frac{W_{L,T}[(V_T - V_F)/V_T]}{W_S} \tag{10.12}$$

Equations (10.11) and (10.12) should give the same result. For the numbers we have cited, we find by Eq. (10.12),

$$x = \frac{3.87[(250-150)/250]}{50} = 0.03$$

Thus the same result does come from Eq. (10.11), which directed our operations, and Eq. (10.12), which evaluated them. It is a mere paper agreement, however. The only real way to determine the actual weight percent of stationary phase in a packing made by the filtration technique is to run a Soxhlet extraction on a portion of the packing [31].

Solvent Stripping

Another way to impregnate the support is through solvent stripping. This is a variation of the slurry technique. The amount of support to be used is weighed out and put into a round-bottomed flask. Enough of the pure solvent that will be used to dissolve the stationary phase is poured over the support to wet and cover it. The weighed amount of stationary phase to be used is then transferred into the round-bottomed flask, usually as washings from the weighing container. Eventually the flask contains all the support, all the stationary phase, and a considerable excess of solvent. The flask is then put onto a rotary stripper (Fig. 10.1) for solvent removal.

In solvent removal by vacuum stripping, the solvent should not be removed too fast, particularly with some of the silicone stationary phases. Too fast a removal

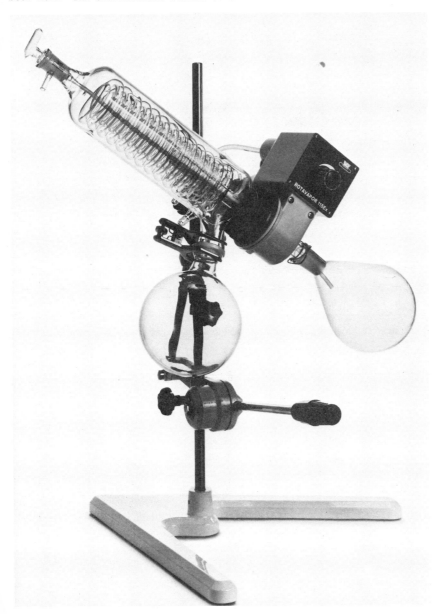

FIGURE 10.1 The rotary stripper is routinely used to remove a volatile solvent from a nonvolatile product. In gas chromatography, the nonvolatile product is the support that has become impregnated with stationary phase as a result of solvent removal by rotary stripping. (Photo courtesy of Brinkmann Instruments, Inc.)

yields an inefficient packing, perhaps caused by uneven deposition. If during stripping the flask is suspended merely in air rather than in a water bath, as is more usual, the solvent will in consequence be stripped off at a desirably low and also automatically self-regulated rate. Initially, the solvent evaporates rapidly and therefore equally rapidly carries away its heat of evaporation. In consequence, the unregulated flask temperature drops sharply at first. At the considerably lower steady-state temperature that is soon reached, the solvent evaporates much more slowly. The lower rate of evaporation is limited by the slow heat transfer into the flask through the film of moisture that quickly condenses on the outside of the flask. Eventually, as most of the solvent becomes vaporized, the flask slowly returns to room temperature.

During the later stages of the stripping, the flask is occasionally removed from the rotary stripper and hit against the heel of the hand so as to break up the cake forming within. The purpose is to achieve a uniform packing. When apparently quite dry, the packing may be further dried, spread out in a pan, and held for an hour or so in an oven.

An early but much-used method of removing the solvent from a slurry was to heat the slurry gently in a hood while gently stirring and mixing the steadily drying packing. This requires constant attention, is somewhat hazardous despite the hood, and fragments the packing particles more than any other method.

Fluidization

Fluidization drying involves a gentle boiling of the packing by a gentle upward current of gas, usually nitrogen, that lifts and tumbles the packing [30]. Fluidized particles, drying and dried, are protected by the gas from mutual abrasion. Also, unless the fines are retained, they float away during the fluidization drying.

Tests of column efficiency and packing reproducibility show that fluidization becomes the more effective as the weight-percent stationary phase decreases--in the tests, from 18 to 3% [32]. But as gas chromatography has matured, weight percents have decreased to and below 3%. Fluidization should increasingly become the drying method of choice.

Except for fluidization drying, all methods of solvent removal and drying tend to abrade the particles of the packing and thereby produce fines. (Of the diatomaceous earths, as we mentioned in the chapter on supports, W is the most friable and fines-producing, G is the least [33].) Fines increase the pressure drop across the column and also decrease the efficiency of a column by increasing the particle size range of the packing [14]. Therefore, to limit the particle size distribution and to remove fines, the dried packing should be resized, preferably just before the column is packed.

Packing the column is simple, straightforward. If the column tubing is metal, it should be washed with acetone until the washings are clean; if glass, silanized just before packing. The packing is poured with gentle lateral tapping into the column through a matching funnel. (Such convenient, outer-diameter-matching funnels with locks to hold them in place during the packing, are commercially available.) The column is tapped or vibrated until no more packing enters. (Attaching the far end of the column to a vacuum pump merely speeds the packing process somewhat--otherwise

Sec. 10.1 The Conventional Packed Column

there is little effect.) The packing is brought to within about 1 cm from the ends of the column. The unfilled ends of the column are then loosely plugged, usually with silanized glass wool pressed in only tightly enough to keep the packing from beginning to run out--the main force holding the packing in place is friction, not the plugs.

C. Practical Column Evaluation

Presently we shall consider the technical evaluation and comparison of conventional and open tubular columns. Here, however, let us consider briefly the practical evaluation of a conventional packed column we may have just packed or purchased.

A conventional packed column should produce at least 250 effective plates per foot. Scott [34] has suggested at least 550. This higher target is attainable enough. Gelpi *et al.* [35], for instance, used several glass columns 1.8 m × 4 mm i.d. holding "3% SE-30, 3% OV-17, and 3% OV-225, all on Gas Chrom Q 100/120 mesh, and 3% OV-17 on Chromosorb W HP, 80/100 mesh... Calculated efficiencies were in the range of 600-800 plates/ft."

In testing a column or an instrument for some target high efficiency, start if possible by having on hand a standard: A column of known, tested quality, with the details of the relevant tests spelled out. But if a new packed or purchased column does not perform as well as desired or expected, and no benchmark column is available, the following parameters should be checked. They are cited in order of decreasing importance or probability of occurrence as causes of an impaired column efficiency: Sample quantity, injection temperature, sample adsorption, carrier gas flow rate, and connections.

Sample Quantity

Columns are quite generally overloaded--too much sample is injected. Also, both the effect and the probability of overload increase as the weight percent of stationary phase decreases; the more efficient the column, the more susceptible it is to performance impairment by overloading. Decrease the amount of component for the peak being used for the efficiency test by two or three successive factors of 10.

Injection Temperature

The injection temperature must be high enough to "flash" the injected sample. The injection temperature required is usually taken to be at least 50°C higher than the boiling point of the highest boiling component injected. Familiarity makes this an insidious source of trouble. *Look up* the boiling points of the components to be injected, and set the injection temperature accordingly.

Sample Adsorption

Determine column efficiency without regard to column inertness, initially. Start by testing with paraffinic hydrocarbons. However, paraffins show sharply decreased solubility in moderately to highly polar stationary phases. This therefore requires injecting unusually heavy paraffins to get a suitable 10 to 20 min retention time. For both reasons, pay even more attention to injecting not too much and to using a high enough injection temperature. Later, test with the substances with which the column is to be used.

Carrier Gas Flow Rate

Flow rates used are usually higher than optimal. Try decreasing the flow rate by 5 to 10 ml/min (as opposed to determining the whole van Deemter curve). The longer the column and the finer the support, the more sensitive the column efficiency is to the particular flow rate used [22].

Connections

Sharp increases or decreases in internal diameter [36, 37] broaden peaks very effectively, decreasing the column efficiency observed. Make any changes in internal diameter, including the column connections, *gradual*.

In general, do not adopt a specific target efficiency for a column or instrument. If the observed efficiency is reasonably good--say, 500 plates per foot--be satisfied, especially if your arbitrarily chosen target is, say, 600 or 700.

If more theoretical plates are required, use a proportionately longer column [22]. If more theoretical plates are required but without a concomitant increase in retention time, use a longer column but decrease the stationary phase loading in inverse proportion. For example, replace a 6-ft 5% column with a 12-ft 2.5% column. If still better performance is required, consider direct injection into a porous-layer open tubular (PLOT) column. Despite the overloading, directly substituting a PLOT column for a conventional packed column is both feasible and rewarding. But this brings us to the consideration of open tubular columns.

10.2 THE OPEN TUBULAR COLUMN

A. History and Characteristics

The open tubular column [38] came into being in 1956 as the physical embodiment of an idea. The idea resulted from a mathematical and experimental study of conventional packed columns, wherein the experimental values for two measurable quantities differed from theory by two orders of magnitude. However, when one of the two quantities, peak width, was measured on an air peak swept from a long, empty tube, then no discrepancy with theoretical prediction was found. Thus there was a striking disagreement with theory in the case of conventional packed columns but none with empty tubes. From this, the idea of an empty but stationary phase-coated tube presented itself. M. Golay [39] reported the development in June, 1957--fairly early in the development of gas chromatography.

From the beginning, open tubular columns yielded separations unprecedented for speed and for the number of theoretical plates. Desty [3, 40-42] and Halasz [43, 44]

Sec. 10.2 The Open Tubular Column

particularly, investigated the characteristics and use of these columns. However, the applications of open tubular columns never matched their capabilities.

Thus as early as 1961 Desty and co-workers complained, "Interest in coated capillary columns has expanded rapidly over the last few years, although it is a little disappointing that their application has not kept abreast of technique development. This is surprising in view of the large increase in scope they offer to gas chromatography particularly in the examination of complex mixtures where conventional packed columns have many limitations" [3]. Although by 1974 the situation had changed enough so that Schomburg et al. [45] could say, "There is little doubt that the future of analytical gas chromatography lies in the use of capillary columns...," they still had to place the full impact of these columns in the future. What Desty had said in 1961 remained essentially true in 1974 (and at this writing, also). Why?

For a long time, over a decade, attempting to use the patented [46, 47] open tubular columns seemed to most prospective users both forbiddingly difficult and relatively unrewarding, despite Desty's marvelous chromatograms--see, for example, Fig. 4.5. The major and complex difficulties of learning to apply open tubular column technology easily, were left largely to the patent holder, Perkin-Elmer. Never licensed to make or use open tubular columns, the other major instrument companies neither minimized these difficulties nor helped solve them. Instead, they developed more and more applications and better and better instruments for packed column gas chromatography.

Despite its explosive growth during the 1950s, packed column chromatography did not lack for further fundamental increases in scope during the 1960s. In 1959, instrumentation for linear temperature programming was introduced [17]. Then, in 1962, derivatization by silylation [18] opened new and much wider areas to gas chromatography. Both techniques were easily understood, readily and widely applied, strikingly beneficial, legally unrestricted, and not *too* new. It was like the 1950s, all over again.

Not until the 1970s did more general interest awaken in open tubular columns. By then the "many limitations" [3] of conventional packed columns had combined with the much improved instrumentation that had meanwhile been developed within open tubular technology. (Also, the original patent [46] was due to expire in 1977.)

Let us now consider the nature, manufacture, and use of open tubular columns.

Compared to the average conventional packed column, an open tubular column is narrower and longer, but it has an open diameter. The two main types of open tubular columns are the wall-coated (WCOT) and the porous-layer (PLOT). The wall-coated column came first, in 1957. The porous-layer column was devised in 1963 [48] and became commercially available under patent [47] about 1965 [49-51]. Each is usually 0.01 or 0.02 in. in inner diameter compared to 0.06 to 0.18 in. for the typical conventional packed column. Also, compared to the usual 6-ft length of a conventional packed column, each tends to be much longer, ranging typically from 100 ft for a PLOT column to 300 ft and longer for the WCOT column. The usual tubing material is stainless steel, but glass received increasing emphasis at least until the advent of fused silica. The September 1975 issue of *Chromatographia* was devoted to papers presented at the First International Symposium on Glass Capillary Chromatography, Hindelang, Germany, May 4 to 7, 1975; see also *Gas Chromatography with Glass Capillary Columns* by Walter Jennings [52]. Fused silica columns appeared in 1979, and are discussed under *Recent Developments*, at the end of this chapter. Nickel columns also began to appear in 1978.

B. The Equation for Height per Theoretical Plate

The equation [53, 54] describing the height per theoretical plate for WCOT columns is analogous to the van Deemter equation for conventional packed columns.

$$H = \frac{B}{\bar{u}} + C_{LP}\bar{u} + C_{G}\bar{u} \tag{10.13}$$

$$H = \frac{2D_G}{\bar{u}} + \frac{k^3}{6(1+k)^2} \frac{r^2}{K^2 D_L} \bar{u} + \frac{1 + 6k + 11k^2}{24(1+k)^2} \frac{r^2}{D_G} \bar{u} \tag{10.14}$$

As with the van Deemter equation, the B term expresses axial diffusion in the gas phase. The C_{LP} and C_G terms express resistance to mass transfer in the liquid and gas phases, respectively. In the detailed expression, D_G and D_L represent gas and liquid diffusivities, respectively; k and K, the partition ratio and partition coefficient, respectively; r, the internal radius of the column; and $\bar{u}$, the average gas velocity.

The minimum height per theoretical plate, approximately equal to the internal diameter [55], is directly proportional to the internal radius (differentiate Eq. (10.14) with respect to $\bar{u}$ to find the minimum, or see Ref. 56). An inner diameter perhaps 1/10 of the 0.25-mm diameter in common use would yield a much more efficient column (and does: see, for example, Ref. 3). However, such an even more minute inner diameter would require sharply higher inlet pressures, even less sample, an even shorter injection time, and even less dead space in the gas chromatograph and the column connections; and be even more likely to get plugged. The trend has been toward the more-convenient, less-demanding larger diameters [57, 58].

Ettre [54] has shown that the C_{LP} term can be reexpressed very closely as

$$C_L = \frac{2}{3} \frac{k}{(1+k)^2} \frac{d_f^2}{D_L} \tag{10.15}$$

Here, d_f, the thickness of the annular coating of the stationary phase on the wall, is expressed explicitly. This reexpression shows that the ways to minimize resistance to mass transfer in the liquid phase are about the same for WCOT columns as they are for conventional packed columns. The film of stationary phase should be as thin as possible. Peaks should be either very close to the air peak (k near zero) or fairly far out on the chromatogram (k quite large), so as to minimize $k/(1+k)^2$.

Sec. 10.2 The Open Tubular Column

For such latter peaks--or substances that spend such a high proportion of time in the stationary phase that they are eluted relatively late in the chromatogram (k quite large)--the resistance to mass transfer in the gas phase becomes increasingly important, ultimately controlling column efficiency. For these peaks, "changes in d_f produce little effect. This has been observed experimentally" [59].

For WCOT columns, resistance to mass transfer in the gas phase is directly proportional to the square of the column radius, inversely proportional to the carrier gas diffusivity. Desty comments:

> It seems worthwhile to emphasize...that the only effective way of decreasing the column HETP is to reduce the column diameter as far as the limits of injection and detector sensitivity will allow. Equally the importance of the low viscosity and high diffusivity of hydrogen in achieving fast performance has not been generally recognized. The optimum gas velocity is nearly three times higher than with most other gases and providing the liquid film thickness is thin the increase in HETP is not very large [60].

This was said in 1965. In 1978, Guiochon came to the same conclusions [61]: "Small diameter capillary columns [with internal diameters less than 0.1 mm]...offer large possibilities of improving current performances...The nature of the carrier gas [suggests that] hydrogen is preferred to helium."

We return to Eq. (10.14). In the C_G term the k coefficient is similar to that for packed columns, namely $[k/(1 + k)]^2$. Here, as k becomes large, the newer expression merely approaches 11/24 quickly rather than unity, slowly. As in the analogous van Deemter term, a minimum C_G (for k near zero) occurs only for peaks near the air peak, long before most actual peaks leave the column. In other words, because in practice most peaks show high k, resistance to mass transfer in the gas phase changes relatively little from one of these peaks to another with either packed or open tubular columns. As mentioned, with these peaks the resistance to mass transfer in the gas phase increasingly controls column efficiency.

C. Manufacture

Dynamic Method

The manufacture of open tubular columns is different from that of packed columns. The wall-coated WCOT column is usually coated by gently driving through it under light gas pressure a solution of the stationary phase in a volatile solvent [62, 63]. This procedure, called the *dynamic method* [62], is not difficult, at least for metal columns: "...the coating and recoating...is a much easier and more rapid procedure than preparing packed columns..." [41]. It also allows columns to be cleaned and recoated easily.

In the dynamic method, the film thickness increases with an increase in any of the following prime variables:

The velocity with which the coating solution is driven through the open tube
The concentration of stationary phase in the coating solution
The viscosity of the solvent in the coating solution [64-68]

The volume of the coating solution ranges from 1/10 [64] or 1/5 [58] of the internal volume of the column to the whole of it (a greater solution volume just wastes time). The solution contains 10 to 20% by volume of the stationary phase, and may also contain a trace of surfactant [68, 69]. The solution is driven through the tube at 2 [64] to 6 cm/sec [58]--in any event, less than 10 cm/sec. When the solution has been completely driven through the coil of tubing, leaving a film of the stationary phase (with a trace of any surfactant) on the wall, the gas is allowed to continue flowing gently overnight, evaporating the rest of the volatile solvent without stripping the stationary phase from the wall.

Static Method

The *static method* calls for slowly heating the solution-filled open tubular column while one end of it is capped [53, 58, 70-73]. The column becomes coated as the solvent slowly evaporates from the coating solution. In the static method, the stationary phase concentration in the coating solution is only 0.5 to 2% by volume, compared to the 10 to 20% concentration used in the dynamic method. The static method is in several ways less convenient; for instance, highly precise thermostatting is required [68]. But it can be made to work reliably, and some prefer it [58].

In the finished WCOT column, the wall coating is perhaps 1 μm thick [74, 75]. In a coated tube initially 250 μm in internal diameter, this leaves over 99% of the original cross-sectional area still open.

D. PLOT Columns

Halasz and Horvath originated the porous-layer open tubular column [47, 48]. In a PLOT column, the open tube is lined by a porous layer--usually a diatomaceous earth-- that is coated with one of the conventional stationary phases. The porous layer may be about 30 μm deep in a tube 0.5 mm in internal diameter [49, 51]. Therefore the PLOT column can carry several times as much stationary phase per unit length as the WCOT column. In consequence, the PLOT column can accommodate a correspondingly large sample. Yet the center of the PLOT column is still open, so the column may be relatively long without requiring an inconveniently large inlet pressure.

The equation [76] describing the height per theoretical plate for PLOT columns is more complex than the one for WCOT columns. However, because it carries essentially the same lessons just discussed with respect to WCOT columns, we shall not display it nor discuss it further. (A few paragraphs hence, we report experimental results on tests of PLOT column efficiency as a function of stationary phase loading.)

PLOT columns can be prepared by either the static or the dynamic method. The static [48, 49] has been increasingly supplanted by the dynamic [58, 77, 78]. Nikelly [77] pioneered a

> ...relatively simple...dynamic coating procedure...[that uses] equipment...usually found in the analytical laboratory [and that] can be followed very easily... Briefly, the technique consists of passing under gas pressure a plug of coating material through coiled capillary tubing. The coating material is a suspension

Sec. 10.2 The Open Tubular Column

of the solid support in a solution of the liquid phase and volatile solvent. The gas flow is continued until all of the solvent is removed leaving a uniform porous layer of liquid phase on the inside walls of the tubing. The analytical properties and applications of the prepared columns (sample capacity, permeability, efficiency, and stability) are approximately the same as those of the commercial PLOT columns...

Nikelly grinds Chromosorb W with an agate or porcelain mortar and pestle, then sizes the product in dry acetone, retaining the fines [77]. "For coating a typical PLOT column (50 ft × 0.02-in. i.d.), about 1.5 grams of suspension are needed. This is prepared by mixing in a small glass-stoppered vial 0.3 grams of solid support and 1.2 grams of liquid phase solution having a concentration of 2 to 10%, usually in chloroform. The tubing (copper, or Type 304 stainless steel) is first flushed with a 1% solution of the stationary phase in chloroform, then coated by passage of the suspension at 10-20 cm/sec, average." (The flushing solution comes from a "filling tube," 4-in. × 1/4-in. o.d., connected just ahead of the tubing to be coated.)

The liquid phase loading or thickness of the porous layer was found to depend more on the percent stationary phase in the coating mixture--and the resultant fluidity of the suspension--than on the velocity of the plug through the tubing.

With PLOT columns as with conventional packed columns, one must decide what stationary phase to use. With the PLOT column, this amounts to deciding what β value --what ratio of gas volume to liquid phase volume--to use. (The gas volume is taken as approximately equal to the inner volume of the tube; the liquid phase volume is calculated from the amount observed to have been deposited within the column [77]. Alternatively, one may calculate β as K/k, finding the partition coefficient K from the literature [41], and measuring a relevant partition ratio k as the ratio of the retention times of a relevant hydrocarbon and air [51]; see also Chap. 2.) Ettre et al. studied what β value to use.

It turns out that the considerations and conclusions concerning stationary phase loading are generally identical for both PLOT and conventional packed columns. With PLOT columns as with conventional packed columns, using a lower loading--a higher β value [51]--increases column efficiency. For most uses of those PLOT columns that are 0.5 mm i.d., as most are, a phase ratio between 55 and 85 is satisfactory. A lower phase ratio, 20 to 30, allows the column to accept proportionately more sample, but decreases column efficiency. A higher phase ratio increases column efficiency but limits even more stringently the already minute sample volume that may be injected without causing unacceptable impairment of column efficiency.

Nikelly makes the practical point that the "phase ratio for [PLOT] columns made with silicone liquid phases is generally higher than 100 because, owing to their viscosity, solutions of such phases are usually used at lower concentrations to permit satisfactory coating rates" [77].

For the solid support in PLOT columns, Blumer [79] suggested using Silanox 101 rather than the fines retained from ground Chromosorb W [77]. This substance is a trimethylsilylated silicon dioxide "with a primary particle size of 7×10^{-7} cm," whereas the particle size of Nikelly's fines is "two to three orders of magnitude" larger [79].

Blumer dispersed 0.4 to 0.65 g Silanox 101 in 10-ml of a methylene chloride solution containing 0.2 to 0.5 g stationary phase. "The suspension is dispersed by brief (30 sec) immersion into an ultrasonic bath."

"Coating columns with Silanox suspension is remarkably uncritical" [79]. Blumer first cleans, then silylates the tubing (stainless steel, Types 304 and 316). Then, like Nikelly, Blumer fills the tubing with the stationary phase solution, then drains it:

> Five- to ten-column volumes of suspension are then added to the filling tube, and the solvent and excess suspension are flushed through the column at the desired ...10-30 cm/second...flow rate... Afterward the gas flow is maintained at 20-50 ml/min, until the odor of the solvent is no longer noticeable.
>
> The ease of drying and conditioning of Silanox coated columns offers a spectacular improvement over conventional PLOT columns... Drying...requires only a few minutes... Conditioning can proceed at 4-6°C/min from 80°C to the upper operating limit of the column [79],

that limit set by the stationary phase.

E. Carrier Gas Flow

Carrier gas volume flow rates for open tubular columns are far lower than those for conventional packed columns. The pressure drops across the usual 0.25- and 0.50-mm i.d. columns are also comparatively very low--so low that flow controllers don't work. The flow rates are determined by pressure controllers operating at the low end of their range, that is, operating with less than usual precision. These uncontrolled flow rates therefore also reflect many operating variables that they should not reflect--temperature programming, for instance. Retention times are in consequence far less reproducible than normal, and require special measures if they are to be used for component identification (see Refs. 80 and 81, for instance).

The optimal volume flow rate for a WCOT column of 0.25 mm i.d. may be 0.5 to 1.0 ml/min [38]; for a PLOT column of 0.50 mm i.d., 3 to 5 ml/min [49].

The flow velocity for a conventional packed column is about 10 cm/sec [14]; for a WCOT column, about 10 cm/sec with nitrogen and 20 cm/sec with helium [40]; and for a PLOT column with a typical β value of 67, about 20 cm/sec with helium [82].

F. Sample Injection

Open tubular columns require specially minute samples, specially injected to achieve plug introduction and an undistorted composition. Desty, Goldup, and Whyman [40] state:

> The importance of introducing a sharp plug of small enough size has been stressed by many authors in connection with conventional packed columns. Keulemans [83] has shown that the maximum sample size above which peak broadening occurs is directly proportional to the product of the plate capacity and the square root of the total number of plates (n). In addition, this small sample, as discussed by

van Deemter [84], must be introduced on to a proportion of the total plates which falls rapidly as the column efficiency increases (according to the reciprocal of the square root of n). It will be appreciated, therefore, that with capillary columns, where plate capacities are of the order of a thousand times lower than packed columns and efficiencies ten or more times higher, the introduction of extremely small samples very sharply becomes of paramount importance.

Overloading a column means injecting a sample size so much larger than the maximum allowable that the height per theoretical plate increases more than 10%. Desty [41] found Keulemans's suggested maximum allowable sample size for packed columns too great for a WCOT column by a factor of 10. He found that a 40-ft column having a 0.5-μm squalane coating could not at room temperature receive more than 0.2 μg (about 0.0003 μℓ) of n-heptane without overloading. But even the smallest microsyringe must deliver perhaps 1,000 times this amount if reasonable reproducibility in sample size is to be achieved. Although PLOT columns can separate components of directly injected samples up to perhaps 0.3 μℓ in size [82], the columns are then grossly overloaded (Fig. 10.2). Thus, proper sample injection into either a WCOT or a PLOT column requires special sample introduction equipment.

The instrumentation used to reduce sample size for open tubular gas chromatography is known as a splitter. The splitter should allow a 50 to 1,000 reduction in sample size, not vitiate the plug characteristics of good gas chromatographic plug injection, be reasonably convenient to use, and not alter the composition of the sample [44, 85, 86].

A splitter design (Fig. 10.3) that has proved satisfactory allows for the following: A heated block into which the sample can be injected and instantaneously vaporized without thermal decomposition; a large flow of carrier gas through the block during sample injection; a separation of the injection point from the column inlet by a long, narrow, sharply convoluted passage within the heated block, the passage design [44] assuring homogeneity of the sample vapor mixture at the time a small part of it enters the column; and a sharp, representative, reproducible sampling of the vapors from the injected material. The reduction in sample size is accomplished by controlling the ratio of flows through two concentric tubes. The vapor band from the sample is swept through the larger tube and sampled by the open end of the smaller tube. The smaller tube leads to (or is) the open tubular column; the larger leads to a vent. The flow rate ratio, and thus the split ratio for reduction of sample size, is controlled by the resistances to flow of the open tubular column and of the vent. Interchangeable needles of small internal diameter control the resistance to flow at the vent.

This arrangement successfully splits the sample without distorting the composition by fractionation. Thus quantitative analysis is, without qualification, not only as feasible with open tubular as with packed columns but also potentially faster because the peaks are narrower [44, 85, 86].

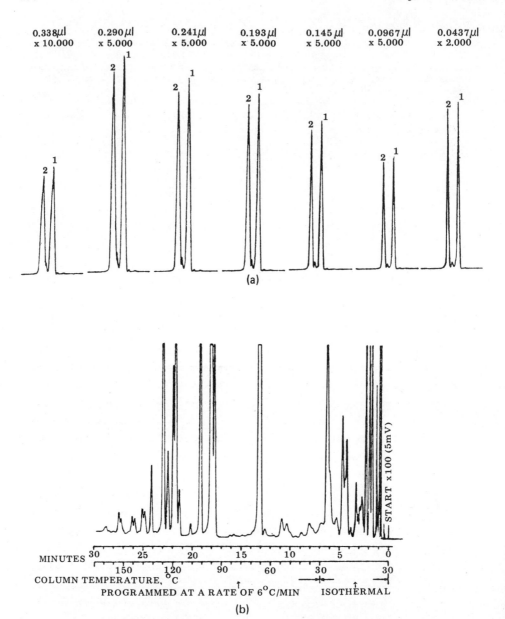

FIGURE 10.2 PLOT columns should be used with sample splitters, but can separate directly injected mixtures, despite the resultant overloading. (a) Chromatograms of a sample containing (1) hexene-1 and (2) n-hexane and an impurity, obtained with various sample sizes. Column: 150 ft. × 0.020 in. i.d. support-coated open tubular, prepared with squalane liquid phase. Column temperature: 100°C. Carrier gas (He) flow rate at column outlet: 2.76 ml/min. The numbers above the peaks refer to the total sample volume entering the column and the recorder attentuation used. (b) Analysis of a gasoline fraction. Column: 50 ft. × 0.020 in. i.d. support-coated open tubular. Liquid phase: DC-550 phenylsilicone oil. Carrier gas flow (He) rate at column outlet: 20 ml/min. Sample volume: 0.2 µl; 5 mv recorder. (Reproduced from Ettre, Purcell, and Billeb, in the Journal of Chromatography, Ref. 82, with permission of the Elsevier Scientific Publishing Company.)

Sec. 10.2 The Open Tubular Column   197

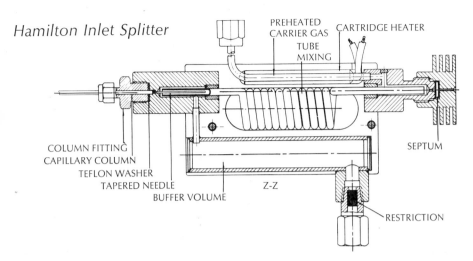

FIGURE 10.3   In a splitter, the vaporized sample is homogenized and then "split" by the ratio of column flow to vent flow. Any connections to the column must be quite free of dead volumes. (Drawing courtesy of the Hamilton Company, Reno, Nevada.)

G. Some Exemplary Chromatograms

Now we come to the results obtainable with open tubular columns. For the best, one need look no farther than the still astonishing work of Desty and his colleagues produced over two decades ago with equipment of their own design and manufacture [3]. Figure 4.5 shows the start of a gas chromatogram exhibiting 300,000 effective plates and made with a column capable of almost a million. Such a column still constitutes a standard for achieving a high number of effective plates. Figure 10.4 depicts a chromatogram completed less than 2 sec after sample injection, but showing a separation of 15 components. This still constitutes a standard for achieving a rapid separation.

If the separations obtained with commercial open tubular columns used in commercial equipment do not match those of Desty et al., nevertheless they are a tremendous improvement over the separations that have been obtained from packed columns (see the improvement, for instance, in Figure 4.6). Such separations with commercial equipment can be very good indeed (Fig. 10.5)[87]. A striking complementary use of computer technology and open tubular gas chromatography is shown in the computer-replotted comparative gas chromatograms made from the extracts of 17 varieties of black pepper (Fig. 10.6)[88]. That work was reported in 1971; the concept was used again in 1978 by Suffet and Glaser [89] for the profiling of part-per-billion trace organic compounds in water.

The design of a gas chromatograph must not limit separations by allowing remixing between the end of the column and the detector. The only design that effectively counters remixing of the column effluent calls for an added gas to sweep as an annulus in the direction of carrier gas flow, over the exit of the column. The added gas

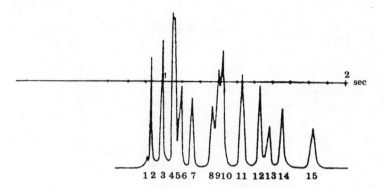

FIGURE 10.4 Fifteen components, separated in less than 2 sec. "Column length: 120 cm. Column diameter: 34.5 μ. Stationary phase: squalane. Carrier gas: hydrogen. Inlet pressure: 200 psig. Key: 1. 2-Methylbutane; 2. n-pentane; 3. 2,2-dimethylbutane; 4. 2,3-dimethylbutane; 5. 2-methylpentane; 6. 3-methylpentane; 7. n-hexane; 8. 2,2-dimethylpentane; 9. 2,4-dimethylpentane; 10. 2,2,3-trimethylpentane; 11. 3,3-dimethylpentane; 12. 2-methylhexane, 2,3-dimethylpentane; 13. 3-methylhexane; 14. 3-ethylpentane; 15. n-heptane." (From Desty, Goldup, and Swanton, Ref. 3. Reprinted with permission of the authors and Academic Press.)

should sweep the column effluent directly into the flame in a flame ionization detector. In any event, the added gas and the column effluent should move into connecting tubing no greater in internal diameter than that of the column.

H. Special Experimental Requirements

The separations achievable by open tubular columns can be vitiated either physically or chemically. Physically, the open tubular column must be accommodated within a gas chromatograph of correct design. Otherwise, extracolumn effects vitiate or ruin the separation of which the column, with its low volume flow rates and sharp peaks, is capable. Chemically, the column itself must merely separate the components of the sample, not adsorb or decompose them. This requires that in the separation of adsorbable or labile components, the open tubular column be made of glass, not metal. We shall consider the physical aspects first, by reviewing a study of extracolumn effects. Then we shall consider the chemical aspects, by reviewing the technology of glass open tubular columns.

Physical: Extracolumn Effects

The column in which gas chromatographic separations are achieved is functionally related and physically connected to the other parts of a gas chromatograph. Each part or cross section of the gas chromatograph introduces its "sample"--its peak of a certain shape--to the next part or cross section; each contributes to peak spreading in a way that should be minimized; each does affect and can limit the performance of the total gas chromatograph.

Sternberg [37] reviewed and studied the extracolumn causes of peak spreading in fixed-flow, isothermal gas chromatography: The "contributions of entering sample bar

Sec. 10.2 The Open Tubular Column

width and shape, connecting tubing [both uniform and changing, in diameter], finite detector volume, and detector and amplifier time constants..." This section reports primarily the language and conclusions of this study. Those conclusions apply to all columns, but it will soon be seen that they bear most heavily on open tubular columns.

Ideal sample introduction is plug injection. (As pointed out in Chap. 3, Sample Introduction, the ideal plug would be a spike: No mixing with the carrier gas, no width, infinitesimal size.) The column changes the incoming plug to an outgoing Gaussian band: The "ideal chromatographic column acts as a Gaussian operator"[37].

Syringe injection of a liquid gives nearly plug injection if the sample introduction port is hot enough and small enough to flash the sample. However, if the sample introduction port is not hot enough, exponential introduction will occur. In this, the sample concentration C_t at any time t following sample injection decreases exponentially as $\exp(-t/\tau)$ from an initial concentration C_0, in accordance with a response time τ characteristic of the vaporization rate:

$$C_t = C_0 \exp(-t/\tau) \tag{10.16}$$

As just mentioned, each part of a gas chromatograph, including the injection syringe, injects its "sample" into the next part, downstream. Any such next part receiving the sample can be considered as a mixing chamber or as a diffusion chamber.

In a mixing chamber, the contents of the chamber are uniform at any given instant. As fresh gas moving at flow rate F enters a mixing chamber of volume V that already contains some sample, the concentration of the sample leaving the chamber falls exponentially in accordance with Eq. (10.16) and at a rate characterized by time constant τ_M, where

$$\tau_M = \frac{V}{F} \tag{10.17}$$

In a diffusion chamber, the contents of the chamber change only by diffusion in and out. The concentration of sample remaining at the entrance to a diffusion chamber after a sample band has passed that entrance also falls exponentially in accordance with Eq. (10.16), and at a rate characterized by a time constant τ_D, here set by a characteristic distance ℓ and the diffusivity D of the contents:

$$\tau_D = \frac{\ell^2}{2D} \tag{10.18}$$

The plate height of a gas chromatographic column is given as the total spreading σ^2 per unit length L of the column:

$$H = \frac{\sigma^2}{L} \tag{10.19}$$

The time-base contribution of a mixing chamber or of a diffusion chamber to the plate height H of the total gas chromatograph is equal to the square of the characteristic time constant τ for that type of chamber:

$$\sigma^2 = \tau^2 \tag{10.20}$$

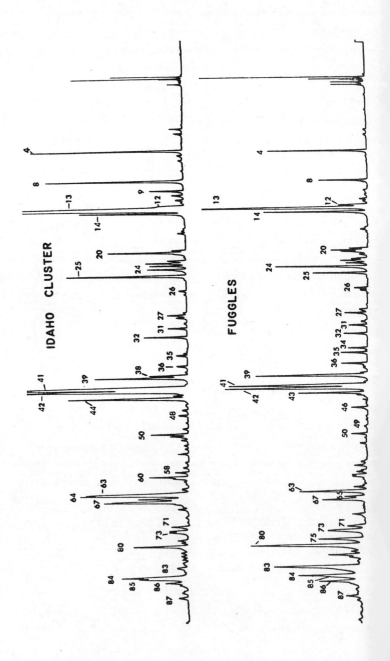

Sec. 10.3 The Open Tubular Column

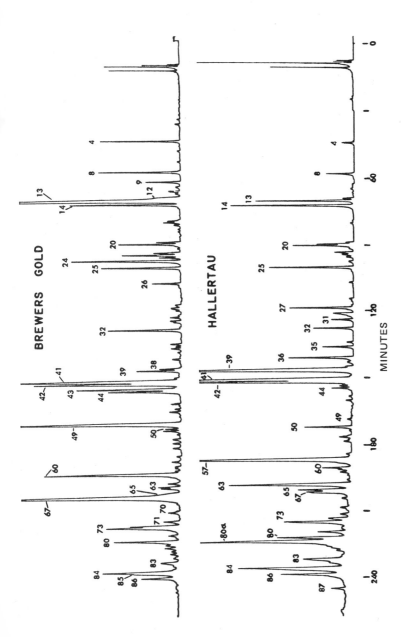

FIGURE 10.5 Excellent separations of these complex, similar mixtures were made with commercial equipment. Note, however, that peaks are identifiable only because the patterns are comparable, not because the retention times are precisely reproducible (try connecting matching peaks with a straight edge). (Reprinted from Buttery, Black, Guadagni, and Kealy, Ref. 87, with permission of the authors and the American Society of Brewing Chemists.)

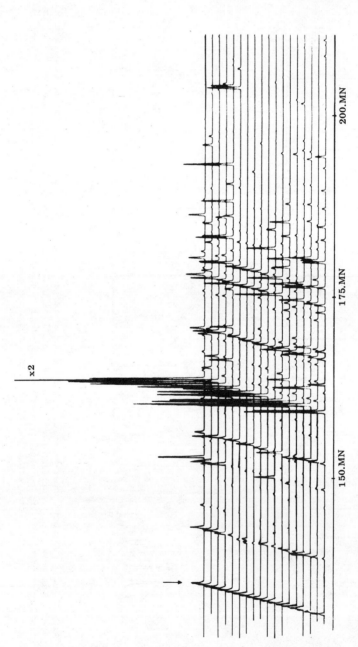

FIGURE 10.6 The extracts of 17 varieties of black pepper were gas chromatographed with open tubular columns. These are segments of the gas chromatograms. For better comparison, the segments were replotted by computer. The segments start with the arrow-indicated internal standard, n-tridecane "flag," at the left. The components are the sesquiterpenes and oxygenated volatiles. (Reproduced with permission of Preston Publications, Inc., from Richards, Russell, and Jennings, Ref. 88, in the Journal of Chromatographic Science.)

Sec. 10.2 The Open Tubular Column

The effect of a diffusion or of a mixing chamber--to put an exponential tail on a nominally Gaussian peak--is the same wherever it occurs. Such chambers can occur anywhere in a gas chromatograph: In the sample introduction system, in the connections to or from the column, or in the detector. Only poor design allows the effect of such chambers to become appreciable.

Symmetrical peak spreading that is negligible occurs during ordinary laminar flow through the nominal lengths of 0.5-mm i.d. connecting tubing found in modern gas chromatographs.

Very effective spreading, albeit symmetrical, occurs from a sharp increase in inner diameter. Such spreading is proportional to the fourth power of the ratio of the inner diameters of the two tubes. The smaller stream tends to retain its spatial integrity as it moves into the larger tube and then to equilibrate laterally. Spreading that is similarly highly effective occurs in a detector that has a signficant internal volume that is not effectively reduced by a purge gas.

Fortunately for gas chromatography, the flame ionization detector has an effective internal volume that can be made essentially negligible. As Desty and coworkers taught [3], the effluent of a column, particularly of an open tubular column, should be swept into the flame by an annulus of hydrogen.

If the thermal conductivity detector has a larger effective internal volume than the flame ionization detector, this is offset (in practice if not in theory) by the generally more easily met technical requirements put to it--peaks from relatively large samples and from conventional packed columns that deliver relatively few theoretical plates per second.

The effective internal volume V_{eff} of a detector, divided by the flow rate F, determines an effective time interval τ that is not a function of the electrical response characteristics of the detector. The plate height contribution σ_T^2 from the effective internal volume of the detector is, again according to Sternberg [37], $\tau^2/12$:

$$\sigma_T^2 = \frac{\tau^2}{12} = \frac{(V_{eff}/F)^2}{12} \tag{10.21}$$

The effective electrical response time of the detector-amplifier combination, according to Schmauch [90], should not be greater than 1/20 of the duration of a given peak at its base. Sternberg [37] finds that for 99% fidelity in peak representation, the ratio of the peak duration at its half-width to the detector-amplifier effective time constant should be about 16; for 95% fidelity, 11.5.

For the evaluation of the equations which are the conclusions of this extensive work (63 pages, 224 equations), Sternberg postulates three peaks. The most important of these is a peak being eluted from an open tubular column 1 min after sample introduction. The column is taken to be 100 ft long by 0.01 in i.d. Such a peak would be not only possible with modern commercially available open tubular columns, but also definitive. This peak would place demands on the equipment more stringent than those from later peaks from the same column or from broader peaks from other columns. These demands would have to be met if the peak were not to be broadened by the extracolumn equipment.

The column in question is considered to exhibit a plate height for this peak of 0.06 cm, or about 5 times the internal radius: A typical value for retained components. This implies that the peak is characterized by 50,000 theoretical plates or, more to the point, that the column is delivering 50,000 theoretical plates per minute (833 theoretical plates per second) to the column exit--as opposed to the detector, or then to the ultimate record, the gas chromatogram.

Certain crucial extracolumn parameters, each taken as fully limiting itself, although in fact the spreading effects of these parameters are additive, follow:

The plug sample volume for that peak could not be larger than 0.058 µℓ of liquid, instantaneously flashed, or larger than 11.5 µℓ of gas.

The connecting tubing could not exceed 0.94 mm (0.04 in) i.d.

Any internal volume having the effect of a mixing chamber could not be larger than 3.25 µℓ.

The time constant of the detector-amplifier-recorder system could not exceed 0.13 sec.

Finally, Sternberg considered the work of Desty et al. [3]. As mentional earlier this work was performed with equipment of their own design and fabrication. This equipment showed on one chromatogram a delivery rate of 2,500 effective plates per second. This required that any mixing chamber volume be restricted to 0.013 µℓ or less, that both the plug sample volume and the detector sensing volume be each 0.045 µℓ or less, and that the detector-amplifier-recorder time constant be 0.0097 sec or less. It is improbable that this many-sided experimental achievement will be surpassed.

10.3 THE GLASS OPEN TUBULAR COLUMN

Glass open tubular columns are absolute prerequisites for effective high-resolution research into the composition of natural mixtures [58, 68, 91-96]. (To glass we now add fused silica, on which columns we comment under Recent Developments, at the end of the chapter.) With metal columns, too many components of such mixtures are adsorbed or decompose.

Most simple mixtures of labile components are still separated by glass conventional packed columns only 2 m long--witness the columns used in the separations reported in the chapter on derivatization (Chap. 12). But separations of the more complex mixtures, including mixtures of biological origin, require glass open tubular columns.

D. H. Desty and his colleagues developed the first machine for drawing glass capillaries for gas chromatography [97, 98]. Adaptations of this device are now commercially available. With them, making glass capillaries is astonishingly simple, and the efficiency-limiting internal diameter [42, 61] is also easily controllable.

The coated-glass open tubular column is frequently connected to auxiliary devices --for example, injectors or tube sections leading to the detector--by shrinkable Teflon tubing [58, 78, 99-101]. Such a shrinkable tube "can easily hold up to 2-3 kg/cm^2 of pressure"[99] or, at atmospheric pressure, be used up to 300°C [78] (1 kg/cm^2 = 14.2 lb/in^2). A good practical complement to the material to follow is the 179-page book, *Gas Chromatography with Glass Capillary Columns*, by Walter Jennings [52]. In particular, the handling of glass capillaries is nicely described there.

Glass open tubular columns have become purchasable. Buying ready-made columns is a route not to be summarily dismissed: "A beginner should start by using one or more purchased columns of guaranteed quality" [102]. And, "Institutes which for years have based their routine work successfully on columns from a specialized manufacturer show no intention of making glass capillaries on their own" [103]. Nevertheless columns, even if purchased, must be judged, tested, handled, and used effectively. Much of the insight for this can be gleaned from examining the processes for coating these columns, as we now do.

A. Adsorption, Apolar Coatings, and Deactivation

The polarity of the stationary phase to be coated markedly affects the adsorptivity and the efficiency of the column. Until the late 1970's, it determined both. Apolar phases can form stable and efficient coats, but at the same time allow a ruinous adsorption of polar solutes onto the glass support [102, 104]. Polar phases prevent adsorption well enough, but tend to form little pools that destroy the efficiency of the column [105].

Apolar coatings, though efficient, must also prevent adsorption. Otherwise, the column affects the composition of the mixture being separated by either subtraction or addition of components, or both. "It can be assumed that columns which exhibit tailing effects with peaks of polar test compounds also show catalytic decomposition, especially at elevated temperatures" [101]. Column stability is also important here. A column nonadsorptive at the moment may not remain so. Sensitive tests for the appearance of adsorption must be applied routinely and frequently to columns that are subjected for any extended periods to relatively high temperatures [101].

Adsorption is detectable by changes in peak shape and retention. Peaks of polar solutes must sensitively reveal such changes, and therefore are used to test for adsorption. Schomburg et al. [101], for example, show the use of a polarity test mixture containing methane, the normal paraffins from C_5 through C_9, n-butanol, benzene, methyl butyrate, toluene, 1-octene, and di-n-butyl ether. The normal paraffins in the mixture act as reference not only for peak shape but also for peak retention, for use in calculating the Kovats retention indices of the polar solutes.

The precision of the Kovats retention indices is highly valuable in detecting that excess retention of polar solutes with respect to the n-paraffins that reveals an ad sorption within the column, an adsorption that mere inspection for tailing might mis

The trouble with apolar coatings is that they do not prevent adsorption onto th wall of the glass capillary. This must therefore be done before coating, by deacti- vation. "The most efficient method of deactivation known is the heat treatment with polar phases according to the method of Aue et al. [our Ref. 106]..." [102]. In the work leading to that method, Aue had attempted "to combine the bonding abilities of silicones with the chromatographic properties of various polyethylene glycols...[u- sually] Carbowax 20M" [106]. The attempt was unsuccessful, but the research led to the discovery that not all the Carbowax 20M could be extracted from a 6% Carbowax 20 on-Chromosorb W packing that had been heated overnight at 260 to 280°C under a gentl stream of nitrogen. Even after exhaustive extraction with hot methanol, the heat- treated packing retained about 0.5% Carbowax 20M. (This packing was described earl: in Sec. 8.5.) The remaining nonextractable Carbowax 20M completely deactivated the Chromosorb W surface, presumably by forming a thin film on that surface. It is the production of just such a film within a glass capillary that is the "most efficient method of deactivation known..." [102].

Such a film is produced only by first coating the glass capillary with a polar phase, in a normal way. This normal coating is then heat-treated at 280°C. This coating must not, of course, form the little pools that not only destroy column ef- ficiency but also leave adsorptive sites uncovered. The formation of such little pools can be prevented by roughening the surface before it is coated, and this must be done. "Consequently, a perfectly deactivated apolar column requires a glass sur face prepared [i.e., roughened] for polar coating" [102].

(For polar stationary liquids, deactivation is not necessary [101].)

B. Deactivation: Procedures and Characteristics

The glass surface within a capillary can be roughened by either etching it or depos iting particles on it. The etching can be done by gaseous HCl, which produces a useful layer of sodium chloride crystals at the same time [107-109]. This layer "i easily produced with suitable density and size of particles; the crystals adhere firmly to the glass surface and they are chemically inert" [102]. Although this ga ous HCl etching process yields efficient columns [58, 68, 110], it may lack reprodu bility and also "is limited to soft glass, from which [essentially] only alkaline columns can be made" [102]. And the sodium chloride layer is soluble in water, whi is a disadvantage. Alternatively, the etching may be done by aqueous HCl.

In a much-cited work that happened to involve aqueous HCl etching, Cronin [111 adapted the Aue heat treatment to producing, for the first time, a stable film of t highly polar Carbowax 20M on a glass capillary. A Pyrex capillary was etched with concentrated HCl (10 N) for 24 hr, then rinsed and dried. The etched capillary was

Sec. 10.3 The Glass Open Tubular Column

coated by passing through it a 2% wt/vol solution of Carbowax 20M in dichloromethane, then drying it with nitrogen. The ends of the nitrogen-filled coated capillary were then sealed and the capillary heated in an oven at 280°C for 16 hr. After the extractable Carbowax 20M was washed out by dichloromethane and methanol, the now deactivated capillary was coated normally with Carbowax 20M. The resulting column "gave excellent performances in respect of both separating power and long-term stability of the polymer film during temperature programming" [111].

Heckman et al. [112] used aqueous potassium hydrogen difluoride for etching both borosilicate and flint glass capillary columns. The resulting satisfactorily roughened surfaces are "already extensively deactivated relative to untreated glass [but are] readily amenable to further deactivation by Cronin's adaptation of Aue's method [or] silanization [by the procedure] of Welsch and coworkers" [112]. (Welsch et al. deactivated the capillary by driving a small portion of pure hexamethyldisilazane [HMDS] through it at about 5 cm/sec, sealing the capillary ends, and holding the sealed capillary at 300°C for 20 hr. The vapor-phase reaction with HMDS under these conditions led "to a very high degree of conversion of the silanol groups," Ref. 113.) The Heckman procedure "comes close to satisfying [the Grob criteria, Ref. 102, for general column preparation]. That is, the method is applicable to both flint glass and borosilicate glass columns, does not require sophisticated equipment or reagents, and is suitable for a wide variety of coating materials" [112].

Schomburg and co-workers [101], having etched both soft and Pyrex capillaries by both gaseous and aqueous techniques, deactivated them by a repetitive adaptation of the Cronin-Aue technique. That is, they laid down a succession of deactivating films, each of a different type, each individually baked on. Only the most thorough deactivation with successive films of Carbowax 20M and Emulphor O under the final stationary phase coat of OV-101 succeeded in yielding undistorted peaks from phenol and from primary n-octylamine, these two solutes having been selected as the probable worst actors with respect to adsorption. However, as they point out, polar-film deactivation increases the effective polarity of what is supposed to be a final apolar coat of OV-101. It was the realization of this undesired polarity interaction that led Welsch and colleagues to employ the HMDS deactivation described in the previous paragraph and to determine that it does not increase the effective polarity of a nominally apolar final coat.

As mentioned, the glass surface within a capillary can be roughened by either etching it or depositing particles on it. A given etching treatment may also produce a usable layer of particles. The layer should stabilize a polar coat and be chromatographically efficient; the particles should be water-insoluble and catalytically inactive.

We have already mentioned that the gaseous HCl etch produces a layer of NaCl particles, and have shown comments on that layer. The Heckman procedure--etching with aqueous hydrogen difluoride, just described--apparently leaves a dense rough layer of K_2SiF_6 crystals that are catalytically inactive and sparingly soluble in water [112].

Grob and Grob [102] have suggested depositing inactive, water-insoluble crystals of $BaCO_3$ by dynamically coating with barium hydroxide solution pushed along by carbon dioxide--a pleasingly clever idea. Let us refer to all these approaches as *particle creation* methods. Then, in contrast, we have *particle injection* methods.

The most prominent particle injection method involves silica gel particles [78, 79, 105, 114-116]. Grob and Grob [102] stated in 1976: "These procedures are... limited by the necessity of using [chromatographically inefficient] thick liquid films... Furthermore, it is our experience that the silica surface is catalytically active with certain liquid phases (*e.g.*, polyglycol) as well as with certain solutes. This activity proves to be a limiting factor for some high-temperature applications." So particle creation methods would seem to be preferred to particle injection methods. However, there seems to be no inherent reason to make this preference permanent. For instance, with regard to chromatographic efficiency, McKeag and Hougen [117] improved the initially unsatisfactory efficiency of several highly polar Silanox [78] columns to a satisfactory level--over 1,000 theoretical plates per meter, plus "high durability"--by dynamically recoating the first Silanox bed with additional liquid phases. And, with regard to catalytic activity, applying Aue deactivation to columns with injected silica particles should make them inert.

C. Special Injection Systems

On-Column Injection into a Packed Precolumn

To protect the labile components to be injected, German et al.[78] devised a special inlet system (Fig. 10.7)--a flow-controlled, replaceable section of a glass conventional packed column held 20 to 50°C above the initial temperature of the column. The sample is injected onto the top 1 cm of packing, which contains 10% SE-30. The remaining 5 cm of the packing contains 1% SE-30. Projecting into the glass wool at the bottom of this more lightly loaded packing is the inlet end of the glass open tubular column. The flow arrangements are such that "the split ratio is solely determined by the precolumn flow controls" [78].

This injection system "proved to be very effective in preventing adsorption or decomposition of sensitive compounds in the initial vaporization step... [It also] provided complete evaporation of the sample prior to the splitting (thus avoiding aerosol splitting), and it traps non-volatile substances which are often present in derivatising biochemical samples..." [78].

Splitless Sampling: The Solvent Effect

Splitless sampling has been described in several papers by Grob et al. [118-120] Two of the more recent of these are particularly instructive [103, 121], as is an even more recent study by Yang et al. [122] of the technique and of their splitless injector, a diagram of which is shown in Fig. 10.8.

In splitless sampling, the sample is injected in company with a much larger quantity of a lower boiling solvent. The quantity of solvent injected--about 5 $\mu \ell$,

Sec. 10.3 The Glass Open Tubular Column

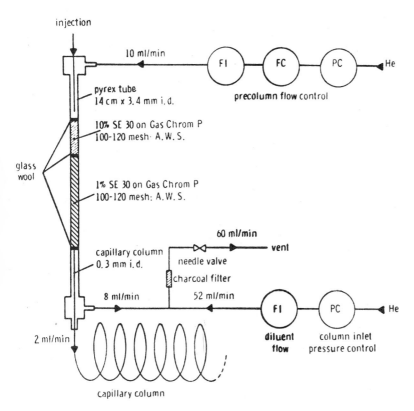

FIGURE 10.7 Capillary column inlet system. The precolumn flow is flow controlled. The flow rate determines the split ratio when the pressure at the head of the capillary column is held constant. The capillary inlet is under pressure control. PC, pressure controller; Fl, flow indicator; FC, flow controller. (Reproduced by permission of Preston Publications, Inc., from German and Horning, Ref. 78, in the Journal of Chromatographic Science.)

with a capillary column--is perhaps 1,000 times larger than normal for a capillary column. The solvent itself is more volatile than the components of the sample, and is injected into a relatively cool column. "The solvent boiling point [should be] 25° below that of the first peak of interest, [and] the initial column temperature should be 15-30° below the boiling point of the solvent" [122].

There may be just one injection, or there may be two. If the sample is a dilute solution (wherein interest is in the trace solutes), then just one injection is made, namely, about 5 to 10 µl of the solution. If the sample is more conveniently handled as a gas, then the injection may be sequential--a first injection of 5 to 10 µl of the solvent, and then some 15 to 30 sec later, the second injection, of perhaps 1 ml of the sample gas.

In either case, *substantially the whole* of the injected material (say, 95% of it) is slowly transferred from the splitless injector into the capillary column. (The completeness of the transfer means that splitless sampling is well suited to

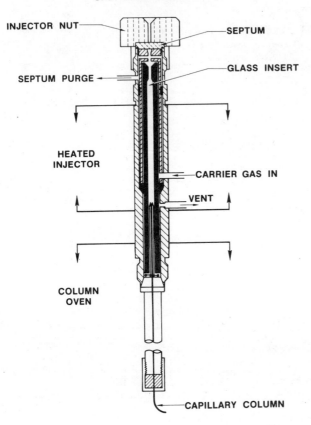

FIGURE 10.8 Cross-sectional view of the Varian Model 3700 splitless capillary injector. Most of what is injected is slowly transferred into the column. The septum purge flow merely prevents contamination from the septum from reaching the column. (Reproduced from Yang, Brown, and Cram, the *Journal of Chromatography*, Ref. 122, with permission of the Elsevier Scientific Publishing Company.)

quantitative analysis. The slowness of the transfer means that a splitless injector need not be at all so hot as a normal one. Splitless injection is therefore far less harsh than normal injection on thermally labile substances.) The transfer from the injector into the column is allowed to continue for at least 40 sec [122]--"as a somewhat arbitrary rule we [take] 60 seconds" [103]. The relatively long duration of the transfer assures that the sampling is representative, *i.e.*, that the compositions of the sample ejected from the syringe and of the sample transferred into the column are the same [122].

On the conclusion of the 60-sec transfer, the injector is quickly purged of the remnants of the sample. This is done by a temporary and large flow of purge gas.

The solvent greatly overloads the capillary column just after injection: "The drastically overloaded solvent peak builds up in the first few meters of the column,

Sec. 10.4 The Evaluation of Column Performance

a condensed liquid layer which is 20-50 times thicker than the...film [of stationary phase]. ...[This] temporarily produced very thick [layer]...causes the solvent to migrate much slower than [normal for] its volatility" [103].

The traces of interst do not overload the column, but the peak width of these traces is initially far from ideal. Suppose the column gas flow (which is identically the injection flow) is 3 ml/min and 1 ml of gas has been injected. The sample is moved into the column in about 20 sec, so each peak is about 20 sec wide. A careful comparison, however, shows "that injection without splitting produces narrower bands for the sample components than does injection with splitting" [121]. This happens because the *solvent effect* [121] sharpens the peaks.

Grob and Grob, on the solvent effect:

> The vapourized material is transferred on to the column essentially as a mixture. In the first stage of separation, the solvent shifts away from the sample components, leaving them on the back slope of its large peak. Thus, the moving vapour plugs of the sample components meet a liquid phase mixed with retained solvent, whereby the concentration of solvent increases rapidly in the direction of migration. Therefore, the front of every plug, in contact with stationary liquid containing more solvent, undergoes much stronger retention than the back of the plug. This effect causes the originally very broad bands of sample components to be condensed to a band width which, under properly selected conditions, may become even smaller than that which can be obtained with stream splitting [121].

With this, we conclude this introductory discussion of splitless sampling. It is clearly the injection technique of choice for the extremely sensitive determination of vaporizable solutes that are less volatile than the solvent in which they are dissolved, for the separation at highest resolution of complex mixtures, for the injection of substances that are thermally labile, for the injection of large volumes of dilute gases (such as environmental air), and for the injection of complex vapor mixtures (*e.g.*, aromas).

Initially, the solvent effect on which splitless sampling depends must be optimized, especially in the choice of solvent. Thereafter, the results are obviously superior. For example, see the truly impressive chromatogram by K. Grob [103] of the middle fraction of cigarette smoke, shown in Fig. 10.9.

10.4 THE EVALUATION OF COLUMN PERFORMANCE

Finding the number n of theoretical plates most simply measures column performance:

$$n = 16 \left(\frac{t_R}{w_b} \right)^2 \qquad (10.22)$$

We assume that the design of the gas chromatograph does not mask and degrade the performance of the column. In this equation, t_R is the total retention time for the peak of a given solute, and w_b is the width of that peak at the baseline.

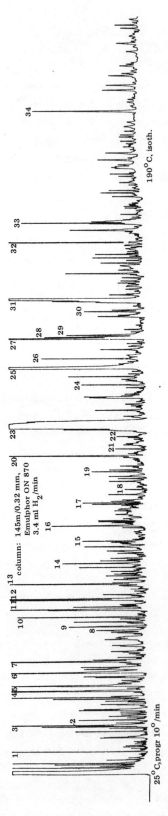

FIGURE 10.9 Natural mixtures such as this--the middle fraction of cigarette smoke--are extremely complex. Notice that the mixture is not fully separated: "...the average number of substances forming a peak is more than three. Thus the actual number of substances involved in the depicted run is at least 3000. Even the highest available separating power is insufficient to manage this task." (Reprinted with permission from Grob, Ref. 103, Pergamon Press, Ltd.)

Sec. 10.4 The Evaluation of Column Performance

We pass from the efficiency of the operation, as measured by n, to the efficiency of the column, as measured by the height h per theoretical plate for a column of length L:

$$h = \frac{L}{n} \qquad (10.23)$$

If the column is very long, as open tubular columns usually are, then the time t_M for the air peak to come through the column may become a sizeable fraction of the total retention time t_R, and the resultant number n of theoretical plates may become speciously high. In this case we calculate a smaller number N of *effective* plates [3, 123, 124] by using the adjusted retention time $t_R' = t_R - t_M$ rather than t_R:

$$N = 16 \left(\frac{t_R - t_M}{w_b} \right)^2$$

or

$$N = 16 \left(\frac{t_R'}{w_b} \right)^2 \qquad (10.24)$$

Note that, as Desty pointed out [3],

$$N = \frac{n}{\left(1 + \frac{t_M}{t_R'}\right)^2} \qquad (10.25)$$

Because $t_R'/t_M = k$, the partition ratio, another expression relating N and n is

$$N = n \left(\frac{k}{1+k} \right)^2 \qquad (10.26)$$

For very early peaks, for instance a peak for which $t_R' = t_M$, the number of effective plates is only one-fourth the number of theoretical plates. For late peaks, the two numbers differ less. For instance, with a peak for which $t_R' = 10 t_M$, $N = (5/6)n$. As k increases, the coefficient of n approaches unity.

Because the number N of effective plates is always smaller than the number n of theoretical plates, the height H per effective plate is always somewhat larger than the height h per theoretical plate, for a column of length L:

$$H = \frac{L}{N} \qquad (10.27)$$

Frequently seen symbols are HETP, for the height h equivalent to a theoretical plate; and HEETP, for the height H equivalent to an effective theoretical plate or, as we use it here, an effective plate.)

Desty et al. suggested that column performance be measured as "the rate of production of effective plates," (N/t_R), where t_R is the unadjusted retention time [3]. This is an adequate measure for many purposes.

Up to this point in this chapter, the evaluation of column performance by parameters such as n, N, and N/t_R suggests an unremitting emphasis on column improvement. Another point of view came from Halasz and Heine [22].

They presented a measure of column performance called the *performance parameter* (PP). The performance parameter measures the ease rather than the extent of column performance. The performance parameter is defined as the product of the minimal column pressure drop Δp_{ne} necessary for a separation and the minimal time t_{ne} also necessary for that separation:

$$PP = \Delta p_{ne} t_{ne} \qquad (10.28)$$

As with the height per theoretical plate, the smaller the performance parameter, the better the column.

Needing only a small pressure drop at once favorably decreases the performance parameter. This happens if the column has a high permeability. High permeability most clearly distinguishes open tubular columns from others. It is the limitation permeability that limits column length, and thus the conveniently available number plates [19, 22, 125].

The permeability of an open tubular column is always much higher than that of conventional packed column. The specific permeability coefficient is equal to $d^2/3$ for open tubular columns of internal diameter d, and is roughly equal to $d_p^2/1000$ for conventional packed columns packed with particles of diameter d_p [21, 22]. For $d_p^2/1000$ to equal $d^2/32$, d_p must roughly equal 5.6d: For the permeability of a conventional packed column to equal that of an open tubular column of internal diameter 0.25 mm, the particle diameter of the packing would have to be about 1.4 mm, or about 14 mesh--a very inefficient packing. When permeability matters, open tubular columns shine. Let us consider some further criteria.

In 1965, Chovin and Guiochon [126] reviewed some measures of column performance that had been devised by other workers. They showed that each measure could be expressed as

$$R_{ij} = \frac{\sqrt{n}}{2} \frac{r_{ij} - 1}{r_{ij} - 2} A \qquad (10.29)$$

differing primarily in the meaning of A. R_{ij} stands for the resolution, or separation per peak width w_b, of two solutes having adjusted retention times t'_{Ri} and t'_{Rj}:

$$R_{ij} = \frac{t'_{Rj} - t'_{Ri}}{mw_b} \qquad (10.30)$$

where m ranges from 1/2 to 4, depending on the choice of the original writer. The relative retention α_{ij} of the solutes equals the ratio t'_{Rj}/t'_{Ri} of the adjusted retention times, expressed so that α_{ij} is always greater than unity; n indicates the number of theoretical plates as defined by Eq. (10.22).

Sec. 10.4 The Evaluation of Column Performance   215

In the opinion of Chovin and Guiochon, the most general of all the measures was that of Said [127], wherein

$$A = 1 - \frac{t_{Ri}}{\overline{t_{Ri} + t_{Rj}}} \qquad (10.31)$$

where t_{Ri} is the unadjusted retention time of solute i and the denominator $\overline{t_{Ri} + t_{Rj}}$ is the average of the two unadjusted retention times.

By means of Said's expression, Chovin and Guiochon compared the performances of various types of columns. They assessed the "cost" of a separation as a linear combination of analysis time, pressure drop, column length, and temperature. (Note that Halasz and Heine [22] would consider a given pressure drop or temperature not as a cost but rather as a limit beyond which a given column or stationary phase, respectively, cannot be conveniently applied. Similarly, column length may tend to involve pressure drop, and thus limit applicability.)

The performances of the various column types, all generally available, were compared twice: First, with all columns theoretically operated at the same temperature; and second, with no such restriction, so that one type of column at one temperature could be compared to a different type of column at another temperature. At a common temperature, the wall-coated open tubular (WCOT) columns clearly surpassed all others, especially for late, high-partition-ratio peaks.

When no common temperature of operation was given, no column type clearly predominated. In such a comparison, and with a given partition ratio, resolution depends more on the relative retention than on the number of theoretical plates, that is, more on selectivity than on efficiency.

Chovin and Guiochon concluded by generally recommending the use of nonadsorptive WCOT columns that contain rather less stationary phase than usual and that are operated at as low a temperature as possible. They also called attention to the great but generally unused potential of short WCOT columns.

Also published in 1965 was a similar, comprehensive study by Struppe [128], which later appeared in a more elaborate form [129]. Struppe used the following expression for resolution R_{ij}:

$$R_{ij} = \frac{1}{2}\sqrt{N_j} \, \frac{\alpha_{ij} - 1}{\alpha_{ij} + (N_j/N_i)} \qquad (10.32)$$

where N is the number of effective plates as defined by Eq. (10.24), and subscripts i and j refer to two just-resolved peaks. (Equation (10.32) is identical to the combination of Eqs. (10.29) and (10.31).)

Struppe proposed, as a measure of column performance that would emphasize speed as well as resolution, a time-dependent resolution [R]:

$$[\underline{R}] = R_{ij}/t_t \qquad (10.33)$$

In Eq. (10.33), R_{ij} corresponds to the R_{ij} of Eq. (10.32), and t_t is the unadjusted retention time of the last peak in the separation. Notice that this new measure does not reduce to the expression of Eq. (10.30), but embodies a deliberate empirical emphasis on separation speed.

In Struppe's survey of published separations, the highest [R] value attained by conventional packed columns was 3.5. Both capillary packed and WCOT columns frequently yielded a value of about 8; significantly, both types of columns exceeded these values for "fast" separations taking less than a minute. In no case were [R] values higher than 13.

The study of Struppe showed that high speed and high resolution are mutually exclusive in gas chromatography. As shown in the last paragraph, below, Halasz and Heine concur [22]. (A similar limitation exists in preparative gas chromatography, where quantity and high resolution each demand a maximum yield index [130] and are mutually exclusive.)

Halasz and Heine [22] compared types of columns by a number of measures. One of these was the measure Desty and colleagues suggested, the delivery rate of effective plates per second [3]. The conventional packed column rarely delivers more than 6 to 10 effective plates per second. The commercial WCOT and PLOT columns may deliver around 20 effective plates per second. Nevertheless, Halasz and Heine point out that "the optimum times of analysis in packed columns are shorter than those in open tubes The assumption that high-speed analysis is achieved *only* in open tubes is thus not correct. If the relative retention is not too small, pressure drop does not introduc any limitation and then high speeds of analysis are obtainable with packed columns" [22].

In general, conclude Halasz and Heine, "Choose the column with the highest possible permeability in which the desired phase ratio is attainable" [22]. If many effective plates are needed, then a high permeability extends the limit that tends to b set by column pressure drop. But if speed is important, then the column must tolerat a high flow rate without losing efficiency: "then $(H/\bar{u})_{opt}$ is the most important parameter" [22].

Further discussion of column types and performances can be found in an excellent paper by D. W. Grant [131] and in a detailed and definitive one by G. Guiochon [132].

RECENT DEVELOPMENTS

Fused silica open tubular columns were introduced in 1979 [133] as the result of research in fiber optics technology at the Hewlett-Packard Corporate Research Laboratories [134]. They are so much more flexible and inert than their glass precursors (the ones discussed in Sec. 10.4) that they have caused a great stir.

The improved flexibility of the fused silica columns derives from their "small outside diameter and thin walls" combined with "the high tensile strength of a pristine silica surface" [134]. Small outside diameter, thin walls: In outer diameter, the fused silica columns are about 0.3 mm *versus* about 0.7 mm for the glass ones; and their walls are about 0.05 mm thick *versus* the 0.23 mm-thick glass walls [135]. High tensile strength: The tensile strength of a straight, 0.3 mm o.d. fused silica tube (of pristine surface) is 25 times that required for the tube to tolerate being bent into a coil 15 cm in diameter [134].

To preserve the initial strength of the fused silica surface, the fused silica tubing is coated as it is drawn. The protective coating is a polyimide that is thermally stable to at least 325°C [135], a temperature adequate for most chromatographic purposes.

Compared to glass capillary columns, coated fused silica tubing is much easier to handle. If the fused silica is drawn straight, as it generally is, it can be not only "conveniently flexed into an appropriate coil" but also "readily inserted by means of graphite ferrules into the usual connections of...injector and detector ports without...difficulty" [135].

The superior inertness of fused silica is ascribed to its low, 0.05 to 1 ppm content of metal oxides [135]. "Better than 99.9% pure," [134] "fused silica is derived from a very high grade, *synthetically* prepared silicon tetrachloride" [135]. "Strong Lewis acids, such as boron, and [chalcogenides] such as calcium and magnesium, are present in...the soft...and [the] borosilicate glasses [that are] used in making conventional glass capillaries, but [such metal oxides] are virtually absent in the fused silica material" [134]. (In contrast, fused quartz comes from natural quartz that may or may not have been chemically purified before its manufacture [135]. Thus fused quartz may or may not be as inert as fused silica.)

The relative inertness of a stationary phase support is revealed, with peaks of polar solutes, by the relative absence of peak tailing and of false increments in the Kovats Retention Index.

Activity in a support increases a given Kovats Retention Index. Increasing inertness decreases an Index toward a theoretical minimum that is approached as activity in the supporting surface approaches zero. Polarity test mixes are used to test inertness.

A proper polarity test mixture contains some components that are basic (such as 2,6-dimethylaniline, the absence of which from a test mixture is considered suspect [135]), some that are acidic (such as 2,6-dimethylphenol, another necessary component), some that are otherwise polar (such as 2-octanol), and some that are quite inert, but serve for calculation of Kovats Retention Indices.

The inertness of fused silica is nicely demonstrated with polarity test mixes. Moreover, in the studies [133, 134] that announced fused silica columns, "the results obtained from the analysis of mixtures of phenols, amines, alcohols, free fatty acids, sulphur-containing compounds, and the like were impressive. Quantitative analysis, where performed, was excellent and the 'tailing' of peaks on nonpolar columns rarely occurred" [135].

Nevertheless, the fused silica surface is not completely inert, but is slightly acid (a characteristic that allows the chromatographing of underivatized free fatty acids [134]). "Although fused silica is inherently less chromatographically active than other glasses, some deactivation is usually required" [134]. This deactivation can be done by an undercoat of Carbowax 20M according to the method of Aue (see Sec. 8.5). Unfortunately, such deactivation falsely elevates the Kovats Retention Index by a participation of the undercoat in peak retention. Also, that undercoat is not stable above 250°C [136]. However, silyl deactivation according to the method of Welsch [113] *is* stable to over 300°C and does not raise the Kovats Retention Index, at least by present standards.

Neither Welsch-deactivated nor untreated fused silica surfaces can be coated successfully with polar stationary phases other than Carbowax 20M [135], at least at this writing. This deficiency is certain to be resolved.

The relative ease of use and the inertness of the fused silica columns have given a substantial impetus to the use of open tubular columns. They should eventually supersede their glass predecessors. However, the impetus from the advent of fused silica columns, despite their advantages, is not likely to change substantially the relative balance of packed *versus* open tubular columns, for the reasons that were cited in the Introduction and for the reasons that follow:

Special mechanical support for the open tubular column is still needed.

Sample splitting is still needed.

Special termination at the detector is still needed.

With the same internal diameter as the glass capillary, the fused silica column offers no particular advantage over its predecessor with respect to column efficiency. (Either could be more efficient, but it is too much trouble.)

No greater amount of sample can be charged to the fused silica open tubular column merely because it is fused silica, therefore the open tubular column remains inferior to the packed column for trace detection.

REFERENCES

1. R. Teranishi, I. Hornstein, P. I. Issenberg, and E. L. Wick, *Flavor Research*, Dekker, New York, 1971.
2. J. M. Ryan and G. L. Dienes, *Drug Cosmetic Ind.* 99, 60 (1966).
3. D. H. Desty, A. Goldup, and W. T. Swanton, in *Gas Chromatography* (N. Brenner, J. E. Callen, and M. D. Weiss, editors), Academic, New York, 1962, pp. 105-136.
4. I. Halasz and C. Horvath, *Anal. Chem.* 36, 1178 (1964).
5. A. Kwantes and G. W. A. Rijnders, in *Gas Chromatography 1958* (D. H. Desty, editor) Academic, New York, 1958, pp. 125-135.
6. A. W. Decora and G. U. Dineen, *Anal. Chem.* 32, 164 (1960).
7. A. B. Callear and R. J. Cvetanovic, *Can. J. Chem.* 33, 1256 (1955).
8. A. M. Filbert, in *Recent Advances in Gas Chromatography* (I. I. Domsky and J. A. Perry, editors), Dekker, New York, 1971, pp. 49-98.

References

9. I. Halasz and C. Horvath, *Anal. Chem. 36*, 2226 (1964).
10. A. T. James and A. J. P. Martin, *Biochem. J. (Proc.) 48*, vii (1951).
11. A. T. James and A. J. P. Martin, *Analyst 77*, 915 (1952).
12. A. J. P. Martin and R. L. M. Synge, *Biochem. J. 35*, 1359 (1941).
13. J. J. van Deemter, F. J. Zuiderweg, and A. Klinkenberg, *Chem. Eng. Sci. 5*, 211 (1956).
14. J. Bohemen and J. H. Purnell, in *Gas Chromatography 1958* (D. H. Desty, editor), Academic, New York, 1958, pp. 6-22; and also Figure 4.
15. J. Bohemen, S. H. Langer, R. H. Perrett, and J. H. Purnell, *J. Chem. Soc. 1960*, 2444 (1960).
16. R. L. Martin, *Anal. Chem. 33*, 347 (1961).
17. A. J. Martin, C. E. Bennett, and F. W. Martinez, Jr., in *Gas Chromatography* (H. J. Noebels, R. F. Wall, and N. Brenner, editors), Academic, New York, 1961, pp. 363-374.
18. C. C. Sweeley, R. Bentley, M. Makita, and W. W. Welss, *J. Am. Chem. Soc. 85*, 2497 (1963).
19. J. A. Rijks, C. A. Cramers, and P. Bocek, *Chromatographia 8*, 482 (1975).
20. C. A. Cramers, J. A. Rijks, and P. Bocek, *Clin. Chim. Acta 34*, 159 (1971).
21. P. C. Carman, *Flow of Gases Through Porous Media*, Butterworths, London, 1956, p. 8.
22. I. Halasz and E. Heine, in *Progress in Gas Chromatography* (J. H. Purnell, editor), Interscience, New York, 1968, pp. 153-208.
23. J. C. Giddings, *Anal. Chem. 36*, 741 (1964).
24. R. H. Kolloff and S. F. Spencer, 16th Mid-America Spectroscopy Symposium, Chicago, Ill., 1965: S. F. Spencer, Technical Paper No. 30, F. and M. Scientific Corp. (now Hewlett-Packard), Avondale, Pa.
25. D. M. Ottenstein, *J. Gas Chromatog. 1*, 11 (1963).
26. Analabs, 80 Republic Drive, North Haven, Conn. 06473, the 10-mesh cut in gas chromatographic supports.
27. R. P. W. Scott, in *Gas Chromatography 1958* (D. H. Desty, editor), Academic, New York, 1958, pp. 189-199.
28. S. J. Hawkes, in *Recent Advances in Gas Chromatography* (I. I. Domsky and J. A. Perry, editors), Dekker, New York, 1971, p. 38.
29. Y. I. Pirogova, M. Y. Shtaerman, and D. A. Vyakhirev, *J. Chromatog. 92*, 249 (1974).
30. Applied Science Laboratories Technical Bulletin 2A, Applied Science Laboratories, P. O. Box 440, State College, Pa. 16801.
31. J. F. Parcher and P. Urone, *J. Gas Chromatog. 2*, 184 (1964).
32. R. F. Kruppa, R. S. Henly, and D. L. Smead, *Anal. Chem. 39*, 851 (1967).
33. Bulletin FF-121, Johns-Manville Corp., Celite Division, Box 5705, Denver, Colo. 80217.
34. R. P. W. Scott, in *Advances in Chromatography*, Vol. 9 (J. C. Giddings and R. A. Keller, editors), Dekker, New York, 1970, pp. 193-214.
35. E. Gelpi, E. Peralta, and J. Segura, *J. Chromatog. Sci. 12*, 701 (1974).
36. S. J. Hawkes, in *Recent Advances in Gas Chromatography* (I. I. Domsky and J. A. Perry, editors), Dekker, New York, 1971, pp. 13-48.
37. J. C. Sternberg, in *Advances in Chromatography*, Vol. 2 (J. C. Giddings and R. A. Keller, editors), Dekker, New York, 1966, pp. 205-270.

38. L. S. Ettre, *Open Tubular Columns in Gas Chromatography*, Plenum, New York, 1965.
39. M. J. E. Golay, in *Gas Chromatography* (V. J. Coates, H. J. Noebels, and I. S. Fagerson, editors), Academic, New York, 1958, pp. 1-13.
40. D. H. Desty, A. Goldup, and B. H. Whyman, *J. Inst. Petrol. 45*, 287 (1959).
41. D. H. Desty and A. Goldup, in *Gas Chromatography 1960* (R. P. W. Scott, editor), Butterworths, Washington, D. C., pp. 162-183.
42. D. H. Desty, in *Advances in Chromatography*, Vol. 1 (J. C. Giddings and R. A. Keller, editors), Dekker, New York, 1965, pp. 199-228.
43. I. Halasz and W. Schneider, *Anal. Chem. 33*, 979 (1961).
44. I. Halasz and W. Schneider, in *Gas Chromatography* (N. Brenner, J. E. Callen, and M. D. Weiss, editors), Academic, New York, 1962, pp. 287-306.
45. G. Schomburg, H. Husmann, and F. Weeke, *J. Chromatog. 99*, 63 (1974).
46. M. J. E. Golay, U.S. Patent 2 920 478 (1960).
47. I. Halasz and C. Horvath, U.S. Patent 3 295 296 (1967).
48. I. Halasz and C. Horvath, *Anal. Chem. 35*, 499 (1963).
49. L. S. Ettre, J. E. Purcell, and S. D. Norem, *J. Gas Chromatog. 3*, 181 (1965).
50. J. E. Purcell and L. S. Ettre, *J. Gas Chromatog. 4*, 23 (1966).
51. L. S. Ettre, J. E. Purcell, and K. Billeb, *Separation Sci. 1*, 777 (1966).
52. W. Jennings, *Gas Chromatography with Glass Capillary Columns*, Academic, New York, 1978; and *Gas Chromatography with Glass Capillary Columns*, 2nd ed., Academic, New York, 1980.
53. M. J. E. Golay, in *Gas Chromatography 1958* (D. H. Desty, editor), Butterworths, London, 1958, pp. 36-55.
54. L. S. Ettre, *Open Tubular Columns in Gas Chromatography*, Plenum, New York, 1965, pp. 13, 14.
55. *Ibid.*, p. 18.
56. *Ibid.*, p. 15.
57. R. Teranishi, I. Hornstein, P. I. Issenberg, and E. L. Wick, *Flavor Research*, Dekker, New York, 1971, p. 95.
58. H. T. Badings, J. J. G. van der Pol, and J. G. Wassink, *Chromatographia 8*, 440 (1975).
59. D. H. Desty and A. Goldup, in *Gas Chromatography 1960* (R. P. W. Scott, editor), Butterworths, Washington, D. C., p. 168.
60. D. H. Desty, in *Advances in Chromatography*, Vol. 1 (J. C. Giddings and R. A. Keller, editors), Dekker, New York, 1965, p. 209.
61. G. Guiochon, *Anal. Chem. 50*, 1812 (1978).
62. G. Dijkstra and J. de Goey, in *Gas Chromatography 1958* (D. H. Desty, editor), Academic, New York, 1958, pp. 56-68.
63. L. S. Ettre, *Open Tubular Columns in Gas Chromatography*, Plenum, New York, 1965, p. 60.
64. R. Kaiser, *Gas Phase Chromatography, Vol. 2: Capillary Chromatography*, Butterworths, London, 1963.
65. D. H. Desty, in *Advances in Chromatography*, Vol. 1 (J. C. Giddings and R. A. Keller, editors), Dekker, New York, 1965, p. 22.
66. M. Novotny, L. Blomberg, and K. D. Bartle, *J. Chromatog. Sci. 8*, 390 (1970).
67. K. D. Bartle, *Anal. Chem. 45*, 1831 (1973).
68. J. Roeraade, *Chromatographia 8*, 511 (1975).

69. W. Averill, in *Gas Chromatography* (N. Brenner, J. E. Callen, and M. D. Weiss, editors), Academic, New York, 1962, pp. 1-6.

70. L. S. Ettre, *Open Tubular Columns in Gas Chromatography*, Plenum, New York, 1965, p. 85.

71. J. Bouche and M. Verzele, *J. Gas Chromatog.* 6, 501 (1968).

72. T. Boogaerts, M. Verstappe, and M. Verzele, *J. Chromatog. Sci.* 10, 217 (1972).

73. G. A. M. F. Rutten and J. A. Luyten, *J. Chromatog.* 74, 177 (1972).

74. L. S. Ettre, *Open Tubular Columns in Gas Chromatography*, Plenum, New York, 1965, p. 26.

75. D. H. Desty, in *Advances in Chromatography*, Vol. 1 (J. C. Giddings and R. A. Keller, editors), Dekker, New York, 1965, p. 208.

76. M. J. E. Golay, *Anal. Chem.* 40, 382 (1968).

77. J. G. Nikelly, *Anal. Chem.* 44, 623 (1972); and *Anal. Chem.* 44, 625 (1972).

78. A. L. German and E. C. Horning, *J. Chromatog. Sci.* 11, 76 (1973).

79. M. Blumer, *Anal. Chem.* 45, 980 (1973).

80. E. Kugler, R. Langlais, W. Halang, and M. Hufschmidt, *Chromatographia* 8, 468 (1975).

81. H. Jaeger, H.-U. Klör, G. Blos, and K. Ditshuneit, *Chromatographia* 8, 507 (1975).

82. L. S. Ettre, J. E. Purcell, and K. Billeb, *J. Chromatog.* 24, 335 (1966).

83. A. I. M. Keulemans, *Gas Chromatography*, 2nd ed., Reinhold, New York, 1959, p. 199.

84. J. J. van Deemter, in *Gas Chromatography 1958* (D. H. Desty, editor), Butterworths, London, 1958, pp. 3-5.

85. L. S. Ettre and W. Averill, *Anal. Chem.* 33, 680 (1961).

86. L. S. Ettre, *Open Tubular Columns in Gas Chromatography*, Plenum, New York, 1965, p. 114.

87. R. G. Buttery, D. R. Black, D. G. Guadagni, and M. P. Kealy, *Amer. Soc. Brewing Chemists' Proc. 1965*, 103; and R. G. Buttery, R. E. Lundin, and L. Ling, *J. Agr. Food Chem.* 15, 58 (1967).

88. H. M. Richards, G. F. Russell, and W. G. Jennings, *J. Chromatog. Sci.* 9, 560 (1971).

89. I. H. Suffet and E. R. Glaser, *J. Chromatog. Sci.* 16, 12 (1978).

90. L. J. Schmauch, *Anal. Chem.* 31, 225 (1959); see also L. J. Schmauch and R. A. Dinerstein, *Anal. Chem.* 32, 343 (1960).

91. K. D. Bartle, L. Bergstedt, M. Novotny, and G. Widmark, *J. Chromatog.* 45, 256 (1969).

92. K. Grob, *Beitr. Tabakforsch.* 3, 403 (1966); *Helv. Chim. Acta* 51, 718 (1968).

93. K. Grob and J. A. Vollmin, *Beitr. Tabakforsch.* 5, 52 (1969).

94. M. Novotny and A. Zlatkis, *Chromatog. Rev.* 14, 1 (1971).

95. D. A. Parker and J. L. Marshall, *Chromatographia* 11, 526 (1978).

96. U. Rapp, U. Schröder, S. Meier, and H. Elmenhorst, *Chromatographia* 8, 482 (1975).

97. D. H. Desty, J. N. Haresnape, and G. Guiochon, *Anal. Chem.* 32, 302 (1960).

98. D. H. Desty, *Chromatographia* 8, 452 (1975).

99. P. Sandra, M. Verzele, and E. Vanluchene, *Chromatographia* 8, 499 (1975).

100. C. A. Cramers and E. A. Verneer, *Chromatographia* 8, 479 (1975).

101. G. Schomburg, H. Husmann, and F. Weeke, *Chromatographia* 10, 580 (1977).

102. K. Grob and G. Grob, *J. Chromatog.* 125, 471 (1976).

103. K. Grob, *Chromatographia 8*, 423 (1975).

104. L. Blomberg, *J. Chromatog. 115*, 365 (1975).

105. A. L. German, C. D. Pfaffenburger, J.-P. Thenot, and E. C. Horning, *Anal. Chem. 45*, 930 (1973).

106. W. A. Aue, C. R. Hastings, and S. Kapila, *J. Chromatog. 77*, 299 (1973).

107. G. Alexander and G. A. M. F. Rutten, *Chromatographia 6*, 231 (1973).

108. G. Alexander, G. Garzo, and G. Palyi, *J. Chromatog. 91*, 25 (1974).

109. G. Alexander and G. A. M. F. Rutten, *J. Chromatog. 99*, 81 (1974).

110. G. Schomburg and H. Husmann, *Chromatographia 8*, 516 (1975).

111. D. A. Cronin, *J. Chromatog. 97*, 263 (1974).

112. R. A. Heckman, C. R. Green, and F. W. Best, *Anal. Chem. 50*, 2157 (1978).

113. T. Welsch, W. Engewald, and C. Klaucke, *Chromatographia 10*, 22 (1977).

114. R. E. Kaiser, *Chromatographia 2*, 34 (1968).

115. J. G. Nikelly and M. Blumer, *Amer. Lab. 6*, 12 (1974).

116. E. Schulte and L. Acker, *Z. Anal. Chem. 268*, 260 (1974).

117. R. G. McKeag and F. W. Hougen, *J. Chromatog. 136*, 308 (1977).

118. K. Grob and G. Grob, *J. Chromatog. Sci. 7*, 584 (1969).

119. K. Grob and G. Grob, *J. Chromatog. Sci. 7*, 587 (1969).

120. K. Grob and G. Grob, *Chromatographia 5*, 3 (1972).

121. K. Grob and K. Grob, Jr., *J. Chromatog. 94*, 53 (1974).

122. F. J. Yang, A. C. Brown, III, and S. P. Cram, *J. Chromatog. 158*, 91 (1978).

123. J. H. Purnell, *J. Chem. Soc. 1960*, 1268.

124. J. H. Purnell, *Nature (Lond.) 184*, 2009 (1959).

125. I. Halasz and E. Heine, in *Advances in Chromatography*, Vol. 4 (J. C. Giddings and R. A. Keller, editors), Dekker, New York, 1967, pp. 207-263.

126. P. Chovin and G. Guiochon, *Bull. Soc. Chim. France 1965*, 3391.

127. A. S. Said, *J. Gas Chromatog. 2*, 60 (1964).

128. H. G. Struppe, *Ber. Bunsenges. f. phys. Chem. 69*, 834 (1965).

129. H. G. Struppe, in *Handbuch der Gas-Chromatographie* (E. Leibnitz and H. G. Struppe, editors), Verlag Chem. GMBH, Weinhein, 1967, pp. 103-157.

130. J. A. Perry, *Chem. and Ind. 1966*, 576.

131. D. W. Grant, *J. Chromatog. 122*, 107 (1976).

132. G. Guiochon, in *Advances in Chromatography*, Vol. 8 (J. C. Giddings and R. A. Keller, editors), Dekker, New York, 1969, pp. 179-270.

133. R. Dandeneau and E. H. Zerenner, *J. High Resoln. Chromatog. Chromatog. Comm. 2*, 351 (1979).

134. R. Dandeneau, P. Bente, T. Rooney, and R. Hiskes, *Amer. Lab. 11*, 61 (1979).

135. S. R. Lipsky, W. J. McMurray, M. Hernandez, J. E. Purcell, and K. A. Billeb, *J. Chromatog. Sci. 18*, 1 (1980).

136. K. Grob, *J. High Resoln. Chromatog. Chromatog. Comm. 2*, 599 (1979).

Chapter 11

THE COLUMN TEMPERATURE

11.1 INTRODUCTION

If the temperature of a column cannot be changed easily, then each wide boiling-range mixture that is to be gas chromatographed must first be fractionated into mixtures, each of narrow enough boiling-range to be isothermally gas chromatographable. But if the temperature of a column can be increased continuously--programmed--after the injection of a sample of a wide boiling-range mixture, then the components can be chromatographed in one pass. Because of this, the advent of temperature programming vastly increased the utility of gas chromatography, at that time already growing explosively.

A generally applicable device for linear temperature programming appeared in 1959 [1]. The obvious convenience of being able to separate high- as well as low-boiling components in the same run changed the primary performance requirements for both gas chromatographs and stationary phases. Thereafter, the prime capability of a gas chromatograph became its ability to increase column temperature at any desired rate and precision, and then to return it sharply to the starting point. The prime capability of a stationary phase became its ability to tolerate high temperatures without distilling or decomposing, in other words, to extend the upper limit available to temperature programming with a given selectivity.

We are about to examine certain aspects of the dependence of column performance on temperature. We describe isothermal gas chromatography, programmed temperature gas chromatography, and the relationship of the two. We close the chapter by noting the finite thermal stability of the stationary phase that ultimately precludes further temperature programming.

11.2 COLUMN BEHAVIOR AS A FUNCTION OF TEMPERATURE

A. Isothermal Conditions

Exponential Retention Time Increase in Homologous Series

The retention times of successive members of a homologous series increase exponentially for a given column, flow rate, and column temperature [2-7]. For example, see Fig. 11.1 [4, 8].

The general expression for this relationship* is

$$\log V_g = c_1 n + c_2 \tag{11.1}$$

Here, V_g is the specific retention volume (defined in the next sentence); n is the number of carbon atoms in each successive member of a homologous series; and c_1 and c_2 are constants. The specific retention volume is defined as the net retention volume V_N per gram of stationary phase of density ρ_L and volume V_L in the column, with the temperature T of actual column use corrected to 0°C [9]:

$$V_g = \frac{V_N}{V_L \rho_L} \frac{273}{T} \tag{11.2}$$

Equation (11.1) is very important. It is not merely empirical but can also be derived. With simplified expressions, we do this. (For a more detailed treatment, see Purnell, Ref. 8.)

For any temperature T, the logarithm of the vapor pressure p° of a pure organic liquid can be calculated from the following equation [10]:

$$\log p° = a - b \log T - c/T \tag{11.3}$$

The constants a, b, and c for numerous organic liquids have been determined and can be found listed in tables. See, for instance, Ref. 11.

Values of p° at various temperatures can be obtained from Eq. (11.3) for a given pure organic liquid having specified constants a, b, and c. These values of p° can then be used to calculate the heat of vaporization ΔH_v of that liquid--should that heat not already have been experimentally determined--by means of the Clausius-Clapeyron equation:

$$\Delta H_v = RT^2 \frac{d \ln p°}{dT} \tag{11.4}$$

The heats of vaporization of the members of various aliphatic homologous series increase smoothly and linearly with the number of carbon atoms per molecule. Examples showing experimentally determined heats are shown in Fig. 11.2 [12].

---

* The notation log is used for common logarithms.

Sec. 11.2 Column Behavior as a Function of Temperature   225

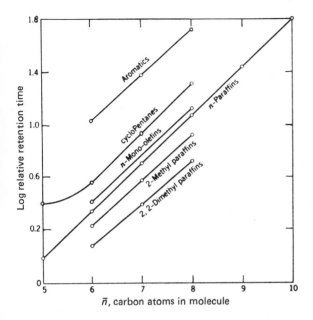

FIGURE 11.1 In general, plotting the logarithm of the retention volumes of the successive members of a homologous series against the carbon numbers of the members yields straight lines. (Reprinted from H. Purnell, *Gas Chromatography*, Ref. 8, with permission of John Wiley and Sons, Inc. Original data obtained by Desty, Ref. 4.)

To return to the Clausius-Clapeyron equation: Another form can be obtained. We note that

$$d\,\frac{1}{T} = d\,T^{-1} = (-1)T^{-2}dT = -\frac{dT}{T^2}$$

and therefore

$$\frac{1}{d(1/T)} = -\frac{T^2}{dT} \tag{11.5}$$

We substitute Eq. (11.5) into Eq. (11.4) in order to obtain this other form of the Clausius-Clapeyron equation that we shall use presently:

$$\frac{d\ln p^\circ}{d(1/T)} = -\frac{\Delta H_v}{R} \tag{11.6}$$

An analogous statement that refers to the solute in dilute solution can be found in chemical thermodynamics texts, for instance, Ref. 13:

$$\frac{d\ln K_H}{d(1/T)} = -\frac{\Delta H_s}{R} \tag{11.7}$$

Here, ΔH_s is the differential molar heat of vaporization of a solute from an infinitely dilute solution, and K_H is Henry's Law constant.

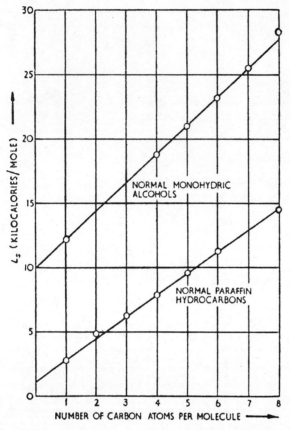

FIGURE 11.2 The heats of vaporization of the successive members of a homologous series also increase linearly with the carbon numbers of the members. The dots are experimental values. (Reproduced with permission of Pergamon Press, Ltd., from E. A. Moelwyn-Hughes, *Physical Chemistry*, 2nd ed., Ref. 12.)

From Eqs. (11.6) and (11.7), we have

$$\frac{d \ln K_H}{d \ln p^\circ} = \frac{\Delta H_s}{\Delta H_v} \tag{11.8}$$

If we let

$$\frac{\Delta H_s}{\Delta H_v} = a$$

then from Eq. (11.8) we have

$$d \ln K_H = a \, d \ln p^\circ$$

We must now point out that Henry's Law constant K_H is essentially the inverse of the partition coefficient K. Henry's Law states that the partial pressure p of a solute equals the mole fraction x of that solute times the Henry's Law constant K_H:

Sec. 11.2 Column Behavior as a Function of Temperature

$$p = K_H x$$

But the partial pressure p of a solute is caused by a weight w_M of that solute in a unit volume of gas or vapor. For our purposes, this can be a unit volume V_M of the mobile phase. Similarly, the mole fraction x of a solute in the nearly pure solvent can be taken as a certain weight w_L of that solute in a unit volume V_L of the stationary phase. Thus

$$K_H = \frac{w_M/V_M}{w_L/V_L}$$

However, as we know, the partition coefficient K is defined as

$$K = \frac{w_L/V_L}{w_M/V_M}$$

Therefore K can be taken as $1/K_H$. This allows us to restate

$$d \ln K_H = a\, d \ln p°$$

as

$$d \ln K = -a\, d \ln p° \tag{11.9}$$

We wish to relate the vapor pressure p° of the pure solute to the retention volume of that solute. We do this by using the definitions of the specific retention volume V_g [Eq. (11.2)] and the net retention volume V_N:

$$V_N = KV_L \tag{11.10}$$

By combining Eqs. (11.2) and (11.10), we can express the partition coefficient K in terms of the specific retention volume V_g:

$$V_g = \frac{KV_L}{V_L \rho_L} \frac{273}{T}$$

$$K = \frac{V_g \rho_L T}{273} \tag{11.11}$$

We take logarithms of this equation:

$$\log K = \log V_g + \log \frac{\rho_L T}{273}$$

$$\log K = \log V_g + c_3 \tag{11.12}$$

Substituting Eq. (11.12) into Eq. (11.9) now relates the logarithm of the specific retention volume V_g to the logarithm of the vapor pressure p° of the pure solute:

$$\log V_g = -a \log p° + c_4 \tag{11.13}$$

The integral of Eq. (11.6) is

$$\ln p° = -\frac{\Delta H_v}{RT} + c_5 \tag{11.14}$$

We are getting closer. We have already seen in Fig. 11.2 the linear variation of the heat of vaporization ΔH_v with carbon number n. Herington [14] pointed out the corresponding empirically observed linear variation of the logarithm of the vapor pressure p° with carbon number for n-alkylbenzenes, n-paraffins, n-fatty acids, and n-alcohols, as shown in Fig. 11.3. But we must still develop the explicit statement that is Eq. (11.1).

If we multiply through Eq. (11.14) by the "a" of Eq. (11.13), and reexpress in common logarithms, we get

$$a \log p° = -\frac{a \Delta H_v}{2.3RT} + c_6 \tag{11.15}$$

But after stating Eq. (11.7) and then Eq. (11.8), we defined a as $\Delta H_s/\Delta H_v$. Therefore, we substitute Eq. (11.15) into Eq. (11.13), and substitute ΔH_s for $a\Delta H_v$. The result expresses the logarithm of the specific retention volume as a linear function of the column temperature T and the heat ΔH_s of solute vaporization from a liquid stationary phase [9, 15]:

$$\log V_g = \frac{\Delta H_s}{2.3RT} + c_7 \tag{11.16}$$

Now, if the heat ΔH_v of solute vaporization from a liquid stationary phase can be taken as equal to the differential molar heat ΔH_s of vaporization of a solute from an infinitely dilute solution, which is really the case only if Raoult's Law holds, then we have

$$\log V_g = \frac{\Delta H_v}{2.3RT} + c_7 \tag{11.17}$$

We must now relate $\log V_g$ to the carbon number n in a homologous series. We do this through Trouton's rule.

According to the approximation known as Trouton's rule [8, 10, 16, 17], the heat of vaporization ΔH_v is directly and linearly proportional to the boiling point T_b:

$$\frac{\Delta H_v}{T_b} = 21 \text{ cal/(mol)(deg)} \tag{11.18}$$

(This approximation suggests that all molecules show about the same increase in entropy, *i.e.*, disorder, on vaporization. It holds, roughly, except for molecules such as water or ethanol that are strongly associated in the liquid state.)

Sec. 11.2 Column Behavior as a Function of Temperature

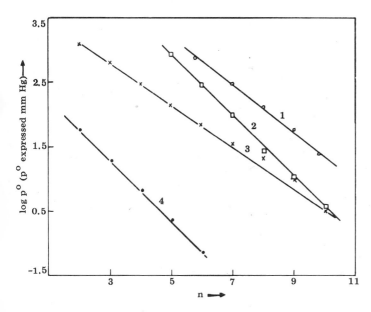

FIGURE 11.3 Plotting the logarithms of the vapor pressures of the successive members of a homologous series against the carbon numbers of the members also yields straight lines. (Reprinted from "The Thermodynamics of Gas-Liquid Chromatography" by E. F. G. Herington, in Ref. 14, with permission of the author and Butterworths Publications Ltd.)

Substituting Eq. (11.18) into Eq. (11.17) yields [18]

$$\log V_g = \frac{c_8 T_b}{RT} + c_7 \tag{11.19}$$

In most cases (as any reader can determine by plotting), the boiling point T_b is a linear function of the carbon number n in a homologous series [8]:

$$T_b = c_9 n + c_{10} \tag{11.20}$$

Substituting Eq. (11.20) into Eq. (11.19) yields

$$\log V_g = \frac{c_{11} n}{RT} + c_{12} \tag{11.21}$$

So, for a given column temperature T, we get the simple, explicit equation that we originally sought to express the relationships shown in Fig. 11.1:

$$\log V_g = c_1 n + c_2 \tag{11.1}$$

The great practical importance of this equation has already been expressed and can be seen demonstrated in Fig. 1.6.

Tendency of Solvent Efficiency to be Better at Lower Temperatures

We distinguish solvent efficiency from column efficiency. Solvent efficiency is a function not of time but of chemical potential, or partial molar free energy [19, 20]. Solvent efficiency can be measured without regard to gas chromatography. Column efficiency, however, is very much a function of time in that it involves diffusional spreading and forced gas flow. Column efficiency is measured with gas chromatographic columns.

We further distinguish selectivity from solvent efficiency, which embraces selectivity. Selectivity is a type of solvent efficiency that depends on thermodynamically nonideal solute behavior in solutions.

A selective stationary phase is one that alters the relative retention of two solutes from that predicted from their boiling points alone. In Chap. 5, expressions were given that describe the solubility forces that produce selectivity. The important dipole-dipole or Keesom force between two molecules, each of which has a dipole moment, is inversely proportional to temperature. On the other hand, the dipole-induced dipole force is temperature-insensitive. Therefore whether solubility-caused selectivity is better at lower column temperatures depends on the case at hand.

Another source of selectivity is adsorption at the gas-liquid interface. Gas-liquid adsorption increases for both nonpolar and polar solutes with increase in stationary phase polarity; it also increases with solute polarity. Thus gas-liquid adsorption becomes increasingly important in attaining selectivity as high selectivity is sought through the use of more-polar stationary phases. Adsorption, including gas-liquid adsorption, is greater, the lower the temperature.

As indicated in Eq. (11.16), the logarithm of the specific retention volume varies linearly with the reciprocal of temperature. Plots of such data are shown in Fig. 11.4 [21]. Although they are nominally straight lines, actually they curve slightly upward. The spacing between adjacent lines tends to increase from left to right, *i.e.*, from higher to lower temperatures. And so on these fairly narrow grounds, two such adjacent components become somewhat easier to separate from each other if the column temperature is lower rather than higher.

Lower column temperatures do improve solvent efficiency, and thus separability [22]. See, for instance, Fig. 11.5 [23]. However, it does not follow that columns should be operated at lower temperatures on these grounds alone. Because a general prediction of the particular components to be separated cannot be made, decreasing the column temperature may merely move two given components closer together.

The only general conclusion regarding solvent efficiency alone is that relative retentions are often temperature-sensitive. Depending on the components involved, using a higher or lower column temperature than the one at hand may well ease a difficult separation. Giddings put it well: "It is possible for two compounds to have identical retention times in an isothermal run at a given temperature, and to have unequal retention times at all other temperatures. In the case of isothermal chromatography the remedy is obvious: operate at a temperature well removed from the intersecting temperature" [25].

Sec. 11.2 Column Behavior as a Function of Temperature

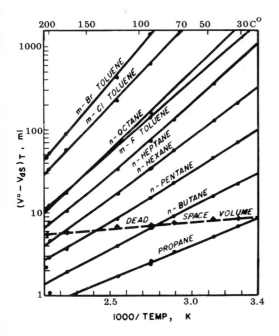

FIGURE 11.4 The logarithm of the net retention volume varies linearly with the reciprocal of temperature. Stationary phase: Apiezon L. (These data were obtained in connection with the study that is described in Sec. 11.2, *Tendency of Efficiency to Improve at Higher Column Temperatures*.) (From *Talanta*, Ref. 21, reprinted with permission of W. E. Harris and H. W. Habgood, and Pergamon Press, Ltd.)

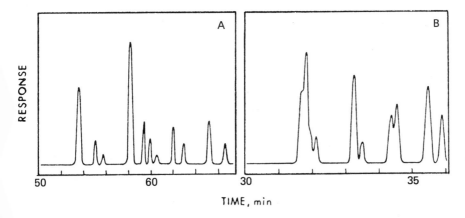

FIGURE 11.5 Lowering the column temperature--in this case from 51°C (B) to 25°C (A)--increases solvent efficiency. The chromatograms show separations of saturated hydrocarbons by a silicone stationary phase in a wall-coated open tubular column. (Reprinted from Harris and Habgood, Ref. 24, *Programmed Temperature Gas Chromatography*, with permission of the authors and John Wiley and Sons, Inc. Original data published by A. G. Polgar, J. J. Holst, and S. Groenings in Ref. 23.)

Column efficiency must be considered along with solvent efficiency. Therefore, we next consider column efficiency as a function of temperature. Then we consider again the difficult separation.

For Each Column, a Best Temperature for Column Efficiency for Each Component

DeWet and Pretorius [26] showed that the height h per theoretical plate varies with the column temperature T as follows:

$$h = A + B_3 T + \frac{C_4}{T} \tag{11.22}$$

The forms of this equation and of the van Deemter equation are similar. As a function of temperature the height per theoretical plate shows a minimum. The column efficiency shows a corresponding maximum.

From the van Deemter equation, Kambara and Kodama [27] derived an equation that linearly relates the logarithm of the number of theoretical plates to the reciprocal of temperature. In support, they showed data (Fig. 11.6) for the elution of two alcohols and two aromatic hydrocarbons from Apiezon L. It can be seen in this figure that with each solute the column yields a sharp maximum number of plates at a temperature characteristic of that solute-column relationship.

We conclude that a given column shows a maximum efficiency for a given solute at a particular temperature characteristic of that solute and column.

For a given column, each solute has a best temperature.

Tendency of Overall Efficiency to Improve at Higher Column Temperatures

Harris and Habgood [21] calculated the expectable changes in the efficiency of a given column (2 m, Apiezon L) for propane, pentane, and octane as a function of column temperatures 30, 100, and 200°C. The results are shown in Fig. 11.7.

The rows of Fig. 11.7 correspond nominally to components but functionally to successive values of the partition ratio. The nominal components are propane, pentane, and octane. The temperature-based variations of the partition ratio with these components are shown in Fig. 11.8. In Fig. 11.7, the parts A, B, and C in the top row reflect the behavior of the propane peak; D, E, and F in the middle row, that of the pentane peak; and G, H, and J in the bottom row, that of the octane peak.

The columns of Fig. 11.7 are not so simply described. They are based on the resistance to mass transfer of the gas phase at 30°C for propane as solute. The transfer resistances at the other temperatures--100 and 200°C--and the behavior of the other solutes--pentane, octane--are calculated from the basic 30°C-propane case: "Parts A, D, and G...are based on the term C_{liq} equal to the term C_{gas} for propane at 30°; parts B, E, and H are based on C_{liq} equal to 10 C_{gas}; and parts C, F, and J are based on C_{liq} equal to 100 C_{gas}, again for propane at 30" [21].

Parts A, D, and G are intended to "illustrate the case of gas resistance being the major contributor to plate height" [21]. (Thus, in these figures, at 30°C, the ratio C_{gas}/C_{liquid} = 1.0 for A, about 4 for D, and about 100 for G; and at 200°C,

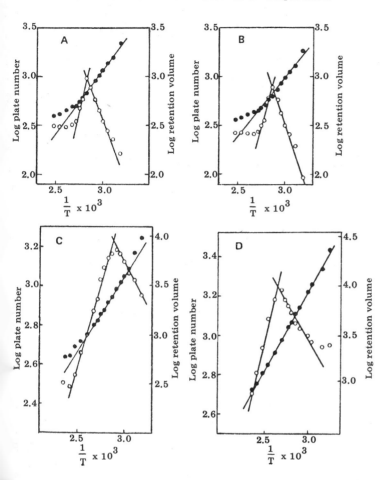

FIGURE 11.6 Each component has its own best temperature that maximizes column efficiency for that component. Here, the empty circles show the logarithm of the number of theoretical plates plotted against the reciprocal of temperature for four components and a given column. The components: (A) isopropyl alcohol, (B) ethyl alcohol, (C) benzene, and (D) toluene. The column: 3 m × 4 mm, containing 10% Apiezon L. The filled circles show the corresponding logarithms of retention times. (Reproduced from T. Kambara and H. Kodama, in the Journal of Chromatography, Ref. 27, with permission of the Elsevier Scientific Publishing Company.)

0 for A, 28 for D, and 42 for G.) The van Deemter curves in these parts show the right branch, the slope of which indicates the resistance to mass transfer, exhibiting lower slopes at higher temperatures. Therefore, at higher temperatures columns such as this would permit higher flow rates at relatively low cost to column efficiencies. In addition, as the temperature increases the minimum decreases in these curves, showing higher efficiency. Further, the minimum moves to higher flow rates, allowing higher plate delivery rate even at highest efficiency: "Under these conditions it is clearly advantageous to work at high temperatures" [21]. "These conditions" would tend to be found with wall-coated open tubular columns, particularly those of larger internal diameters, or with conventional packed columns of very low loadings.

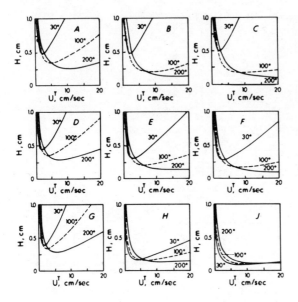

		30°		100°		200°	
Chart	Compound	c_{liq}	c_{gas}	c_{liq}	c_{gas}	c_{liq}	c_{gas}
A	propane	50	50	4.7	39	0.5	20
B	propane	91	9.1	8.6	7.1	0.8	3.7
C	propane	99	1.0	9.4	0.8	0.9	0.4
D	pentane	20	78	5.8	48	1.0	28
E	pentane	35	14	10	8.7	1.8	5.1
F	pentane	38	1.6	12	0.9	2.0	0.5
G	octane	1.0	109	1.3	71	1.0	42
H	octane	1.9	20	2.4	13	1.8	7.6
J	octane	2.1	2.2	2.6	1.4	2.0	0.8

c_{liq} and c_{gas} given in sec × 10^3

FIGURE 11.7 These van Deemter plots are discussed in detail in Sec. 11.2A, *Tendency of Overall Efficiency to Improve at Higher Column Temperatures*. They explore column efficiency as a function of partition ratio, temperature, and the liquid-to-gas ratio of resistances to mass transfer. Note also the importance of the magnitude of resistance to mass transfer: The lower the resistance to mass transfer, the lower the slope of the right branch, the more efficient the column, and the wider the range of the higher gas velocities. (From *Talanta*, Ref. 21, reprinted with permission of W. E. Harris and H. W. Habgood, and Pergamon Press, Ltd.)

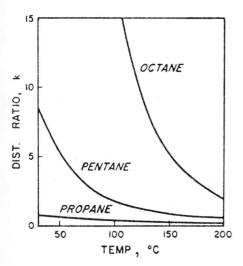

FIGURE 11.8 Data from Fig. 11.4 are replotted here to show the variation of partition ratio with temperature for propane, pentane, and n-octane. Both the partition coefficient and the partition ratio vary exponentially with the reciprocal of the absolute temperature. (From *Talanta*, Ref. 21, reprinted with permission of W. E. Harris and H. W. Habgood, Pergamon Press, Inc.)

The behavior shown in the right column is roughly the converse of that shown in the left. Resistance to mass transfer in the stationary phase here greatly exceeds that in the gas phase, at least at 30°C. (C_{liquid}/C_{gas} = 99 in part C, about 24 in part F, although only unity in part J.) However, as the temperature increases, the diffusivity of the liquid phase increases exponentially [28]. Primarily because of this, the resistance to mass transfer of the liquid phase decreases sharply with increasing temperature.

Higher temperature sharply improves column efficiency for very early peaks, *i.e.*, k about 1, Fig. 11.8. This improvement can be seen in part C of Fig. 11.7. With later peaks, however (part J, k = 50 to 100 at 30°, 20 to 30 at 100°), where in this example the resistances are about the same--and *low*--at all temperatures, the van Deemter curves are, in consequence, about the same for all temperatures and here are also nearly ideal. The minima are consistently lower, corresponding to higher column efficiency. The right branches are very nearly horizontal, permitting almost any increase in flow rate, and thus in plate delivery, without cost to higher column efficiency.

(Note that these favorable conditions require not merely a balance of resistances, but also a balance of *low* resistances. Compare the 30°C curve of part A with any of those in part J, for instance. Both have a balance of resistances, but the resistances involved in part J are only about 2% of those in part A. One never increases the resistance to mass transfer of one phase in order to achieve a balance with that of the other phase.)

Experimental data [29-31] support the conclusion of Harris and Habgood that "column efficiency generally should increase with increasing temperature" [21].

Separability versus *Column Temperature*

Separability involves column efficiency, solvent efficiency, and the particular components to be separated. Inadequate separability, that is, a difficult separation, occurs when two similar components cannot quite be separated by a given column. We have already seen that a given column shows a maximum column efficiency--a maximum number of plates--for a given component at a *best* temperature that is characteristic for that component and column. Does each separation, and thus any difficult separation, also have a *best* temperature?

Giddings found that, in theory, isothermal separability "is increased by lowering the temperature" [22, 32]. This suggests a preferred temperature direction rather than a best temperature.

Scott [30] studied the effect of temperature on the performance of WCOT columns. He concluded:

> It may be theoretically deduced that any given column would have an optimum operating temperature to give the maximum resolution between any two given substances. If the retention ratio of the pair of substances concerned remains constant or increases with operating temperature, the best resolution will be obtained on a column with a thick film of stationary phase and low [ratio of gas volume to liquid volume], at a comparatively high temperature.
>
> If the retention ratio of the two substances to be separated decreases with increase in temperature, then the best resolution would be obtained on columns with thinner films and higher [gas-to-liquid volume ratios] operated at lower temperatures.

Whether the temperature should be higher or lower depends on the two components to be separated. See Fig. 11.9, for instance [33, 34]. Each component has a best temperature for column efficiency, but for the pair that temperature may also be the very one at which the two retention times coincide. In that case, as Giddings recommended, some other temperature, "well-removed," should be used [25].

Does isothermal operation maximize separability, or is separability enhanced by temperature programming? In their book, *Programmed Temperature Gas Chromatography*, [35], Harris and Habgood conclude, "When the principal concern is separation of two closely spaced peaks, the best resolution probably will be obtained from isothermal operation. As [a temperature program rate] approaches zero...resolution approaches the value it would have at the initial [lowest] temperature [of the program]..."

Halasz and Heine agree: "Programmed gas chromatography [of] the temperature and/or the inlet pressure always leads to shorter times of analysis but to *worse* resolution than is achievable with optimized constant temperature or pressure operation" (their italics) [36]. (At the end of the next section, we point out that programmed temperature operation does *not* always lead to shorter times of analysis.

Sec. 11.2 Column Behavior as a Function of Temperature 237

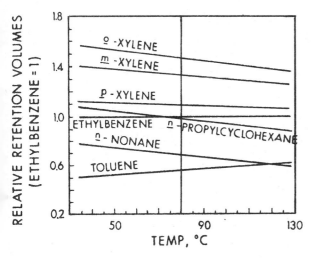

FIGURE 11.9 Each component has a best temperature for column efficiency, and solvent efficiency increases with decreasing temperature. Nevertheless, separability depends on the case at hand. Consider ethylbenzene and n-propylcyclohexane at 75°C, or toluene and n-nonane at 120°C. The separability of these pairs increase as the temperatures cited are moved away from, and are impossible at those temperatures. (Reprinted with permission from J. V. Mortimer and P. L. Gent, Ref. 33. Copyright by the American Chemical Society.)

B. Programmed Temperature Gas Chromatography (PTGC)

History, Characteristics, Effects on Instrumentation

The best description of PTGC is the just-mentioned book of W. E. Harris and H. W. Habgood, *Programmed Temperature Gas Chromatography* [24].

The PTGC concept was introduced in 1952 [37]. Six years later, simple automatic instrumentation for PTGC became commercially available and the concept acquired its name. The first review of PTGC marked the practical advent of the technique and the birth of a major company in gas chromatography [1].

In PTGC, the column temperature is raised some given number of degrees per minute during elution of a sample. From front-panel controls on almost any gas chromatograph that is equipped for PTGC, the operator can preset the following constants for a given run: An initial temperature, a corresponding initial period during which the initial temperature will obtain, a rate of temperature rise, a maximum temperature at which the programming will stop, and a final period during which the maximum temperature will obtain.

Figure 11.10 shows the striking differences between isothermal and programmed temperature gas chromatograms [24, 38]. In isothermal gas chromatography, peak width increases linearly with retention time [39] and, as Eq. (11.1) describes, retention time increases exponentially with carbon number in a homologous series. In contrast,

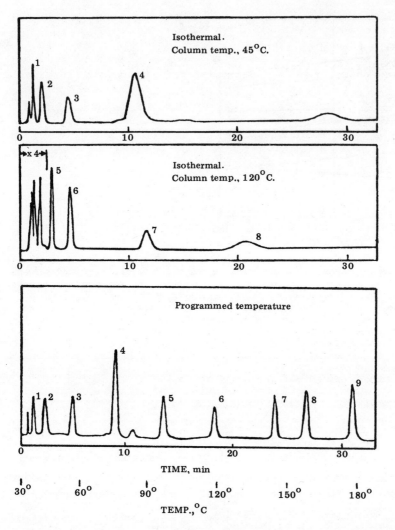

FIGURE 11.10 In PTGC, peak widths are constant and retention times increase only linearly with carbon number in a homologous series. With a sample of any appreciabl boiling range, such as the one shown here--boiling range 226°C--PTGC is necessary. Peaks 1 through 6, n-C_3 through n-C_8; 7, $CHBr_3$; 8, m-chlorotoluene; and 9, m-bromotoluene. (Reprinted from Harris and Habgood, Ref. 24, *Programmed Temperature Gas Chromatography*, 1966, with permission of the authors and John Wiley and Sons, Inc. Data originally published by the authors in Ref. 38.)

in PTGC peak width is constant and retention times rise only linearly with carbon number in a homologous series [1]. PTGC is necessary if the sample to be separated displays any appreciable boiling range; PTGC is necessary if survey work is to be at all efficient.

PTGC caused basic and lasting changes in gas chromatographic instrumentation. 1959, most gas chromatographs contained only one column and one detector--a thermal conductivity detector, often housed in the column oven. But thermal conductivity detectors are highly sensitive to both the wall temperature of the detector and the carrier gas flow rate through the detector. For PTGC, the detector had to be thermally isolated in a separate oven with individual temperature control.

(Consider, for the moment, a thermally isolated detector that is supplied with the constant flow rate we discuss in the next paragraph. Such a detector is divorced from the source of the vapors it detects. Thus PTGC has no effect whatsoever on detector response. Similarly, to the extent that quantitative analysis depends on detector response, PTGC has no effect on quantitative analysis.)

PTGC also brought about direct control of carrier gas flow rate. Before PTGC, carrier gas flow was adjusted by pressure control. However, because gas viscosity and therefore column pressure drop increase with increasing temperature, the carrier gas flow tends to decrease during PTGC. Such a decrease is quickly reflected in the deviating baseline signal from a single, flow-sensitive detector. To hold the carrier gas flow constant during PTGC, and thus also the retention times (from run to run) and particularly the baseline (during each run), flow controllers were introduced. They became standard, expectable components in gas chromatographs.

Stationary phase volatility was also involved. The stationary phases so valued in 1959 for their high selectivity are volatile, so volatile that by today's standards they are useless. With programmed temperature, the easily detectable and exponentially increasing concentrations of vapors from these phases would quickly send a baseline off-scale. Detecting and recording a PTGC gas chromatogram with the combination of such phases and one-detector, one-column equipment was almost impossible. A better use of the thermal conductivity detector helped to answer this problem and also alleviated the stringency of flow-control requirements.

The reference elements of a thermal conductivity detector are identical to the sensing elements, and so can serve equally well as detectors. Gas chromatographs were redesigned to accommodate two columns in the same column oven. The thermal conductivity detector was redesigned to serve as two electrically opposed, equally sensitive detectors, one for each column. With the effluents from two matched columns going to equal and opposing sides of the same detector bridge, changes in stationary phase vapor concentration became in theory undetectable and in practice far less important. Detecting and recording a PTGC gas chromatogram became possible, despite stationary phase volatility.

Thus, as mentioned earlier, the prime characteristics of both gas chromatographs and stationary phases became those relevant to the demands of programmed temperature. The gas chromatograph had to be able to bring about sharp but well-controlled changes in column temperature, and yet record a gas chromatogram; the stationary phases had to withstand these changes without decomposing or vaporizing to a prohibitive degree, and yet retain the selectivity relevant to each.

To permit rapid changes in column temperature, high-velocity, high-power, low-mass column ovens were introduced. PTGC or no, the column oven must allow easy column installation and interchange, therefore air had to remain the heat-exchange medium. In the newer designs, however, the air would be moved in much greater quantity and velocity, by and over much more powerful fans and heaters. Further, the column oven would become low-mass: A light inner aluminum shell of low thermal mass, insulated from the more substantial walls of the outer oven.

(Note: PTGC does not always lead to shorter times of analysis. Any routine analysis that involves temperature change requires not only the heating during the analysis but also the subsequent cooling and stabilization before the next analysis. That cooling and stabilization can easily take longer than the separation itself. Separations that require no more than five to ten minutes can be most quickly performed isothermally.)

PTGC in More Detail: Terminology

In the simplest case of PTGC, column heating begins on sample injection. The column temperature is then the *initial temperature* T_0. On injection, the column temperature begins to increase at r degrees per minute. Throughout, the carrier gas flow rate F is constant. The r/F ratio is the *program*. A given component is eluted from the column at a corresponding column temperature T_r known as the *retention temperature*.

The retention temperature T_r is defined as the temperature at which the apex of a given component peak emerges from a given column. However, as studies we are about to cite demonstrate, the retention temperature is more than just an observation with a name.

In PTGC, the retention volume V is related by the following approximation to the experimental observations and parameters that we have just named:

$$V = (T_r - T_0)\frac{F}{r} \qquad (11.23)$$

Thus when the column gas traverses a certain number $(T_r - T_0)$ of degrees under a given r/F program a component showing a PTGC retention volume V is eluted:

$$\frac{r \text{ (degrees/time)}}{F \text{ (volume/time)}} = \frac{(T_r - T_0) \text{ degrees}}{V \text{ volume}}$$

For a given column, the retention temperature T_r is a predictable function of the r/F program. The retention temperature cannot be predicted from column temperature alone, nor from any relationship between the heating rate r and the time from injection to elution. The retention temperature varies with the flow rate F as well as with the heating rate r.

With a given column, the retention temperature of a component not only is dependent on both r and F but also is determined by the ratio of r to F, the r/F program. Fig. 11.11, developed by Harris and Habgood [38, 40], clearly demonstrates this.

The ordinate of Fig. 11.11 is the r/F program. The abscissa is the retention temperature T_r.

Sec. 11.2 Column Behavior as a Function of Temperature 241

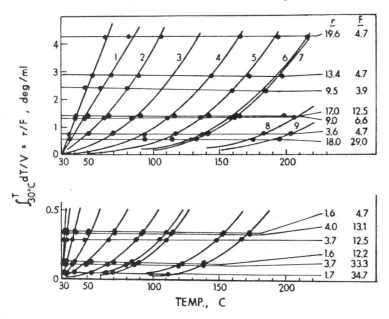

FIGURE 11.11 The retention temperature (abscissa) varies with the program r/F (ordinate). Changing only either the heating rate r or the flow rate F changes the retention temperature; changing both, so as to hold r/F constant, leaves the retention temperature unchanged. The solid lines are calculated from isothermal data; the points are experimental. Components 1 through 6, $n-C_3$ through $n-C_8$; 7, m-fluorotoluene; 8, m-chlorotoluene; and 9, m-bromotoluene. Stationary phase: Apiezon L. (Reprinted from W. E. Harris and H. W. Habgood, Ref. 24, *Programmed Temperature Gas Chromatography*, 1966, with permission of the authors and John Wiley and Sons, Inc. Original data published by the authors in Ref. 38.)

The solid lines in Fig. 11.11 are theoretical, calculated from isothermal data. The dots are experimentally observed retention temperatures. The numbers of the curves correspond to components. The heating rates r and the carrier flow rates F of the various r/F programs are shown at the right of the figure.

It can be seen that if either the heating rate or the flow rate is changed by itself, the retention temperature of a given component changes. However, if both the heating rate and the flow rate are changed so as to hold the r/F program constant, the retention temperature of the component does not change.

(The converse of this directly concerns PTGC reproducibility. If a retention temperature is to be precisely reproduced, both the heating rate r and the flow rate must be reproducible *from other settings* with corresponding precision, and also, from a practical point of view, with corresponding ease. At this writing, both temperature and heating rate are precisely and easily reproducible, but flow rate is not. In Sec. 11.2B, *The Retention Index in PTGC*, we comment further on flow control.)

Although the retention temperature can be predicted from isothermal data [41], it pertains only to PTGC. There is, however, a temperature that explicitly relates PTGC to isothermal gas chromatography: The *significant temperature* T'.

The concept of the significant temperature T' was formulated by Giddings [42]. The significant temperature T' is related to the program r/F and to the retention temperature T_r. In his assumptions, Giddings assumed first that the heating rate r would be constant from T_0 to T_r, and second that the flow rate F would not vary rapidly with change of temperature (it does not). During the development of his thesis, Giddings commented, "We will find for many purposes that a programmed temperature process may be considered equivalent to an isothermal process, providing the latter is run at T'. For the preceding reasons the temperature T' is called the 'significant temperature' for programmed runs."

A column programmed from T_0 to T_r by a program r/F yields about the same resolution as the same column operated not only isothermally but also optimally at the significant temperature T' [22]. A column optimized for isothermal gas chromatography is therefore also optimized for PTGC. The optimal conditions include such facets as, for a packed column, low loading and small particle size and size distribution.

Giddings commented that the significant temperature T' is "ordinarily about 40°C below the retention temperature" T_r [42, 43]. This comment in no sense constitutes a definition. Many samples benefit from temperature programming over less than a 40° range. The definition of significant temperature does not really allow assigning any given number of degrees as the specific temperature difference between the significant and retention temperatures.

Column Operation and Performance in PTGC

PTGC column requirements. We have just seen that column specifications are about the same for isothermal gas chromatography as for PTGC. Given a stable stationary phase, a good isothermal column is equally good for PTGC.

Sample introduction for trace detection. PTGC can ease markedly the requirement for sample introduction as a plug. If the initial column temperature T_0 is far below the retention temperature T_r at which the solute will be eluted, then each incoming solute molecule is immobilized at the head of the column largely without regard to the time of its arrival. All the solute molecules, which may have been entering the column over quite a period of time relative to the normal duration of sample injection now appear in one narrow band at the head of the column--just as with plug introduction. The cool column integrates the arriving solutes.

Somewhat more explicitly, the following equation is approximately true for a given solute in a gas chromatographic column:

$$\log V = c \frac{1,000}{T} \qquad (11.24)$$

where V is the retention volume, T is the absolute temperature, and c is approximately constant for a given solute [44]. With Apiezon L as stationary phase, c can be observed to be about 1 for n-butane as solute, about 2 for n-octane, and about 2.3 for m-bromotoluene [44].

Sec. 11.2 Column Behavior as a Function of Temperature

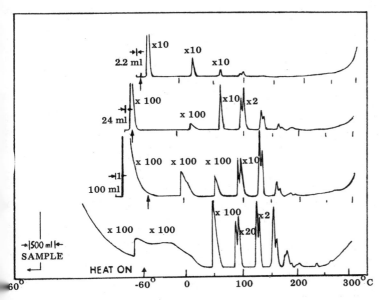

FIGURE 11.12 Injecting a solute at a column temperature far below that at which the solute will be eluted not only eases or effectively cancels the need for plug introduction but also improves trace detectability by stripping from greatly extended samples. Here, high-boiling traces were stripped by an alumina column at −60°C from natural gas. Note the enlarged samples, up to 500 ml in size. (Reproduced with permission of the National Research Council of Canada from the *Canadian Journal of Chemistry*, 43, 1560-1568, 1965, Ref. 31.)

For absolute temperatures T_0 and T_r, corresponding retention volumes V_0 and V_r, and c assumed constant for a given solute,

$$\log \frac{V_0}{V_r} = c \left[\frac{1,000}{T_0} - \frac{1,000}{T_r} \right] \tag{11.25}$$

If we arbitrarily take c as 2, then for sample introduction at 0°C and elution at 100°C, log (V_0/V_r) equals about 2, and thus V_0 about 100 V_r. This is the measure of permissible sample dilution. In other words, under these conditions the usual amount of n-octane could be injected, but it could exist in the sample at 1% of its normal concentration, and 100 times the normal volume of sample could be injected during an almost indeterminate period. The n-octane peak eluted at 100°C would not change shape or decrease in peak height [45].

This easing of requirement for plug introduction and increase in trace detectability brought about by extended sample size without increased peak width are illustrated in the PTGC chromatograms shown in Fig. 11.12 [46].

The initial temperature T_0. The low column temperature that can be used especially for trace concentration during the injection of unusually dilute samples is not to be considered a proper initial temperature T_0. Having the initial temperature so unusually low does not affect the ultimate separability of the sample components, but it can waste time if the column is programmed throughout [47]. After such an injection, the operator should discontinuously raise the column temperature up to a proper initial temperature, rather than program it to the initial temperature.

Otherwise, the initial temperature should be lower rather than higher. We have already mentioned that separability increases with decreasing temperature, as the separation function of Giddings suggests [22, 32]. If the initial temperature is favorably low, then during the components' initially slow movement, once the program has commenced "...very effective separation is taking place in the initial part of the column" [47]. In any event, the initial temperature and the program should be set low enough that the first components of interest take several--say, 5--times as long to be eluted as the air peak [48].

The program r/F. In accordance with previous remarks, for best separability and resolution the initial temperature T_0 should be relatively low [32] and the heating rate r and program r/F should be minimal [35, 49]. Increasing the heating rate r "usually leads to an impaired separation" [49, 50].

With very long columns such as open tubular, too high a heating rate r may cause the components eventually to exist primarily in the mobile phase. In such a case "...the latter part of the column may actually hinder separation...Long capillary columns should therefore be run at high flow velocities or low heating rates, or both" [47].

On the other hand, increasing the heating rate r reduces the time per analysis if a corresponding degree of resolution may be sacrificed.

Column efficiency. Habgood and Harris [38, 51] point out that the number of theoretical plates developed during a PTGC run can be measured from a corresponding isothermal chromatogram in which

$$n = 16 \left(\frac{V_{T_r}}{w_b} \right)^2 \tag{11.26}$$

Here, V_{T_r} is the retention volume of the component measured isothermally with the column held at the retention temperature T_r. The width w_b is measured conventionall Testing Eq. (11.26), Said found good agreement between theory and experiment [52]. The plate height for PTGC is about that determined isothermally at the retention temperature [38]. This "is not unexpected...the greater part of the band movement takes place just below the retention temperature" [53].

Resolution. For isothermal gas chromatography, Ober [54] suggested measuring the resolution of two components as the difference $V_j - V_i$ between the two retention volumes of these components, divided by the average $(V_i + V_j)/2$ of the two. Harris and Habgood [55] refer to this as the "intrinsic resolution" R_I. The isothermal

Sec. 11.2 Column Behavior as a Function of Temperature

intrinsic resolution is

$$R_{IT} = \frac{V_j - V_i}{(V_i + V_j)/2} \qquad (11.27)$$

An expression [56] equivalent to Eq. (11.27) is

$$R_{ij} = R_I \frac{\sqrt{n_{av}}}{4} \qquad (11.28)$$

Here the observed resolution R_{ij} is defined conventionally as the peak separation $V_j - V_i$ divided by the average peak width $(w_{bi} + w_{bj})/2$; and n_{av} is the average number of theoretical plates shown by the two peaks. Equation (11.28) expresses this observed resolution R_{ij} as the product of the intrinsic resolution R_I, which expresses solvent efficiency, and the theoretical plate function $\sqrt{n_{av}}/4$, which expresses the contribution of column efficiency.

The corresponding expression for PTGC [57] replaces the retention volumes V_i and V_j by the program r/F and the retention temperatures T_{Ri} and T_{Rj}:

$$R_I = \frac{1}{(r/F)} \frac{T_{Rj} - T_{Ri}}{[(V_{T_{Ri}} + V_{T_{Rj}})/2]} \qquad (11.29)$$

Notice that the intrinsic resolution is inversely related to the program r/F.

Programmed intrinsic resolution R_I increases with decreasing initial temperature T_0 [58].

What can we now conclude about programmed *versus* isothermal resolution? Giddings said, "The heating process may be considered merely as a mechanism for obtaining a range of temperatures in proper sequence with no direct effect upon the separability obtainable...The degree of separation of neighboring peaks...is equal for programmed temperature and isothermal chromatography providing the latter is operated at or near the significant temperature, T'...[Given] a temperature T', the optimum conditions for this temperature must be obtained" [22].

How closely this statement holds depends on whether a given separation has an optimum temperature. If it does not, then a succession of column temperatures will do as well as one. However, if an optimum temperature does exist, or if the separation improves the lower the column temperature, then only an isothermal separation will produce maximum resolution.

Harris and Habgood said, "...within the restriction of certain assumptions-- closely spaced peaks, retention volumes large compared with dead-space volume, low initial temperature, and retention temperature signficantly higher than initial temperature--intrinsic resolution for PTGC is equal to isothermal intrinsic resolution at the significant temperature" [55]. But in their concluding remarks they also said, as we have seen, "When the principal concern is separation of two closely spaced peaks the best resolution probably will be obtained from isothermal operation. As [a temperature program rate] approaches zero...resolution approaches the value it would have at the initial [lowest] temperature [of the program]..." [35].

In sum, we can view isothermal gas chromatography as the limiting case of PTGC for maximized resolution. PTGC resolution increases with decreasing heating rate and initial temperature. Therefore, in the limit, PTGC resolution approaches but does not equal the resolution of optimized isothermal gas chromatography conducted at the designated significant temperature.

The Retention Index in PTGC

The PTGC retention index $I_{PT(i)}$ of a substance i can be determined from the retention temperature T_{Ri} of that substance and the bracketing retention temperatures T_{Rn-C_z} and $T_{Rn-C_{z+1}}$ of the relevant normal paraffins $n-C_z$ and $n-C_{z+1}$:

$$I_{PT(i)} = 100 \frac{T_{Ri} - T_{Rn-C_z}}{T_{Rn-C_{z+1}} - T_{Rn-C_z}} + 100\,z \qquad (11.30)$$

The isothermal Kovats retention indices tend by the properties of logarithms to cancel out variations of experimental parameters such as column length and loading, and flow rate. These indices are conveniently determinable, experimentally stable, adequately precise. The isothermal indices have become standard tools that will probably become steadily more useful.

In contrast, a PTGC index determined according to Eq. (11.30) does not cancel out variations in experimental parameters. For instance, a retention temperature corresponds to a given program r/F. Precisely reproducing a retention temperature demands precisely reproducing both the heating rate r *and the flow rate* F. But precisely reproducing a flow rate from another flow setting is an almost insuperable experimental difficulty.

Quite generally, a given flow rate cannot within any reasonable time be reproduced within 0.5% of its value, once the setting on the flow controller has been changed by 5 or 10 ml/min. Indeed, the flow rates for open tubular columns are not even directly controlled--the flow controllers are bypassed for such columns. Such programmed chromatograms cannot be reproduced well, and therefore require many internal standards--see, for instance, Figs. 10.5 and 10.6.

PTGC retention indices that correspond to the isothermal Kovats retention indices have not been and are not likely to become usefully standard tools in gas chromatography. Watts and Kekwick [59] suggest a relative retention temperature--the ratio of the retention temperature of a given substance to that of a standard. Whether this becomes widely accepted remains to be seen.

Sec. 11.3 Stationary Phase Stability as a Function of Temperature

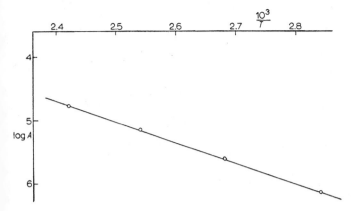

FIGURE 11.13 The logarithm of the stationary phase vapor pressure varies linearly with the reciprocal of the absolute temperature. Stationary phase: di-n-propyl phthalate. (Reproduced with permission of the Elsevier Scientific Publishing Company from N. Petsev and C. Dimitrov, Ref. 61, in the *Journal of Chromatography*.)

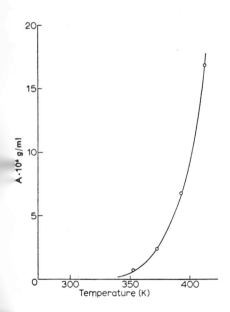

FIGURE 11.14 The vapor pressure of a stationary phase increases exponentially with linear increase in temperature. With a "thermally stable" stationary phase, this increase becomes detectable only at satisfactorily high temperatures. Stationary phase: di-n-propyl phthalate. (Reproduced with permission of the Elsevier Scientific Publishing Company from N. Petsev and C. Dimitrov, Ref. 61, in the *Journal of Chromatography*.)

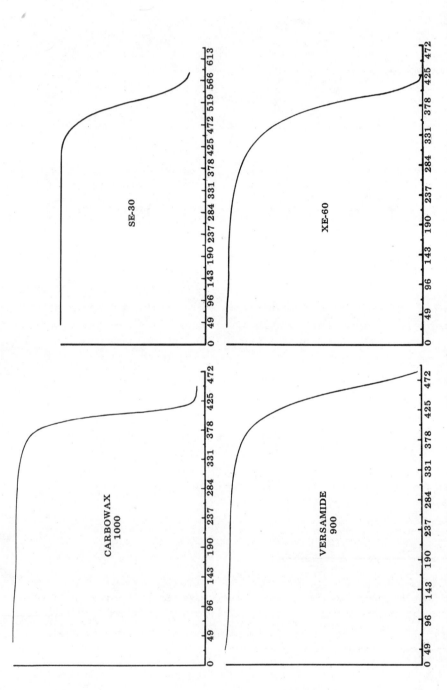

FIGURE 11.15 All stationary phases in gas-liquid chromatography show substantial weight loss with increasing temperature. Each sample weight is plotted along the ordinate; the sample temperature is plotted along the abscissa. Each sample eventually shows almost quantitative weight loss with increasing temperature. (Reproduced with permission of Preston Publications, Inc., from R. W. McKinney, J. F. Light, and R. L. Jordan, Ref. 63, in the Journal of Gas Chroma-

11.3 STATIONARY PHASE STABILITY AS A FUNCTION OF TEMPERATURE

Before temperature programming, a stationary phase was valued for its selectivity, but after for its thermal stability. (Now, given an acceptably adequate thermal stability in the modern stationary phase, high selectivity is once again being sought.)

We distinguish newly made columns from "conditioned" columns. A newly made column may show anomalously high concentrations of detectable vapors. These are removed by conditioning the column. A column is conditioned by holding it at its maximum operating temperature for some minimum period while carrier gas is passed through it. *During conditioning, the exit of a column is not attached to the gas chromatograph.*

Ideally, the stationary phase is stripped of its volatiles by the manufacturer [60]. Ideally, a newly made column needs no conditioning, but *always test*. Raise the column temperature very slowly from a relatively low level to the maximum operating temperature, while the detector is set at high sensitivity; watch to see if the baseline is adequately stable.

Let us assume we have a conditioned column. For a given temperature, the amount of stationary phase carried per unit time from a conditioned column is directly proportional to the carrier gas flow rate [61, 62]. In general, the vapor pressure of a conditioned stationary phase in a gas chromatographic column is described by Eq. (11.14)(see Fig. 11.13)[61]. The more familiar display of this behavior is shown in Fig. 11.14: the baseline surges exponentially upward as the column temperature is increased. "Thermal stability" means that this surge is held to within a satisfactory limit until a satisfactorily high temperature is reached.

An exponential increase in vapors from the stationary phase resulting from a linear increase in column temperature may be expected. The details differ from phase to phase, but the phenomenon is general (see Fig. 11.15)[63], unless the stationary phase is chemically bound. (Then, the maximum operating temperature must usually be observed even more carefully, although this situation may finally be changing. See the discussion of the bound stationary phase in Chap. 8, Supports, and in *Recent Developments* at the end of Chap. 8.)

In short, each column has a maximum operating temperature. Beyond that temperature, temperature programming may not be further applied except at hazard to column and instrument. At that temperature, the behavior described by Eq. (11.1) and shown in Fig. 11.1 again sets in. Pressure programming [64], which means automatically and progressively increasing the carrier gas flow rate, alleviates the difficulty when the maximum operating temperature is reached. But the better answer, the one that has been adopted, is to produce and use stationary phases that have satisfactorily high maximum operating temperatures.

REFERENCES

1. A. J. Martin, C. E. Bennett, and F. W. Martinez, Jr., in *Gas Chromatography* (H. J. Noebels, R. F. Wall, and N. Brenner, editors), Academic, New York, 1961, pp. 363-374.
2. N. H. Ray, *J. Appl. Chem.* **4**, 21 (1954).
3. J. D. Ogilvie, M. C. Simmons, and G. P. Hinds, *Anal. Chem.* **30**, 25 (1958).
4. D. H. Desty and B. H. F. Whyman, *Anal. Chem.* **29**, 320 (1957).
5. B. M. Craig and N. L. Murty, *Can. J. Chem.* **36**, 1297 (1958).
6. J. C. Hawke, R. P. Hansen, and P. B. Shorland, *J. Chromatog.* **2**, 547 (1959).
7. E. R. Adlard, in *Vapour Phase Chromatography* (D. H. Desty, editor), Butterworths, London, 1957, pp. 98-114.
8. H. Purnell, *Gas Chromatography*, Wiley, New York, 1962, Chapter 10.
9. A. B. Littlewood, C. S. G. Phillips, and D. T. Price, *J. Chem. Soc. 1955*, 1450 (1955).
10. E. A. Moelwyn-Hughes, *Physical Chemistry*, 2nd ed., Pergamon, London, 1961.
11. *Ibid.*, Chapter XVI, Table 9, p. 701.
12. *Ibid.*, Chapter XVI, Figure 1, p. 703.
13. F. T. Wall, *Chemical Thermodynamics*, 2nd ed., W. H. Freeman, San Francisco, Calif., 1965, pp. 355-357.
14. E. F. G. Herington, in *Vapour Phase Chromatography* (D. H. Desty, editor), Butterworths, London, 1957, pp. 5-14.
15. H. Purnell, *Gas Chromatography*, Wiley, New York, 1962.
16. F. Daniels, *Outlines of Physical Chemistry*, Wiley, New York, 1958, Chap. IX, Equation 20, p. 182.
17. J. H. Hildebrand and R. L. Scott, *The Solubility of Nonelectrolytes*, 3rd ed., Reinhold, New York, 1950, pp. 77-79.
18. D. W. Grant and G. A. Vaughan, *J. Appl. Chem.* **6**, 145 (1956).
19. F. T. Wall, *Chemical Thermodynamics*, 2nd ed., W. H. Freeman, San Francisco, Calif., 1965, p. 188.
20. E. A. Moelwyn-Hughes, *Physical Chemistry*, 2nd ed., Pergamon, London, 1961, p. 261.
21. W. E. Harris and H. W. Habgood, *Talanta* **11**, 115 (1964), and also Figure 3.
22. J. C. Giddings, in *Gas Chromatography* (N. Brenner, J. E. Callen, and M. D. Weiss editors), Academic, New York, 1962, p. 69.
23. A. G. Polgar, J. J. Holst, and S. Groenings, *Anal. Chem.* **34**, 1226 (1962).
24. W. E. Harris and H. W. Habgood, *Programmed Temperature Gas Chromatography*, Wiley, New York, 1966.
25. J. C. Giddings, in *Gas Chromatography* (N. Brenner, J. E. Callen, and M. D. Weiss editors), Academic, New York, 1962, p. 73.
26. W. J. DeWet and V. Pretorius, *Anal. Chem.* **30**, 325 (1958).
27. T. Kambara and H. Kodama, *J. Chromatog.* **17**, 66 (1965).
28. E. A. Moelwyn-Hughes, *Physical Chemistry*, 2nd ed., Pergamon, London, 1961, p. 715.
29. J. J. Duffield and L. B. Rogers, *Anal. Chem.* **32**, 340 (1960).
30. R. P. W. Scott, *J. Inst. Petrol.* **47**, 284 (1961).
31. L. Hollingshead, H. W. Habgood, and W. E. Harris, *Can. J. Chem.* **43**, 1560 (1965)

References

32. J. C. Giddings, *Anal. Chem. 32*, 1707 (1960).
33. J. V. Mortimer and P. L. Gent, *Anal. Chem. 36*, 754 (1964).
34. W. E. Harris and H. W. Habgood, *Programmed Temperature Gas Chromatography*, Wiley, New York, 1966, Figure 5.10.
35. *Ibid.*, p. 138.
36. I. Halasz and E. Heine, in *Progress in Gas Chromatography* (J. H. Purnell, editor), Interscience, New York, 1968, p. 202.
37. J. H. Griffiths, D. H. James, and C. S. G. Phillips, *Analyst 77*, 897 (1952).
38. H. W. Habgood and W. E. Harris, *Anal. Chem. 32*, 450 (1960).
39. S. Dal Nogare and R. S. Juvet, Jr., *Gas-Liquid Chromatography*, Interscience, New York, 1962, p. 63.
40. W. E. Harris and H. W. Habgood, *Programmed Temperature Gas Chromatography*, Wiley, New York, 1966, Figure 3.05.
41. *Ibid.*, Section 3.03, Section 5.04.
42. J. C. Giddings, in *Gas Chromatography* (N. Brenner, J. E. Callen, and M. D. Weiss, editors), Academic, New York, 1962, pp. 57-77.
43. W. E. Harris and H. W. Habgood, *Programmed Temperature Gas Chromatography*, Wiley, New York, 1966, p. 104.
44. *Ibid.*, Figure 3.03.
45. *Ibid.*, p. 119.
46. *Ibid.*, Figure 9.22.
47. J. C. Giddings, in *Gas Chromatography* (N. Brenner, J. E. Callen, and M. D. Weiss, editors), Academic, New York, 1962, p. 71.
48. *Ibid.*, p. 59.
49. *Ibid.*, p. 70.
50. S. Dal Nogare and W. E. Langlois, *Anal. Chem. 32*, 767 (1960).
51. W. E. Harris and H. W. Habgood, *Programmed Temperature Gas Chromatography*, Wiley, New York, 1966, Equation 5.07.
52. A. S. Said, in *Gas Chromatography* (N. Brenner, J. E. Callen, and M. D. Weiss, editors), Academic, New York, 1962, pp. 79-90.
53. W. E. Harris and H. W. Habgood, *Programmed Temperature Gas Chromatography*, Wiley, New York, 1966, p. 115.
54. S. S. Ober, in *Gas Chromatography* (V. J. Coates, H. J. Noebels, and I. S. Fagerson, editors), Academic, New York, 1958, pp. 41-50.
55. W. E. Harris and H. W. Habgood, *Programmed Temperature Gas Chromatography*, Wiley, New York, 1966, Section 5.05.
56. *Ibid.*, Equation 5.27, p. 122.
57. *Ibid.*, Equation 5.29, p. 123.
58. J. F. Fryer, H. W. Habgood, and W. E. Harris, *Anal. Chem. 32*, 1515 (1961).
59. R. B. Watts and R. G. O. Kekwick, *J. Chromatog. 88*, 165 (1974).
60. C. R. Trash, *J. Chromatog. Sci. 11*, 196 (1973).
61. N. Petsev and C. Dimitrov, *J. Chromatog. 30*, 332 (1967), and also Figures 2 and 3.
62. N. Petsev and C. Dimitrov, *J. Chromatog. 34*, 310 (1968).
63. R. W. McKinney, J. F. Light, and R. L. Jordan, *J. Gas Chromatog. 6*, 97 (1968), and also Figures 2, 8, 10, and 14.
64. L. S. Ettre, L. Mazor, and J. Takacs, in *Advances in Chromatography*, Vol. 8 (J. C. Giddings and R. A. Keller, editors), Dekker, New York, 1969, pp. 271-326.

Chapter 12

DERIVATIZATION

12.1 INTRODUCTION *

Compared to its precursor, the gas chromatographic derivative should be more volatile, separable, stable, or detectable.

In an ideal derivatization, a given precursor is derivatized quantitatively and quickly. Then, without further processing, the reaction mixture can be injected directly into the gas chromatograph. The corresponding derivative, stable and preservable in the reaction mixture, is also stable during gas chromatography, yields a symmetrical peak, and can be separated from similar derivatives.

If a given substance contains two or more functional groups that no one reagent can successfully derivatize, a sequenced derivatization is usually used. The characteristics of an ideal sequenced derivatization are described in Sec. 12.7B. Such a derivatization was first developed by Horning, Moss, and Horning [1] for the catecholamines.

This chapter is a selective review of current optimum practices in certain areas of derivatization for gas chromatography. The areas are primarily those of various biologically active substances. It will be seen that each area has its own characteristics and, therefore, its own peculiarities of derivatization practice.

This derivatization review also appears elsewhere in abbreviated form [2]. Other reviews of derivatization for gas chromatography appeared in 1967 [3], 1974 (in Japanese) [4], 1975 [5], 1976 [6], and 1977 [7].

12.2 ABBREVIATIONS

BSA   N,O-Bis(trimethylsilyl)acetamide
BSTFA   Bis(trimethylsilyl)trifluoroacetamide

---

* In collaboration with Charles A. Feit.

CMDMS	Chloromethyldimethylsilyl-
DAM	Diazomethane (CH_2N_2)
DIMAP	Dimethylamino-dimethyl phosphine
DMF	N,N-Dimethylformamide
DMF-DMA	N,N-Dimethylformamide-dimethylacetal
DMSO	Dimethylsulfoxide
DNBS	2,4-Dinitrobenzenesulfonic acid
DNP	2,4-Dinitrophenyl-
DNPH	2,4-Dintrophenylhydrazine
HFB	Heptafluorobutyryl
HFBA	Heptafluorobutyric anhydride
HMDS	Hexamethyldisilazane
IMDMS	Iodomethyldimethylsilyl-
MBTFA	N-Methyl-bis(trifluoro)acetamide
MO	Methyl oxime
MOPEG	3-Methoxy-4-hydroxyphenylglycol
MPA	N-Methyl-phenethylamine
MSTFA	N-Methyl-N-trimethylsilyl-trifluoroacetamide
NBB	n-Butyl boronate
PA	Phenethylamine
PFB	Pentafluorobenzene (either pentafluorobenzoyl- or -benzyl)
PFP	Pentafluoropropionyl-
PFPA	Pentafluoropropionic anhydride
PGB	Prostaglandin B
$RB(OH)_2$	A boronic acid with radical R
TCE	2,2,2-Trichloroethanol
TFA	Trifluoroacetyl-
TFAA	Trifluoroacetic anhydride
TMAH	Trimethylanilinium hydroxide
TMCS	Trimethylchlorosilane
TMH	Tetramethylammonium hydroxide
TMS	Trimethylsilyl-
TMSDEA	Trimethylsilyldiethylamine
TSIM	Trimethylsilylimidazole

12.3 COMPARISON OF COMMON REAGENTS AND DERIVATIVES

Presented here for reference, the following comparisons among and comments on the most common reagents have been gleaned from the works reviewed for this chapter.

A. Silylating Reagents

In this chapter, the term "silyl" means a group centered on a trisubstituted silicon. The common silyl group is the trimethylsilyl, $-Si(CH_3)_3$, abbreviated TMS.

"Silylation," as used here, indicates the replacement of an active hydrogen by a silyl group. (An early and massive review of silylation was the 1968 book of Pierce [8].)

If it is applicable, silylation is the preferred method of derivatization.

At this writing, the silylating reagents of choice are bis(trimethylsilyl)acetamide (BSA), trimethylsilylimidazole (TSIM), and bis(trimethylsilyl)trifluoroacetamide (BSTFA). (For the structures and abbreviations of these and other silylating reagents see Fig. 12.1.) Of these BSTFA is by far the most effective. BSTFA silylates, for instance, 10 times faster than BSA [9, 10]. BSTFA is usable where BSA is not, for instance, in the silylation of homovanillic acid and vanillyl mandelic acid [11], purines and pyrimidines [12], intact glucuronides [13], shikimic and quinic acids [10] and neuraminic acid [14].

TSIM is less generally effective than either BSA or BSTFA, but excels in silylating highly hindered hydroxyls [15-17]. TSIM is also the reagent of choice for silylation in the presence of water [18, 19] as, for instance, in the silylation of syrups.

BSA-TMCS is much more effective than HMDS-TMCS, but BSA (and BSA-TMCS) may be erratic [18]. For example, BSA may enolize and silylate keto groups that are not affected by either TSIM or BSTFA [20]. Also, although BSA has been found less effective than TSIM for silylation in the presence of water [18], trace water can catalyze silylation by BSA [1, 21].

B. Perfluoro Derivatives

The most common perfluoro derivatizing groups for sensitizing substances to detection by electron capture are the trifluoroacetal (TFA), the pentafluoropropionyl (PFP), the perfluorobenzene (PFB)--either the pentafluorobenzoyl or the pentafluorobenzyl--and the heptafluorobutyryl (HFB). (For the structure, names, and abbreviations of these groups, see Fig. 12.2.) These groups have repeatedly been compared [22-36]. For stability to hydrolysis and high temperatures, ease of overall use, and the degree of imparted sensitivity to detection by electron capture, the perfluorobenzene derivative seems most preferable [32].

The HFB group is stable toward water at pH≤6, but decomposes with aqueous ammonia [25]. It is also thermally unstable [29]. Further, the anhydride HFBA is not easily removed from a reaction mixture by evaporation, although washing the mixture is quite effective, if less convenient [33]. Although in favorable structures the HFB group imparts more sensitivity for detection by electron capture than either the TFA or PFP groups, this is highly dependent on the structure derivatized (indeed for cyclohexylamine there is no improvement at all [37]).

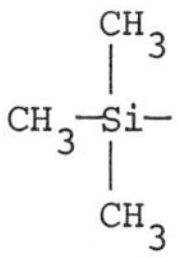

Sec. 12.3 Comparison of Common Reagents and Derivatives

Trimethylsilyl (TMS)

(TMS)Cl
Trimethyl-
chlorosilane
(TMCS)

(TMS)$_2$NH
Hexamethyl-
disilazane
(HMDS)

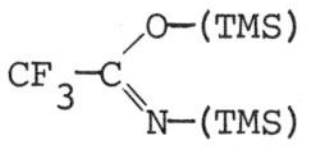

Bis(trimethylsilyl)-
trifluoroacetamide
(BSTFA)

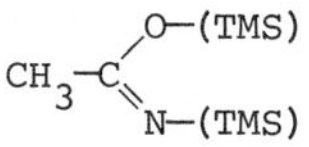

N,O-Bis(trimethylsilyl)-
acetamide
(BSA)

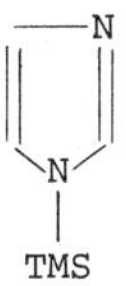

Trimethylsilyl-
imidazole
(TSIM)

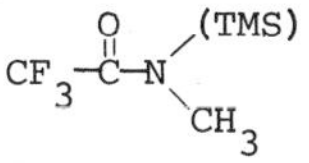

N-methyl-N-trimethylsilyl-
trifluoroacetamide
(MSTFA)

FIGURE 12.1 The trimethylsilyl (TMS) group and some common trimethylsilylating reagents.

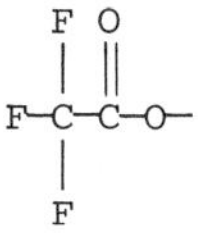

Trifluoroacetyl
(TFA)

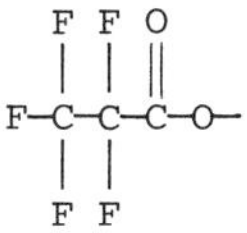

Pentafluoro-
propionyl
(PFP)

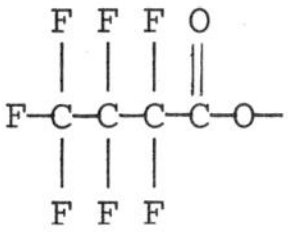

Heptafluoro-
butyryl
(HFB)

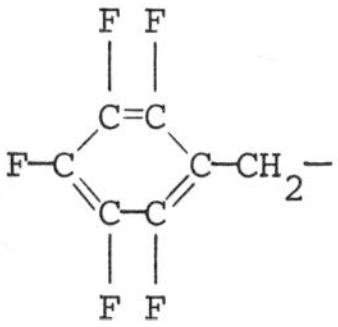

Pentafluorobenzyl
(PFB)

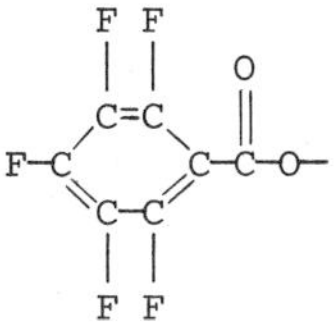

Pentafluorobenzoyl
(PFB)

FIGURE 12.2 Some common perfluoro derivative groups.

12.4 HYDROXY COMPOUNDS

A. Alcohols

Flame Ionization Detection

Typically fast, mild, and convenient, silylation is the most widely used method of derivatizing alcohols. See, for example, quantitative silylations using BSTFA in the determination, in one case, of vitamin D_2 (30 min, room temperature)[38] and, in another case, of the components of a polyethylene terephthalate prepolymer mixture--ethylene and diethylene glycols, mono-(2-hydroxy-ethyl)- and bis(2-hydroxyethyl) terephthalates, and terephthalic acid--(10 min, 80°C)[39]. In each case, portions of the reaction mixture itself were injected into the gas chromatograph.

Although the usual silyl derivative is the trimethylsilyl, the higher trialkyls may fractionate component groups better. For instance, the tri-n-butylsilyl derivatives separate the dihydroxy- from the monohydroxycannabinoids better than the trimethylsilyl derivatives [40]. (The conditions of derivatization--here, 30 min at room temperature--remain unchanged with increase in alkyl chain length.)

The promising boronic acids, $RB(OH)_2$, form cyclic boronates with pairs of hydroxyls. The reaction, for practical purposes instantaneous at room temperature, can occur either in aqueous solution or with the dried sample. The reaction mixture is injectable. (For stability in long-term storage, the derivatives should be dried.) In one illustrative application, the n-butyl boronates of sorbitol, mannitol, and galactitol were formed and used in the assay of sorbitol by gas-liquid chromatography [41].

Compared to silyl derivatives, alditol acetate derivatives are much more trouble to prepare. This is primarily because of the first step, a long reduction followed by a repetitive removal of the reducing agent, $NaBH_4$. (A complete alditol acetate derivatization [42] is described in detail at the end of Sec. 12.10 C, Amino Sugars.) However, the alditol acetates are stable and yield only one derivative per derivatized component (one sugar, silylated, can yield several silyl derivatives--see the discussion of carbohydrate derivatization in Sec. 12.10, Carbohydrates). A given acetate derivative takes over twice as long as the corresponding trimethylsilyl derivative to be eluted [43]. Except for the longer elution times and the lack of separation "of the 1,2(2,3)- and 1,3-diacyl-glycerols when run as the acetates," Myher and Kuksis [43] found the separation of the trimethylsilyl and the alditol acetate derivatives of the natural diacylglycerols "equally effective."

Electron Capture Detection

Braestrup [44] compared the TFA, HFB, and PFP derivatives of 3-methoxy-4-hydroxy phenylglycol (MOPEG). "Pentafluoropropionyl was chosen because of high sensitivity towards [electron capture detection] and ease of preparation, as previously described in detail by others" [22, 23]. The preparation required contact of MOPEG with pentafluoropropionic anhydride (PFPA) for 15 min at room temperature, followed by removal

Sec. 12.4 Hydroxy Compounds

of the excess PFPA by a stream of nitrogen at room temperature. The sample was injected after being taken up in ethyl acetate.

Gas Chromatography-Mass Spectrometry

The Braestrup study just cited involved both electron capture and mass fragmentographic detection of the PFP derivative (mass spectrometry, including mass fragmentography, is discussed in Chap. 13, Qualitative Analysis). "These two quantification methods give virtually identical results...the [mass spectrometer] gives more exact specificity by measuring one or more characteristic mass fragments of the compound at its [gas-liquid chromatographic] retention time...the mass fragmentographic technique must be regarded as conclusive evidence for the identity and purity of the MOPEG [as] measured by [gas-liquid chromatographic-electron capture detection] procedure alone" [44].

B. Phenols

Flame Ionization Detection

Amines, stable as salts but unstable as free bases, are best derivatized from the salts. This also eliminates one preliminary extraction. Accordingly, and using BSA for 20 min at room temperature, Smith and Stocklinski were able to silylate the phenolic dihydroxy tertiary amine, apomorphine, as its HCl salt (the salt was actually the hydrochloride hemihydrate)[45].

Trimethylsilylimidazole (TSIM) is not a reagent of choice for the silylation of nonhindered hydroxyls. However, in tests of all the silylating reagents, TSIM most quickly silylated hindered hydroxyls [16, 17](see more in Sec. 12.11 Steroids). It has been found that trifluoroacetic acid (TFA)(5% vol/vol) speeds the hindered-hydroxyl silylation of any reagent [16]. In this, TFA is more effective, volume for volume, than TMCS. TFA speeds hindered-hydroxyl silylation by the slower reagents more than by the faster; it speeds silylation by TSIM only slightly. But, TFA-promoted or no, TSIM was still found the most effective reagent for the silylation of hindered hydroxyls [16].

Flame Photometric Detection

Nonhindered phenols can form diethyl phosphate esters. These esters can be selectively and sensitively detected, down to 10 pg of the ester of thymol, by the flame photometric detector. Electron capture detection of thymol pentafluorobenzoate is only 5 times more sensitive, and is less selective [46].

Electron Capture Detection

Among seven derivatives of thymol prepared and tested by McCallum and Armstrong [47] for sensitization to electron capture (EC), the pentafluoroaryl derivatives were judged best on grounds of reactivity, sensitivity, and cost. The pentafluorobenzoate was found about 7 times more sensitive to EC detection than the heptafluorobutyrate (and about 10,000 times more sensitive to EC detection than the underivatized thymol to flame ionization detection).

Ehrsson et al. [25] found that the aqueous stability of the heptafluorobutyric (HFB) ester of p-tert-butylphenol is "very good when the aqueous phase has pH≤6.0, while a slight decomposition is evident after one week with alkaline media. Aqueous ammonia gives rapid decomposition of the HFB ester... The HFB ester of p-tert-butylphenol decomposed quantitatively on stationary phases containing free alcoholic groups such as Carbowax 20M and Amine 220. On a polyester phase [neopentylglycol succinate] no sign of decomposition could be seen when about 1 µg of p-tert-butylphenol was chromatographed."

Seiber and co-workers [48] chose ethers over esters as phenol derivatives, because of the hydrolytic instability of the esters. The "stability and relative ease of formation...excellent electron capture responses and...relatively short retention times" of the pentafluorobenzyl ethers, for example, have "led to their use for rapid qualitative analysis of the phenolic fraction from charcoal samples of river water." Kawahara [26] and Brötell and colleagues [27] also chose the ethers.

Kawahara [26] pointed out that, whereas the trifluoroacetate of phenol "hydrolyses in the presence of water" and "is detectable only to about one microgram," the sensitivity to detection by electron capture of the water-stable phenyl pentafluorobenzyl ether is 1,000 times better than that of the phenol trifluoroacetate.

Brötell and co-workers [27], also seeking both stability in aqueous solution and EC sensitivity, used the pentafluorobenzyl ether for detection of the drug pentazocine. Adsorption losses were minimized by deactivating the column with desipramine both before and during the injection of the sample. The "minimum detectable amount" was then found to be about 3 pg of PFB-pentazocine.

C. Antibiotics

The *Quantitative Gas Chromatography of Antibiotics* has been reviewed by Margosis [49]. Most gas chromatographic antibiotics first require derivatization. The usual derivative is trimethylsilyl; the detector used is generally flame ionization.

In the determination of erythromycin [50], the silylation reagent is BSA-TMCS-TSIM (5:5:2, vol/vol), mixed in that order and then added to an equal volume of pyridine that contains an internal standard, 1,3-Dimyristin. One milliliter of this reagent is used to silylate 10 mg of the erythromycin powder (24 hr, 75°C). During the development of this procedure, the derivatives were examined by gas chromatograph mass spectrometry to make sure that all the active hydrogens of the hydroxy groups had been silylated.

Obtaining just one derivative from a given parent molecule, particularly an antibiotic molecule, is by no means always either straightforward or successful [51]. The spectinomycin molecule, for example, has "two secondary amines, two secondary hydroxyls, one tertiary hydroxyl, and one enolizable carbonyl" [52]. With such a molecule the number of derivatives obtained markedly reflects the conditions of silylation. It was eventually established that a single derivative is obtained from spectinomycin by silylating with hexamethyldisilazane for 1 hr at room temperature [52].

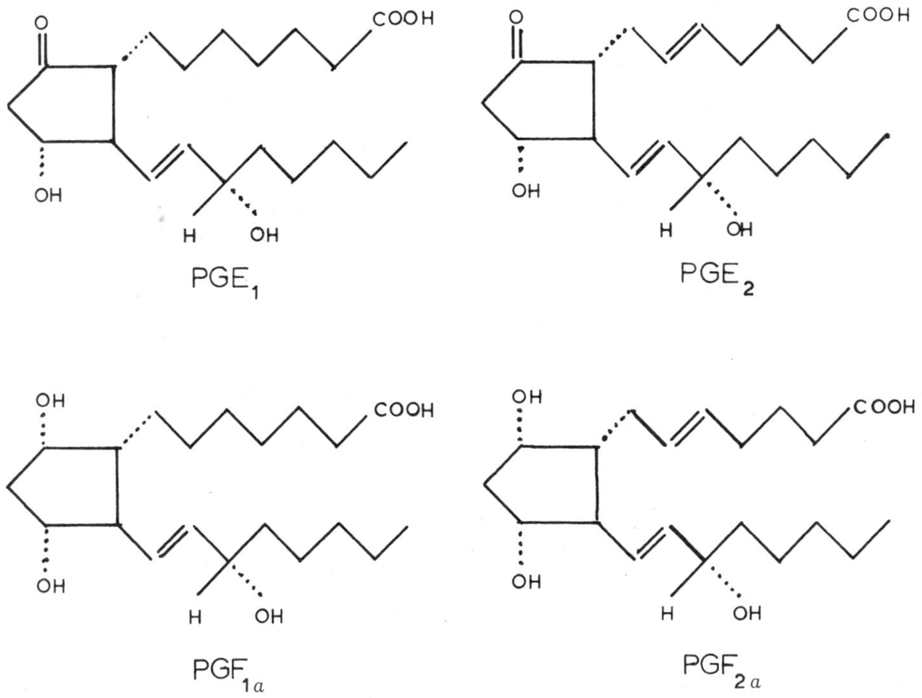

Sec. 12.4 Hydroxy Compounds

FIGURE 12.3 The E_1, E_2, $F_{1\alpha}$, and $F_{2\alpha}$ prostaglandins.

D. Prostaglandins

A review [53] shows that the functional-group complexity and trace concentration of prostaglandins principally complicate applying gas chromatography to prostaglandin analysis. Thus in such analyses we see, almost as prerequisites, multiple derivatization, ultraselective and ultrasensitive detectors, and the use of internal standards. (See Fig. 12.3 for the structures of some prostaglandins.)

Flame Ionization Detection

Derivatives suitable for a gas chromatograph-mass spectrometer (GC-MS) combination can be detected by a flame ionization detector (FID). However, if used at all in prostaglandin analysis--as in the two cases to be cited next [54, 55]--the FID is only an adjunct to such a GC-MS combination.

Preparing derivatives of A, B, E, and F prostaglandins, Vane and Horning [54] sequentially derivatized first the ketone groups as methoxime (MO) (CH_3ONH_2 in pyridine, 2 to 3 hr, 60°C) and then the hydroxylic and carboxylic acid groups as trimethylsilyl ethers and esters, respectively (BSTFA, 1 to 2 hr, room temperature).

With a mixture of the E_1, E_2, $F_{1\alpha}$, and $F_{2\alpha}$ prostaglandins, Pace-Asciak and Wolfe [55] also first formed the MO group: "This converted the E prostaglandins without altering the F..." Then, using n-butyl boronic acid (2 min, 60°C), they "converted the F prostaglandins to the 9,11-n-butylboronate [NBB] derivatives without altering the methoxime derivatives of the E prostaglandins." Finally, after drying the derivatives again, they converted the "remaining [hydroxyl and carbonyl] functional groups...to the [TMS] derivatives."

An interesting facet of this paper by Pace-Asciak and Wolfe is their use of dimethoxypropane. "A disadvantage of the NBB-TMS derivatives is their ease of solvolysis." To remove the water formed with the NBB derivatives, they used dimethoxypropane as reaction solvent. "In the presence of water dimethoxypropane is hydrolyzed to acetone and methanol which are removed with nitrogen when the reaction is terminated" [55]. (Note: the dimethoxypropane-water reaction is both instantaneous and endothermic--the reaction cools the substance it dries. Such a reaction is thus also ideal for drying heat-labile substances [56].)

Electron Capture Detection

Using "the inherent electron-capturing properties of prostaglandin B (PGB) compounds" for the electron capture detection of prostaglandins "is limited to the PGB compounds or to prostaglandins that can be readily and quantitatively converted to the PGB's..." [29]. Of the common derivatives, the

thermal instability of the heptafluorobutyrate derivative, resulting in the formation of multiple GLC peaks, renders it unsuitable...Since all naturally occurring prostaglandins contain a C-1 carboxyl group that has been shown to undergo a facile reaction with benzylic halides, the C_1-pentafluorobenzyl ester of prostaglandin $F_{2\alpha}$ ($PGF_{2\alpha}$) was prepared to examine its electron-capture detection properties...

$PGF_{2\alpha}$ was treated with a solution of pentafluorobenzyl bromide in acetonitrile... and a solution of diisopropylethylamine in acetonitrile...The reaction mixture was heated at 40° for 5 min and then evaporated to dryness under nitrogen. This esterified product was treated with BSA at room temperature for 5 min. The reaction mixture was then injected...the electron-capturing GLC response [to the product] was found to be linear over the concentration range of 0.2-5.6 ng $PGF_{2\alpha}$ [29].

Gas Chromatography-Mass Spectrometry

From freshly prepared diazomethane, Nicosia and Galli [57] prepared prostaglandin methyl esters, then silylated the nitrogen-dried esters with TSIM in piperidine: "The reaction mixture was injected after a few seconds...in all cases, a single derivative with excellent [GC] properties" was obtained. (However, "due to the great instability of PGA's to alkali, very stringent conditions are required, such as...silanized glassware and freshly silanized columns...") "The estimation of nanogram levels of prostaglandins was carried out using...multiple ion detection."

Also from freshly prepared diazomethane, Kelly [58] methyl-esterified prostaglandin $F_{2\alpha}$ (10 min, room temperature). Using butyl boronic acid in acetone/benzene (2:1, vol/vol), he then converted the dried ester to the NBB (20 min, 60°C), after which the product was dried again under vacuum. Last, the product was silylated (BSTFA, 2 hr, room temperature), then dried again under vacuum. For stability, BSTFA was added again and the glass reaction tube sealed in a flame. The product could then be kept 10 days at room temperature, longer at -15°C. Later, "the total contents of the tube were injected."

Sec. 12.5 Carbonyls

The NBB ester was used partly to distinguish between cis- and trans-diols and partly to secure a relatively high molecular m/e value for the derivative: "In general, the higher the mass monitored, the lower the chance of ions from other compounds interfering in the measurements" [58].

The selectivity of mass spectrometric detection can be increased still further by the use of isotopically altered derivatives used as internal standards. The heavier ions from such derivatives are likely to be completely free from interference from other ions. Samuelsson, Hamberg, and Sweeley [59], using CD_3ONH_2, prepared and mixed a trideuterated prostaglandin methoxime with an unaltered prostaglandin methoxime. These were injected as the trimethylsilyl derivatives. Simultaneously monitoring both a normal ion (m/e 470) and the corresponding but heavier (in those circumstances, essentially unique) ion (m/e 473) from the normal and trideuterated derivatives, respectively, enabled a quantitative determination of 3 ng of the derivatized PGE_1.

Similarly, the original prostaglandin itself can be prepared in a deuterated form, then mixed in known weight ratio with the sample for normal extraction and derivatization procedures [60].

12.5 CARBONYLS

Most carbonyls do not require derivatization. When required, carbonyl GC derivatives have been either hydrazones or oximes.

Papa and Turner [61] reviewed carbonyl derivatization *via* 2,4-dinitrophenylhydrazine (DNPH), a standard method in wet qualitative organic analysis. In 1972, Kaalio *et al.* [62] applied the approach to flavor carbonyls, using both flame ionization (10 to 10,000 ng) and electron capture (20 to 500 pg) detection.

Fales and Luukkainen [63] criticized DNPH derivatives as potentially unstable if exposed to light and air. They suggested instead the O-methyloxime (MO) derivatives: "...stable, easily crystallized derivatives, that survive operations such as treatment with acetic anhydride or trimethylsilyl chloride and hexamethyldisilazane." They formed the methoximes by bringing carbonyls into contact with methoxyamine hydrochloride in pyridine. The reaction mixture was allowed to stand overnight at room temperature. The sample was then dried of excess pyridine, taken up in benzene, and injected.

Once formed, MO derivatives can be variously derivatized. One method is silylation. Using BSTFA-1% TMCS for 3 hr at 75°C, for instance, Butts [64] silylated cytidine and deoxycytidine.

Vogh [65] reviewed carbonyl derivatization in 1971 and concluded that oximes would be the derivatives of choice for carbonyls in automobile exhausts. For stability, oximes should make contact only with glass, he found.

If left underivatized, carbonyl keto groups can be enolized and silylated by prolonged silylation [20]. At room temperature, either BSA or HMDS:TMCS (2:1, vol/vol) monosilylated acetovanillone in 12 min, but disilylated it partly in 10 hr and completely in 24 hr. This disilylation proceeded without regard to reagent excess, heating, or solvent-catalyst tried. Neither TSIM nor BSTFA was found to produce such disilylation.

In 1974, Korolczuk and co-workers [66] substituted the phenylhydrazones for the 2,4-dinitrophenylhydrazones: "...preparation of the derivatives is simpler, [faster] ...separation from aqueous solution by extraction with diethyl ether is also very easy, rapid, and quantitative." The phenylhydrazones do not require glass columns, as do the 2,4-dinitrophenylhydrazones, nor such high column temperatures.

12.6 ACIDS

A. Carboxylic Acids

Flame Ionization Detection

For gas chromatography, carboxylic acids are quite generally derivatized, most frequently to the methyl esters. (More on methylation is given in Sec. 12.6B, Fatty Acid Mixtures.)

If conducted at -60°C with dioxane as solvent for both sample and reagent, methylation by diazomethane (DAM) gives complete conversion for both monobasic and dibasic acids [67]. DAM, which is unstable, is best prepared on the spot, frequently by decomposing N-methyl-N-nitroso-p-toluene sulphonamide with KOH [67]. Special microapparatus for the derivatization has been devised [68].

By a quantitative, fast (less than 10 min), but extremely mild reaction, Greeley [69] used an organic base, phenyltrimethylammonium hydroxide; a polar solvent system, 80% N,N-dimethylacetamide-20% methanol (the particular composition is not critical); and 1-iodobutane to convert stearic acid to butyl stearate (99.4%). To keep the reaction time short, the solvent must be anhydrous; and the acid-base reaction must take place before the alkyl iodide is added. West [70] applied this approach to make methyl esters.

Under adequately anhydrous conditions (for this, use a 100:1 molar excess of the silylating reagent), carboxylic acids can be silylated readily enough. BSTFA is at least 10 times faster than BSA and, in contrast to BSA, can fully silylate either shikimic acid or quinic acid in 3 min at room temperature [9, 10].

To use the phosphorus-detecting selectivity and sensitivity of the thermionic detector [71], Schulz and Vilceanu [72] esterified carboxylic acids with α-hydroxyphosphonic acid ester (I) ($R_1 = R_2 = CH_3$; $R_3 = R_4 = H$) in the presence of dicyclohexylcarbodiimide for 60 min at 50°C to produce O,O'-dialkyl-α-hydroxyphosphonic acid esters (II). After being cooled, the solution could be injected into the gas chromatograph. Nanogram quantities of acids such as butyric, valeric, and caproic were easily detectable. If the baseline was stabilized by backflushing the solvent, quantities down to 10 pg were also detectable.

$$\text{I:} \quad \begin{array}{c} R_1O \\ R_2O \end{array} P(=O) - C(R_3)(R_4) - OH \qquad \text{II:} \quad \begin{array}{c} R_1O \\ R_2O \end{array} P(=O) - C(R_3)(R_4) - O - C(=O) - R_5$$

I II

Electron Capture Detection

Smith and Tsai [73] esterified 10 carboxylic acids with 10% 2,2,2-trichloroethanol (TCE) in trifluoroacetic anhydride (TFAA) (10 min, 100°C). The TFAA was then removed by evaporation and the TCE by passing it over a short silica-gel column (20 g). Except for hippuric and p-nitrobenzoic acids, conversion was quantitative; 0.1 to 1.0 ng were detectable.

B. Fatty Acid Mixtures

Of the methods recommended [74] in 1975 for the esterification of fatty acids of higher [75-78], "both high and low" [79], and "low and medium" [80] molecular weights, the standardized BF_3-methanol method [77] is probably the most widely accepted and certified procedure. It avoids isomerization, "quantitatively converts the fatty acids to methyl esters in ten minutes, (and) almost completely eliminates the objectionable loss of lower fatty acids in the water layers..." [74]. A variation gently cleaves cerebrosides to fatty acid methyl esters [81].

An alternative, elegant method [82] was developed for the methylation of the fatty acids in small amounts of tissue (3 to 5 mg) or serum (5 to 100 µℓ). After alkali hydrolysis, acidification, and hexane extraction, the fatty acids are extracted into 10 µℓ of aqueous α,α,α-trifluoro-m-tolyl ammonium hydroxide. Mixed with methyl propionate within a syringe, 1 µℓ of this is sandwich-injected into a gas chromatograph, where methylation occurs in the injection block. Fatty acid distributions so determined are statistically equivalent with those determined by other esterification procedures.

N,N-Dimethylformamide quantitatively converts acids to esters of choice either in 10 min at 60°C or on sandwich-injection [83].

Free fatty acids in dried extract of plasma react with N,N-carbonyldiimidazole and methanol to form methyl esters. "The reactions go to completion within a few minutes at room temperatures...Alcohols may...be chosen...to make many esters" [84]. The reaction is gentle enough not to transesterify triglycerides and cholesteryl esters.

Not unduly volatile with the shorter-chain fatty acids, benzyl ester derivatives of short-, medium-, and long-chain (C_1-C_{20}) fatty acids can be prepared quickly, easily, and quantitatively by reaction with phenyldiazomethane [85, 86]. They furnish distinctive fragments for gas chromatography-mass spectrometry.

Electron Capture Detection

To determine as little as 2 to 5 mg C_2-C_8 aliphatic acids in biological specimens, Alley and co-workers [87] esterified the acids with TCE (32 min at room temperature). They used HFBA as catalyst [88, 89]. The excess TCE was then esterified with palmitic acid, whereupon palmitic ester and the HFBA were removed by washing. The procedure can be modified for application to long-chain fatty acids [87].

C. Phenolic, Hydroxy, and Keto Acids

Flame Ionization Detection

Heating dried extracts of homovanillic acid (3-methoxy-4-hydroxy-phenyl acetic acid) and vanillyl mandelic acid (3-methoxy-4-hydroxy-phenyl-β-hydroxy acetic acid) with an excess of BSTFA:TMCS (9:1, vol/vol) for 20 min at 70°C quantitatively silylated the acid, phenol, and hydroxy groups [90]. (A variety of other derivatization methods had earlier been tried unsuccessfully [11].) The reaction mixture could be injected directly and could, given moisture exclusion by effective closure and the presence of excess BSTFA, "be stored for long periods of time." In other hands [91], and with similar acids (o-hydroxy-phenylacetic acid, β-phenyl lactic acid, phenyl pyruvic acid, and p-hydroxyphenyl pyruvic acid), heating was not found necessary because the BSTFA "gave rapid, for all practical purposes, instantaneous, reaction." In still other hands [92], "the keto acid (phenylpyruvic) was first converted... (NH_2OH)...to an oxime," then "all acid, alcohol, phenolic, and enolic groups quantitatively converted in one step" with BSTFA:TMCS (100:1, vol/vol; 15-30 min, 70°C) "to the TMS ester, ether, and enol-ether derivatives." Prior conversion to an oxime can be done under aqueous conditions and stabilizes the keto acid during drying and subsequent silylation [93].

Electron Capture Detection

Pentafluoropropionic anhydride (PFPA) and a halogenated alcohol (in 1974, 2,2,3,3,3-pentafluoro-1-propanol [30]; later, 1-chloro-1,1,3,3,3-pentafluoro-2-propanol) [94, 95] were used (15 min, 75°C) to esterify the carbonyl groups of the phenolic acid metabolites of biogenic amines. Following evaporative removal of these reagents, the hydroxyl and indole groups, which the first step could not quantitatively convert, were then acylated with more PFPA (5 min, 75°C). The excess PFPA was again removed by evaporation (it was for this facile evaporative removal that PFPA was preferred over HFBA). Down to 5 pg of the derivatives were detectable [30].

D. Sulfonic Acids

Examples of aromatic and aliphatic sulfonic acids (p-toluene-sulfonic acid, anthraquinone-2-sulfonic acid, 3,4-dideoxy-4-C-sulfo-D-hexosulose and its 2,4-dinitrophenylosazone, and L-cysteic acid, or their corresponding pyridinium salts) were converted to their trimethylsilyl derivatives by shaking with TSIM for 10-15 min at room temperature. The TSIM "reacted with both hydroxyl and sulfonate groups smoothly at room temperature in the absence of a solvent. Furthermore, the versatility of the technique allowed sequential derivatization of other functional groups, such as amine, by addition of 1-(trifluoroacetyl)imidazole, if required" [96].

"Volatile derivatives of aminobenzene sulfonic acids were easily formed by heating the sulfonic acid with a mixture of PCl_5 and PCl_3 (45 min, 95°C). The product ($-SO_2Cl$ and $-NH-POCl_2$) and reagents are soluble in benzene and the sample can be injected directly..." [97].

12.7 AMINES

From the point of view of derivatization for gas chromatography, biologically important amines are divided into simple amines and hydroxyamines. Primary and secondary amines are usually treated as one group. As a rule, when secondary amines are derivatized it is for sensitization to detection by electron capture. Tertiary amines are as yet always handled separately.

Hydroxyamines are handled inclusively in the laboratory, the practice as a rule not differentiating between, for instance, catecholamines and indoleamines. Also, the trend is increasingly toward ideal sequential derivatization, as brought out in Sec. 12.7B, Catecholamines and Indoleamines.

A. Simple Amines

Mixtures of the biologically important amines often include both primary and secondary amines. The secondary amines are much harder to derivatize than the primary. The few reagents and reaction conditions that can successfully derivatize secondary amines in practice are more than adequate for full N,N-derivatization of the primary. Often, the secondary amine group is merely left intact unless sensitization for electron capture detection is required. Also, derivatizing only a single hydrogen on a primary amine is often considered satisfactory.

Tertiary amines differ from primary and secondary amines in at least two respects relevant to this discussion. First, without derivatization, the tertiary amines are more readily separable by gas chromatography than either secondary amines or N-derivatized primary amines. Second, derivatization of tertiary amines requires a more radical change of molecular structure, not just the replacement of one or both N-hydrogens that is characteristic of and adequate for the primary and secondary amines. Nevertheless, tertiary amines are derivatized for gas chromatography to yield either more certain identification or more sensitive detection. Both will be discussed.

Flame Ionization Detection

Primary amines can be not only fully and quantitatively but also easily derivatized by N,N-dimethylformamide-dimethylacetal (DMF-DMA). (The reaction is shown in Fig. 12.4.) The use of this reagent with primary amines to yield one N-dimethylaminomethylene group per primary amine group was reviewed in 1970 [98] and illustrated in 1972 [99] and 1975 [100]. In the first example [99], quantitative on-column derivatizations of cyclooctylamine and phenethylamine were achieved regardless of which one

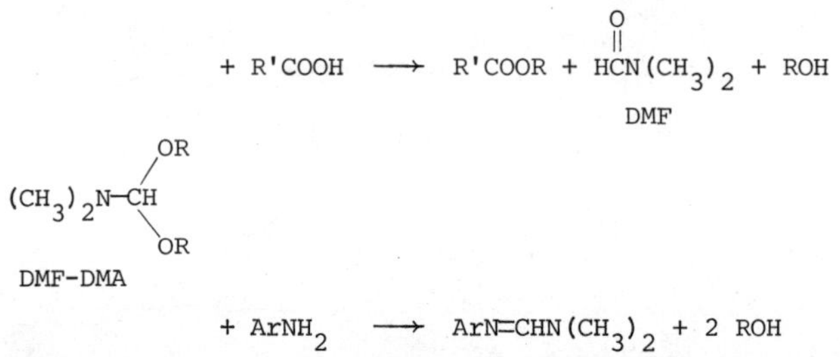

$$(CH_3)_2N-CH(OR)_2 \quad \text{DMF-DMA}$$

$$+ R'COOH \longrightarrow R'COOR + HCN(CH_3)_2 \text{ (DMF)} + ROH$$

$$+ ArNH_2 \longrightarrow ArN=CHN(CH_3)_2 + 2 ROH$$

FIGURE 12.4 Alkylation with N,N-dimethylformamide dimethylacetal (DMF-DMA).

of the eight common solvents tested was being used. In the second example [100], each of the three diamines--hexamethylenediamine, bis(4-aminocyclohexyl)methane, and 2,2-bis(4-aminocyclohexyl)propane--was converted quantitatively to the N-dimethylaminomethylene derivative by 10 min heating at 60°C. The first diamine formed a single derivative; the latter diamines produced the relevant stereoisomers.

As already mentioned, the difficulty of amine derivatization increases in the order, primary-N, primary-N,N, and secondary-N. Frequently, or perhaps usually, achieving only the primary-N derivative yields satisfactory results. For instance, the TFA derivatives were found adequately volatile in the determination by GLC of the polyamines putrescine $[NH_2(CH_2)_4NH_2]$, spermidine $[NH_2(CH_2)_4NH(CH_2)_3NH_2]$, and spermine $[NH_2(CH_2)_3NH(CH_2)_4NH(CH_2)_3NH_2]$. For the derivatization, the cleaned-up and dried samples were taken up in equal portions of CH_3CN and TFAA, sonicated 1 min, and heated at 100°C for 5 min. The CH_3CN and TFAA were then removed by N_2 evaporation at room temperature. Finally, the derivatives were redissolved in ethyl ether that contained phenanthrene as an internal standard [28].

Similarly, only one TMS group is usually put onto a primary amine group, and the secondary amines are usually left intact. In a study with BSA, it was found that the secondary amines could be trimethylsilylated, but in the practice adopted they were

Sec. 12.7 Amines

not, although conditions were severe enough (acetonitrile solvent, TMCS "catalyst," 2 hr, 60°C) to convert all the primary amines to the N,N-di(TMS) derivatives [1].

Usually, tertiary amines are derivatized specifically for increased sensitivity to detection by electron capture. Street [101], however, to secure "a further criterion for identification purposes", used the Hofmann degradation reaction (the reaction is shown in Fig. 12.5) to produce one or more tertiary amine derivatives having retention times different from (and usually less than) that of the parent. After alkylation to produce a quaternary ammonium halide that was then dried and redissolved, the halide was converted to the hydroxide by moist silver oxide. On injection into the gas chromatograph, the hydroxide decomposed to the derivative or derivatives-- usually one or more related olefins. (The Hofmann degradation reaction has been much used in the derivatization of barbiturates. See Sec. 12.9A, Barbiturates.)

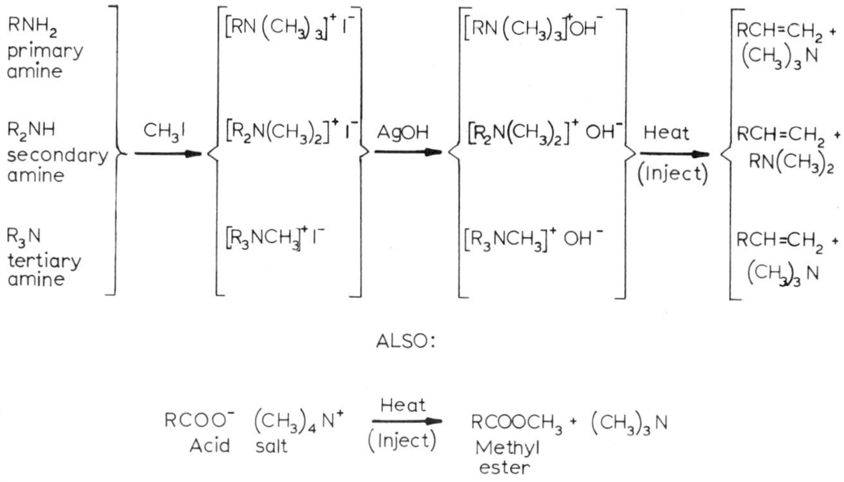

$$RNH_2 \text{ primary amine} \quad R_2NH \text{ secondary amine} \quad R_3N \text{ tertiary amine} \xrightarrow{CH_3I} \begin{cases} [RN(CH_3)_3]^+ I^- \\ [R_2N(CH_3)_2]^+ I^- \\ [R_3NCH_3]^+ I^- \end{cases} \xrightarrow{AgOH} \begin{cases} [RN(CH_3)_3]^+ OH^- \\ [R_2N(CH_3)_2]^+ OH^- \\ [R_3NCH_3]^+ OH^- \end{cases} \xrightarrow[\text{(Inject)}]{\text{Heat}} \begin{cases} RCH=CH_2 + (CH_3)_3N \\ RCH=CH_2 + RN(CH_3)_2 \\ RCH=CH_2 + (CH_3)_3N \end{cases}$$

ALSO:

$$RCOO^- \; (CH_3)_4N^+ \xrightarrow[\text{(Inject)}]{\text{Heat}} RCOOCH_3 + (CH_3)_3N$$
Acid salt → Methyl ester

FIGURE 12.5 The Hofmann degradation and flash alkylation in gas chromatography.

Electron Capture Detection

The sensitivity of electron capture detection varies widely with both the sensitizing group and the substances derivatized [31-33, 37, 102]. With respect to phenethylamine, the TFA, PFP, and HFB derivatives were found to be more sensitive than the parent amine by factors of 200, 2,000, and 10,000, respectively [31]. Despite the increasing chain lengths of the derivatizing groups, the peaks of the derivatives not only are symmetrical but also have nearly identical retention times [31, 103]. In one illustrative application [102], HFB-imidazole was used to acylate the drug protryptiline, a secondary amine (15 min, 60°C). The smallest detectable amount of the HFB-protryptiline was 0.05 ng, corresponding to a concentration in the original sample of 10 ng/ml.

Of 10 perfluorobenzene reagents tested for the derivatization of phenethylamine (PA) and N-methyl-phenethylamine (MPA), "the best derivatizing reagent for secondary amines was pentafluorobenzaldehyde and pentafluorobenzoyl chloride... The sensitivity of detection [of PFB-PA and -MPA] by the electron capture detector is approximately 2×10^3 that obtained with a flame ionization detector. Amounts of the amines down to 10 pg could easily be quantified" [32].

Cummins [37] found that the relative ECD sensitivities of the PFB and HFB derivatives of six amines differ widely, as shown in Table 12.1. Cummins noted that in general the "heptafluorobutyryl derivatives of aliphatic and nonaromatic amines are very insensitive towards ECD." Indeed, HFB-cyclohexylamine shows no more sensitivity with ECD than the underivatized cyclohexylamine does with FID.

TABLE 12.1 The Relative Sensitivities to Electron Capture Detection of Some PFB and HFB Amine Derivatives

Amine	PF benzamide	HFB amide
Cyclopropylamine	1,000	15
Cyclohexylamine	500	1
Aniline	133	670
Benzylamine	153	25
Dibenzylamine	23	83
Heptylamine	670	5

Walle and Ehrsson [33] studied various aspects of acylation by HFBA. They emphasized the importance of completely removing the excess HFBA--which need originally be only in 25% excess--before injection. Evaporation does not remove the HFBA completely but does lead to the partial loss of some derivatives. However, extracting the benzene reaction mixture with aqueous ammonia completely removes the HFBA but does not hydrolyze or extract the derivatives (note, however, as mentioned earlier, that aqueous ammonia rapidly decomposes the HFB ester of p-tert-butylphenol [25]).

Comparing several ECD-sensitizing reagents, Matin and Rowland [104] concluded that pentafluorobenzoyl chloride is suitable for either primary or secondary amines, alone. But if the two occur together and the primary might interfere with the determination of the secondary (because the PFB derivative of a primary amine is 10 to 100 times more sensitive than the PFB derivative of the corresponding secondary amine then the HFB derivative is probably best.

With no active hydrogens, tertiary amines must be derivatized differently from primary and secondary ones. For instance, Walle [105], using TFAA, trifluoroacylated the rings of tertiary aromatic amines to form trifluoroacetophenones, detectable by electron capture down to less than 1 pg.

Sec. 12.7 Amines

A tertiary amine can be sensitized for electron capture detection via a quaternary ammonium intermediate, given that the tertiary nitrogen is not too sterically hindered. By the end of 1975, this approach had succeeded only with N,N-dimethyl tertiary amines. In each example cited below, the salient initial reactant was a chloroformate.

In the first example, the tertiary amine was deaminated, and the chloroformate chlorine sensitized the bulky residue to detection by electron capture. N,N-Dimethyl-dibenze[b,f]thiepin-10-methyl, treated for 30 min at 60°C with ethyl chloroformate, was determined down to a level corresponding to a concentration of about 2.5 ng/ml of blood, serum, or urine (the method requires 1 to 3 ml of sample)[106].

With no deamination, two routes have been used, both through the carbamate that results from treatment with the chloroformate. In the first route, the carbamate was reduced to a secondary amine that was then ECD-sensitized by the pentafluorobenzyl derivative [107-109]. In the second route, the carbamate itself was ECD-sensitized by treatment with PFB-chloroformate [110]. However, not all tertiary amines react so favorably--methadone, for instance, deaminates to an ECD-insensitive product--nor are the reaction conditions without complications [110].

B. Catecholamines and Indoleamines

The structures and names of some common catecholamines are shown in Fig. 12.6.

Flame Ionization and Mass Fragmentographic Detection

Modern practice in the application of gas chromatography to the separation of hydroxyamines began with an analytical method developed by Horning, Moss, and Horning [1]. In that method, TSIM was used to silylate the hydroxyl groups under conditions that would not affect the amine hydrogens present (acetonitrile, 2 to 3 hr, 60°C). Then, BSA:TMCS (2:1, vol/vol) was added directly to the reaction mixture. Under the conditions employed (another 2 to 3 hr at 60°C), the BSA-TMCS affected neither the O-TMS ethers resulting from the first derivatization nor the secondary amines present. It did fully convert the primary amines to the N,N-di(TMS) derivatives. The reaction products were stable for several days in the final reaction mixture, which could be directly injected. All resultant eight peaks were symmetrical and fully separated. The method was the first *ideal sequenced derivatization*.

An ideal sequenced derivatization is seen to be characterized by a related sequence of operations and set of properties:

    The different groups that are to be derivatized in sequence cannot be really successfully derivatized by just a single reagent. More than one derivatization is necessary.
    Each reagent yields one clean derivatization. A given reagent affects neither prior nor subsequent derivatizations.

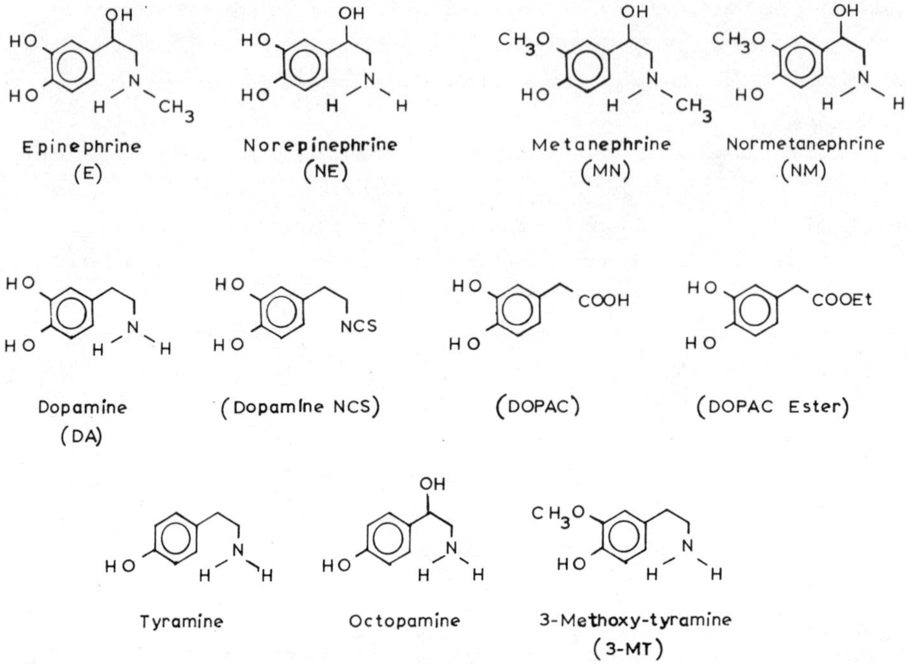

FIGURE 12.6 Some catecholamine structures, their names, and abbreviations.

    Reaction mixtures require no processing between successive derivatizations. Each
       succeeding reagent is added directly to the reaction mixture resulting from
       the previous derivatization.
The reaction products are stable and preservable in the reaction mixture.
The final reaction mixture is directly injectable into the gas chromatograph.
The derivatives are stable during gas chromatography, yield symmetrical peaks
       without undue delay, and are mutually separable.

    Phenolalkylamines as bases are highly sensitive to loss through oxidation, but
not as salts. Donike [111, 112], working directly with the salts in an ideal
sequenced derivatization, first silylated and then N-trifluoroacylated them. The derivatized products, effectively protected against moisture in several ways, were reported as being stable in the reaction mixture and preservable for months. The silylating reagent was N-methyl-N-trimethylsilyl-trifluoroacetamide (MSTFA); the acylating reagent was N-methyl-bis(trifluoro)acetamide (MBTFA). The reaction product from each reagent is the volatile, catalytically inactive N-methyl-trifluoroacetamide.

    Because of the favorable cracking patterns of the N-TFA-O-TMS derivatives, and given the loss-preventing, complementary facets of the overall Donike procedure, femtomol quantities of the phenol-alkylamines, so derivatized, are detectable by mass fragmentography. The derivatives are not suitable for detection by electron capture.

In a very similar ideal sequenced derivatization, the amine or amine hydrochloride sample (about 1 mg in 1 ml methanol) was taken to dryness, then warmed at 80°C for 5 min with 50 µl of BSTFA. "After the addition of 5 microliters MBTFA, the sample can immediately be injected into the gas chromatograph" [113].

Lovelady and Foster [114] dissolved isolated catecholamine residues in 0.1 ml BSA and added 0.1 ml each of tetrahydrofuran and TFAA (mix, 10 min, 50°C). Down to 0.1 pg of the O-TMS, N-TFA derivatives could be detected by FID.

Using PFPA in large excess, Willner et al. [115] derivatized phenethylamine, phenylethanolamine, and the corresponding d_4 and d_3 internal standard replicates (45 min, 70°C). The excess PFPA was then evaporated by dry N_2. Concentrations down to 5 pmol/g tissue could be determined by mass fragmentography.

Electron Capture Detection

Pentafluoropropionate (PFP) and heptafluorobutyrate (HFB) have been generally preferred over trifluoroacetates as the perfluoro derivatives for phenolic amines. Both Anggård and Sedvall [22] in 1969 and Karoum et al. [23] in 1972 found the PFP derivative preferable to the HFB one. Similarly, Gelpi et al. [116] in 1974, using reaction times (from 3 min to over 4 hr) and temperatures (from room to 100°C) relevant to the substance at hand, chose pentafluoropropionic anhydride (PFPA) for catecholamines, indoleamines, and the alcoholic metabolites of these. After the reaction, excess PFPA and solvent were removed by drying under a stream of helium. After being redissolved in benzene, the sample was ready for injection. The ECD/FID factors-of-sensitivity increase for the PFP derivatives of metanephrine and normetanephrine were 166 and 95, respectively.

Edwards and Blau [117] used a tissue-extraction procedure and the reagents 2,4-dinitrobenzenesulfonic acid and BSA, in that order, to make the N-dinitrophenyl (DNP), O-TMS derivatives of certain hydroxyamines (no secondary amines, no catecholamines). Subsequently, the further addition of more BSA to the benzene solution of the derivatives protected them "against hydrolysis even when the solution was shaken with dilute acids or bases." The ECD/FID factor-of-sensitivity increase was about 20, with minimum detectable amounts of 0.1 ng DNP-phenylethylamines and as little as 8 pg DNP-diethylamine, corresponding to "as little as 20 ng/g tissue" for the phenylethylamines.

Perfluorobenzaldehyde and BSA were used to form the N-pentafluorobenzylimine, O-TMS derivatives of catechol primary amines. Directly injectable solutions of the derivatives were shown to be preservable without loss at room temperature for weeks. ECD peaks corresponded to 1.8 ng amine, or 180 µg amine per gram of tissue [118].

C. Nitrosoamines

Brooks and colleagues [119] showed that "some of the biologically important nitrosoamines form electron capturing derivatives when reacted with fluorinated anhydrides in the presence of pyridine." The reaction is long and multistep. The derivative is 6,000 times more sensitive to ECD than to FID; a quantity of 17 ng is detectable.

D. Phosphoryl Amines

Karlsson [120] used BSTFA:TMCS:pyridine (10:2:5, vol/vol) to silylate O-phosphorylethanolamine, O-phosphorylserine, and O-phosphorylthreonine. Detection was by mass spectrometry.

E. Guanido Compounds

Using hexafluoroacetylacetone, Erdtmansky and Roehl [121] simultaneously extracted, derivatized, and sensitized to ECD drugs and metabolites containing the guanido group $[R-C(=NH)-NH_2]$. Unit picogram quantities of the derivatives were found detectable.

12.8 AMINO ACIDS

Flame Ionization Detection

A successful method for the quantitative gas chromatographic analysis of the 20 natural protein amino acids should apply to any one or equally well to all at once (this review does not treat derivatives for protein-sequencing). Preferably in a single step, the derivatization should encompass all of the dozen different functional groups among these acids. Separation of the derivatives should be complete, average but 1 or 2 min per acid, and admit automatic integration. The final figures should reflect the starting composition reliably, accurately, and acceptably precisely.

Gehrke, his co-workers, and his graduate students have for over a decade vigorously and inventively pursued the development of such a method (see Refs. 122-124, for instance).

After the first 5 years of this pursuit, they could state their "most important ...criteria...for a suitable volatile derivative [of any amino acid]: (a) no rearrangements or structural alteration should occur during formation; (b) derivatization reaction should be 95 to 100% complete; (c) no sample loss on concentration; (d) stability with respect to time; (e) derivative must have increased volatility; and (f) there must be little or no reactivity of the derivative with the substrate and/or support phase [123]. Also, for "single column separation" of acceptable duration, the derivatives had to be mutually separable by a conventional packed column of conventional length.

By the end of 1975, silylation had not become, at least yet, the method of choice for amino acid derivatization. Repeatedly, otherwise successful approaches were found somehow wanting. For instance, BSA gave "sharp single peaks for all the amino acids except arginine, which showed signs of decomposition on the column" [125]. Similarl

Sec. 12.8 Amino Acids   273

BSTFA was specifically invented for the amino acid problem [123, 124], but not finally used for it by the inventors [126, 127]. Nor had silylation comparison studies [128, 129] yielded a fully satisfactory amino acid silylating reagent by 1971 (see, however, Ref. 130).

These comments of silylation apply to other promising derivatizations, some much used, some very new. The 2,4-dinitrophenyl derivatives of amino acids have been thoroughly studied--Rosmus and Deyl [131] refer to over 1,000 papers and at least nine reviews--but the applications are replete with exceptions. Similar reservations apply to the phenyl thiohydantoins [132]. Promising new derivatives--the oxazalidinone [133, 134], the N-heptafluorobutyryl n-butyl ester [135, 136]--have not undergone the exhaustive testing on which silylation foundered.

Thus, in 1969 Gehrke et al. concluded, "The derivative of final choice was the N-trifluoroacetyal [N-TFA] butyl ester" [123]. That conclusion seems still valid at this writing.

In 1971 [124], 1972 [137], and 1974 [122], Gehrke and co-workers described in detail the effective but elaborate systems of analysis involving the N-TFA, n-butyl derivative. Both Casagrande [138] and Raulin et al. [139] showed that that derivative allows an additional 15 to 18 nonprotein amino acids and some amino sugars to be separated in the presence of the 20 protein amino acids, given certain adjustments (Raulin used a 6 ft column, 0.325% ethylene glycol adipate on 80/100 Chromosorb G). The derivatization has been used [140, 141] for the analysis of nanogram and picogram quantities of the amino acids. (Moodie [142] treats the N-TFA n-butyl derivatives with ethoxyformic anhydride (20 min, 150°C ; or 45 min at room temperature) to solve the problem of the instability of the N-TFA group on histidine.) The N-TFA, n-butyl ester derivatization is shown in Fig. 12.7.

As Gehrke et al. wrote in 1969, "There is still need of further simplification of the N-TFA n-butyl derivatization reaction, for a method which will lend itself to

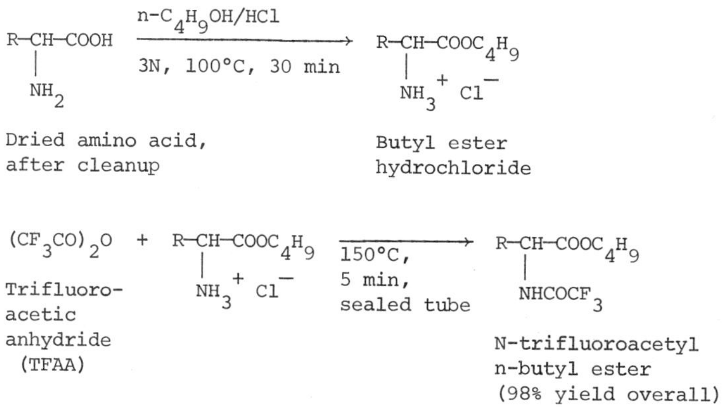

$$\text{R-CH(NH}_2\text{)-COOH} \xrightarrow[\text{3N, 100°C, 30 min}]{n-C_4H_9OH/HCl} \text{R-CH(NH}_3^+ Cl^-\text{)-COOC}_4H_9$$

Dried amino acid,   Butyl ester
after cleanup   hydrochloride

$$(CF_3CO)_2O + \text{R-CH(NH}_3^+ Cl^-\text{)-COOC}_4H_9 \xrightarrow[\text{5 min, sealed tube}]{150°C,} \text{R-CH(NHCOCF}_3\text{)-COOC}_4H_9$$

Trifluoro-   N-trifluoroacetyl
acetic   n-butyl ester
anhydride   (98% yield overall)
(TFAA)

FIGURE 12.7 The N-TFA, n-butyl reaction for amino acid derivatization.

automation, and for a derivative that will be simpler in certain of its chromatographic aspects; as well as for entirely new derivatization and chromatographic approaches" [123].

Electron Capture Detection

Iodine-containing amino acids are naturally sensitive to detection by electron capture and can be detected in picogram quantities. For this purpose, Nihei et al. [143] in two steps converted triiodothyronine and thyroxine to their methyl-N,O-dipivalyl derivatives.

Also for detection by electron capture, Wilk and Orlowski converted L-pyrrolidone carboxylic acid [144] and pyrrolidone carboxylate and γ-glutaryl amino acids [145] to their N-pentafluoroacyl esters with pentafluoropropanol and pentafluoropropionic anhydride (PFPA)(1:4, vol/vol)(15 min, 75°C, dry with N_2; add more PFPA, 5 min, 75°C, redry with N_2).

Similarly, the oxazolidinone [133, 134] and N-HFB ester derivatives [135, 136] are used for sensitizing amino acids to detection by electron capture.

12.9 AMIDES AND IMIDES

A. Barbiturates

"The barbiturates do not give stable derivatives with the common silylating reagents; however, they may be readily methylated to dimethylbarbiturates [that] are more volatile than the unmethylated compounds and can be gas chromatographed with little or no adsorption"[146]. A number of related structures are shown in Fig. 12.8.

In the derivatization of barbiturates for gas chromatography, the barbiturates have as a rule been alkylated by flash alkylation. This is an adaptation of the Hofmann degradation (shown in Fig. 12.5). The adaptation was introduced in 1963 by Robb and Westbrook [147].

MacGee [148], reviewing flash alkylation in gas chromatography, pointed out that the most-used modern reagents for flash alkylation were demonstrated in 1881 by Hofmann as alkylating agents. These reagents are two: What is now generally known as trimethylanilinium hydroxide (TMAH); and tetramethylammonium hydroxide (TMH), which was used by Robb and Westbrook [147].

Solow et al. [149], reporting in 1974 on a method that had been applied successfully in her laboratory to thousands of samples [148], added TMH to the toluene extract of buffered serum, which extract contained three internal standards. After vortex-mixing and centrifuging the mixture, they took 1 µl from the bottom TMH phase and injected it. Total processing time was about 15 min; all processing was done at room temperature. The injection port temperature was held at 350°C "for maximum yield of the methyl esters... 280°C...tends to produce residual compounds from incomplete vaporization of the TMH." Seven antiepileptic drugs could be measured in the ensuing gas chromatography.

Sec. 12.9 Amides, Imides

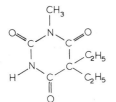

Mephobarbital
5-ethyl-1-methyl-
5-phenylbarbituric acid

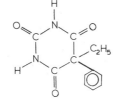

Phenobarbital
5-ethyl-5-
phenylbarbituric acid

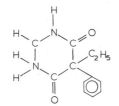

Metharbital
5,5-diethyl-
1-methyl-barbituric acid

Primidone
5-ethyl-hexadro-4,6-
dioxo-5-phenyl-pyrimidine

FIGURE 12.8 Some barbiturate and barbiturate-related structures and names.

Flash alkylation can be extended by means of the heavier tetraalkylammonium hydroxides, to C_6 in one case [150]. Both ethyl [148] and butyl [151] have been substituted for the unusable methyl in the separation and determination of methylphenobarbitone and phenobarbitone, for instance, and in similar applications [152].

In 1969, TMAH was introduced as a substitute for TMH because "a quaternary ammonium base [that] would produce a better leaving group than trimethylamine...should [in consequence allow a] shorter reaction time and milder conditions for thermal decomposition of its salts" (Brochmann-Hanssen and Oke, Ref. 146). These workers found TMAH superior "for flash-heater alkylation of barbiturates, xanthine bases, and phenolic alkaloids." Their injection port temperatures: 250°C for barbiturates, 275°C for xanthine and phenolic alkaloids. Barrett [153] and Kananen et al. [154] are among those who agreed, using TMAH for their analyses of Dilantin and the barbiturates, respectively. As with TMH, N-alkylation with groups heavier than methyl can be accomplished with the corresponding tetraalkylanilinium hydroxide [155].

The MacGee review [148] shows in both presentation and floor discussion that the results of flash alkylation reflect operational parameters such as injection rate: "...the rate of injection of the sample into the vaporizer influenced the solvent response and the peak heights...slow injection yielded...more complete heating of the sample...a more efficient reaction...better elution patterns than...rapid;" on injection port material: "When you have a metal injection port with metal columns, the reaction goes much more easily" (Kupferberg); and port-to-packing gap: "...for most reproducible methylation...we need a gap of 1.5 cm" (both Baylis and MacGee).

Thus one worker concluded, "The main drawback of...flash alkylation with quaternary ammonium hydroxides...is the influence of flash heater geometry and of the velocity of injection on the completeness of derivatization" [156].

In at least one case, operational parameters have been found to affect not only the efficiency but also the very nature of an important flash alkylation, and this only in 1974. In the TMAH flash alkylation of phenobarbital, a major product can be not only N,N-dimethylphenobarbital, as expected, but also N-methyl-α-phenyl-butyramide (MPB)[157]: "Two parameters affecting its formation are the concentration of TMAH and the time phenobarbital is in TMAH prior to GLC injection." Indeed, with a high TMAH concentration, "a better basis for the quantitative determination of phenobarbital" can be had with MPB than with the N,N-dimethylphenobarbital.

In response to such untoward characteristics of flash alkylation, other approaches to barbiturate derivatization have been brought forward.

In 1973, Dünges suggested "refluxing an acetone solution of free barbituric acids and of methyl iodide in the presence of potassium carbonate as condensating agent" [156]. The refluxing of 5 to 25 μl is done with a microrefluxer [158] for about 30 min. A portion of the supernatant solution is then injected.

Also in 1973, Venturella et al. [159] found that DMF-DMA reacts "quantitatively and reproducibly with glutethimide, phenobarbital, hexobarbital, secobarbital, amobarbital, aprobarbital, and pentobarbital to form the corresponding acetals (see the reaction in Fig. 12.9). The compound to be derivatized is dissolved in DMF-DMA, then warmed 10 min on a steambath. After dilution with internal standard solution and dimethylformamide, the solution (3 μl) is injected into the gas chromatograph.

In 1974, Greeley [35] suggested converting a to-be-derivatized acidic compound such as a barbiturate to a soluble salt with an organic base such as TMAH or, preferably, TMH in a highly polar solvent system. A primary alkyl iodide is then added. This yields the corresponding alkyl derivative at room temperature within 10 min. The particular alkyl is a matter of choice. The approach has been adopted in preference to flash alkylation for the derivatization of xanthines [160].

In 1974, Horning et al. sequentially derivatized a number of anticonvulsant drugs, as follows: After isolation, and depending on the drug in question, the drug

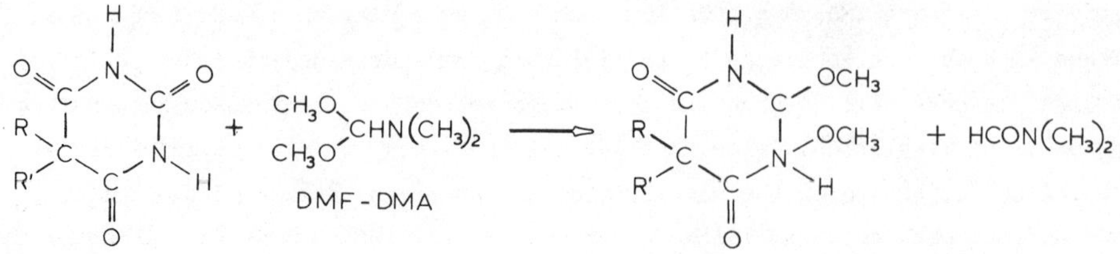

FIGURE 12.9 Barbiturate acetal formation with dimethylformamide dimethylacetal [159]

Other Amides and Imides

Åhrsson [34] reported that the PFB derivatives of seven secondary amides tested take much longer to prepare than the trifluoroacetates but are far more stable, particularly to hydrolysis. Both sensitize to electron capture.

Saccharin: With sodium carbonate present, methylation of the imide saccharin by methyl iodide in dimethylsulfoxide gives just one derivative rather than the two resulting from methylation by diazomethane [156, 160-162]. Alternatively, saccharin can be trimethylated by BSA [163].

Ureas

Using TFAA, Evans [36] converted monoureas and symmetrically disubstituted ureas to their trifluoroacetates (2 to 3 min, room temperature). Products are solid, stable for weeks if protected from moisture, and yield symmetrical peaks.

Cyclophosphamide

Cyclophosphamide (I) and isophosphamide, an internal standard for I, were converted to their mono-trifluoroacetates by treatment for 20 min at 70°C with trifluoroacetic anhydride:ethyl acetate (2:1, vol/vol) [164].

12.10 CARBOHYDRATES

Carbohydrate derivatization for gas chromatography has been extensively reviewed by Pierce [8] (1968, 70 pages on silylation), Sloneker [165] (1968, 48 pages), and Dutton [166] (1973, 150 pages).

By 1967, the "almost universal derivatization procedure...for carbohydrate analysis by GC" [8] was that of Sweeley et al. [167]: 10 mg carbohydrate, 1.3 ml reagent--HMDS:TMCS:pyridine (2:1:10, vol/vol)--room temperature, shake 30 sec, let stand 5 min; inject.

The Sweeley procedure seemed to require that the sample be rigorously dried for silylation. It does not. Drying prevents anomerization in water solution; but water does not prevent silylation. Water silylates immediately to hexamethyldisiloxane (see Ref. 168, for example), thus its presence merely requires more reagent [169].

Wet or aqueous sample had long been silylated. In 1958 Langer and co-workers silylated water-alcohol mixtures [168], and in 1967 Bentley and Botlock silylated 2% aqueous sugar solutions [170], both using HMDS-TMCS.

Brobst and Lott [171], citing the delays of drying and the inconsistent yields of trioses and tetraoses with the Sweeley formulation, substituted trifluoroacetic acid (TFA) for the TMCS in that formulation and used a sequential procedure: Dissol in pyridine about 60 mg solids syrup containing up to 50% water; add 0.9 ml HMDS; add 0.1 ml TFA; shake vigorously 15 sec, let stand 15 min at room temperature. (Ho ever, according to Bentley and Botlock [170], this procedure only incompletely silylates glucose.) Investigating silylation in the presence of water and looking f a reagent that could be premixed, Brittain and colleagues [18, 19] found that TSIM in pyridine "is excellent and reliable for wet or dry sugars...its action is smooth and rapid."

TSIM--a mild, selective, but effective reagent--has also been used to silylate all the hydroxyl groups of N- and O-acylated sialic acids without eliminating either the O-acetyl or the O-glycolyl groups in the process (15 min, room temperature)[172

Derivatization ideally yields one peak per component. Getting more than one pe per component may be caused by faulty derivatization or by anomerization. With respect to faulty derivatization, incomplete silylation, for instance, may produce several peaks per component [170]. Alternatively, too harsh conditions or silylatin reagents may cleave linkages [13] or enolize and silylate ester groups [173], in eac case making the derivatization and subsequent analysis less clear.

Anomerization causes the "main defect" of glycose derivatization procedures: "the production of several products from a single glycose owing to the formation of anomeric derivatives of possible furanose and pyranose ring forms" [174]. "Thus, fo a suitable derivative (that is, one produced by a method of derivatization that does not introduce additional peaks), it is theoretically possible to obtain a peak for t derivative of each anomer of both the pyranose and furanose forms, in addition to th of the acyclic, free carbonyl form" [175].

To restate: A single glycose can exist in any distribution among five forms. One of these five is acyclic. The four other possibilities are the two stereoisomer i.e., anomers, of each of the two possible cyclic forms (the furanose, or five-membered acetal ring; and the pyranose, or six-membered acetal ring). Anomerization the change in solution from one form toward an eventual equilibrium distribution of all the anomeric forms, can occur during solution for derivatization. It is anomerization that principally complicates glycose derivatization.

(The α- to β-$_D$-glucose anomerization rate is essentially zero in dry dimethylsulfoxide (DMSO) and remains low in DMSO-water mixtures. With respect to water, the "α-$_D$-glucose anomerization in 50% pyridine is about 48 times faster than in 50% DMSO The rate for 50% dioxane is similar to that for 50% DMSO [176].)

"Several investigators eliminated the problem of multiple peaks by reduction of the anomeric center. Acylation of the resultant alditols, usually with acetic anhydride and subsequent GLC, resulted in a single peak for each sugar" [177]. (Primarily because of the lengthy reduction, alditol acetates tend to be more trouble to prepare than silyl derivatives, but they are less trouble to preserve. A descriptio of alditol acetate preparation is given at the end of Sec. 12.10C, Amino Sugars.)

Although prior reduction of the anomeric center may eliminate multiple peaks, it can also destroy molecular identity. For example, if an aldose is reduced, "It becomes difficult to determine the end of the carbon chain from which a given ion fragment arises" [178]. Converting a sugar--in the example, glucose--to the oxime before either silylating or acetylating the glucose hydroxyls "effectively labels the carbonyl carbon" without requiring the use of a deuterated reagent [178].

Again, such reduction would in another example destroy the family identity of components from among the aldonic, hexuronic, and hexaric acids. Therefore, in that case, the components were silylated [179]. Earlier reasoning of this type led to oxidizing neutral glycoses to their corresponding aldonic acids, which were then silylated. Although the oxidation itself was lengthy and multistep, the overall result was the ideal one: "Each glycose in a mixture...converted to a single derivative" [174].

A. Glucuronides

Derivatization allows gas chromatography to be applied to intact rather than hydrolyzed glucuronides: "The usual hydrolytic procedures...are time-consuming, require large amounts of the conjugate, and may be destructive to biochemically significant aglycons" [13]. Derivatizing the β-$_D$-glucuronic acid conjugates of 1-naphthol and androsterone, Ehrsson et al. [180] first methylated the carboxylic groups (diazomethane, 15 min, 20°C; N_2 drying), then trifluoroacetylated the alcoholic groups (TFAA, 15 min 20°C). The derivatives were stable only in the continued presence of the TFAA. A "rapid... trimethylsilylation procedure [for] identifying urinary glucuronides...useful for microgram samples" was used by Mrochek and Rainey for intact glucuronides that had been isolated from urine and dried (BSTFA-TMCS, then vacuum drying)[13].

B. Sugar Phosphates

Sugar phosphates are usually trimethylsilylated (e.g., BSA-TMCS, 1 hour, room temperature [181]). Harvey and Horning [182] derivatized 43 sugar phosphates for GC-MS study; BSTFA:TMCS:CH_3CN (2:1:2, vol/vol), reacting with the free phosphate or its metal or cyclohexylammonium salt for 10 min at 80°C was used to trimethylsilylate all the phosphates except the pentose- and hexose-1-phosphates. These were trimethylsilylated by TSIM-CH_3CN for 5 min at room temperature: "Substituted oxime derivatives were prepared [CH_3ONH_2·HCl:sugar phosphate (1:1, wt/wt) in pyridine, 1 hr, room temperature] from several of the sugar phosphates in an attempt to reduce the number of gas chromatographic peaks (due to α- and β-furanose and -pyranose structures) and to increase the stability."

C. Amino Sugars

Amino sugars were among the first carbohydrates to be silylated for gas chromatography [14, 167]. The primary amino groups were monoacetylated and then the hydroxyl groups were trimethylsilylated by HMDS-TMCS (15 min, room temperature)[14]. When BSTFA became available, it was used in 1968 to silylate N-acetyl neuraminic acid (BSTFA:CH_3C (3:10, vol/vol), 2 hr, 125°C)[183].

Porter [42] succinctly reviewed the current analysis of glycoproteins for carbohydrates, particularly amino sugars. Methanolysis is nondestructive but produces multiple peaks that complicate quantitation. Reduction followed by acetylation yiel a single peak for each monosaccharide, but the preceding resin hydrolysis, devised s as not to destroy neutral sugars, does not release the amino sugars from the resin. Also, amino sugars tend to undergo thermal decomposition if separately derivatized a alditol acetates. Porter suggested that, after the resin hydrolysis of glycoproteir sugars, the "resin bound glucosamine and galactosamine [be] deaminated with $NaNO_2$ to the neutral 2,5-anhydrohexoses. Hexoses and 2,5-anhydrohexoses are then quantitated as the corresponding alditol acetates." The procedure was applied to several different glycoproteins. Thus the alditol acetate derivatives, introduced in 1961 [184] a applied since then (Refs. 177, 185, for example), continue to be used at this writir

The derivatization of glycoses to alditol acetates is illustrated by the procedure just cited [42]. In it, the hexoses and 2,5-anhydrohexoses are reduced in water solution ($NaBH_4$, 1 hr, room temperature). Then, the "excess $NaBH_4$ is decompos by...glacial acetic acid," the samples are dried, and the borate "removed as volati trimethylborate" by four additions of methanol-HCl "with concentration to dryness... after each addition. For acetylation, ...pyridine and...acetic anhydride are added. [In sealed containers,] the samples are heated at 100°C for 15 min...removed, mixed. [and heated at 100°C] for an additional 15 min"[42]. This reaction mixture may be injected into the gas chromatograph.

12.11 STEROIDS

Early studies are reviewed in *Gas Chromatographic Determination of Hormonal Steroids* edited by F. Polvani, M. Surace, and M. Luisi [186].

Flame Ionization and Mass Spectrometric Detection

Without the prior blocking of the steroid keto group, "enol TMS ethers may in some cases be the principal products of silylation of steroids which contain reactive keto groups" [187]. "When reactive ketone groups are present (this is true for many human steroids), and when catalyzed reaction conditions are employed, the derivative of choice are MO-TMS derivatives" [15].

The MO-TMS derivative, first presented in 1966 by Gardiner and Horning [188], b become standard in steroid derivatization for gas chromatogaphy. In 1974, for instance, it was used in studies reported from Canada [189], France [190], and Sweden [191].

With the MO-TMS derivative, the keto groups are converted to methyl oximes by reaction with methoxyamine in pyridine (3 hr, 60°C [189]; or overnight at room tempe ature [191]). The hydroxyls are then converted to trimethylsilyl groups by reaction with one or another silylating reagent added to either the methoxime reaction solution or the dried methoxime.

The steroid methoxime, introduced by Fales and Luukkainen [63], is relatively stable and has good GC properties. The methoxime is "the usual derivative for steroid ketonic groups which are relatively unhindered (for example, 3, 16, 17, and 20-keto steroids)" [192]. (See Fig. 12.10 for steroid numbering and some steroid structures and names.)

The methoxime-steroid ketone reaction may yield two isomers: "The isomers are presumably geometric isomers of the syn/anti type by analogy with the well-known reaction of oxime formation. In some instances isomers are not found; the course of the reaction apparently depends upon steric interrelationships...the major C_{19} and C_{20} human urinary steroids do not form isomeric MO and MO-TMS derivatives. The 11-keto group does not react to form a MO derivative, and 17- and 20-keto groups lead to only one MO isomer. Compounds with a 3-keto-Δ^4, 3-keto-5α-H, and 16-keto structure [give] isomers" [193]. In 1974, Delaforge et al. commented: "The separation of 3-MO-Δ-4-steroid isomers has not yet been observed on an apolar column but only on polar packed columns... The separation of syn/anti 3-MO-5α isomers is increased on capillary column compared to a packed column. For all the other positions of the methoxime group, including those at 3 in the 5β series, 6, 16, 17, and 20, only one peak is observed. These properties allow rapid identification of a 3-keto-Δ-4- or 5α-steroid by GC alone" [190].

Analogous to the methyloximes are the butyl- [194], pentyl- [194], and benzyl- [195] oximes: "They are...prepared in the same way...[and are] eluted much later than the corresponding MO derivatives. This effect makes it easy to distinguish steroids with a reactive keto group from closely related steroids containing hydroxyl groups alone, or with an unreactive ketone (11-one) group" [195].

(The reaction mixture from methoxyamine hydrochloride contains hydrogen chloride which, if injected, has untoward consequences for the oximes. Any of these oximes can, on injection, undergo an undesired Beckmann fission to the nitrile: "The fission reaction...is not a purely thermal reaction...[but] evidently requires an appropriate surface which is generated easily by hydrogen chloride." If testing with the fission-reactive androsterone MO-TMS derivative shows such active sites to be present, "a fresh upper portion of column packing is required... The injection of solutions containing hydrogen chloride should be avoided" [196].

For gas chromatography, the steroid hydroxyl groups "are usually converted to trimethylsilyl (TMS) ethers" [192]. "Steroid hydroxyl groups may be classified as unhindered, moderately hindered, or highly hindered, on the basis of the conditions needed for derivative formation... Three [correspondingly] useful conditions have been defined. These are (a) BSA at room temperature, (b) BSA:TMCS or BSTFA:TMCS [2:1, vol/vol] at room temperature or at 60°C, and (c) TSIM:BSA:TMCS [3:3:2, vol/vol] at 60-80°, [respectively]" [15]. (Note: with condition (c), use MO-TMS derivatives [15].)

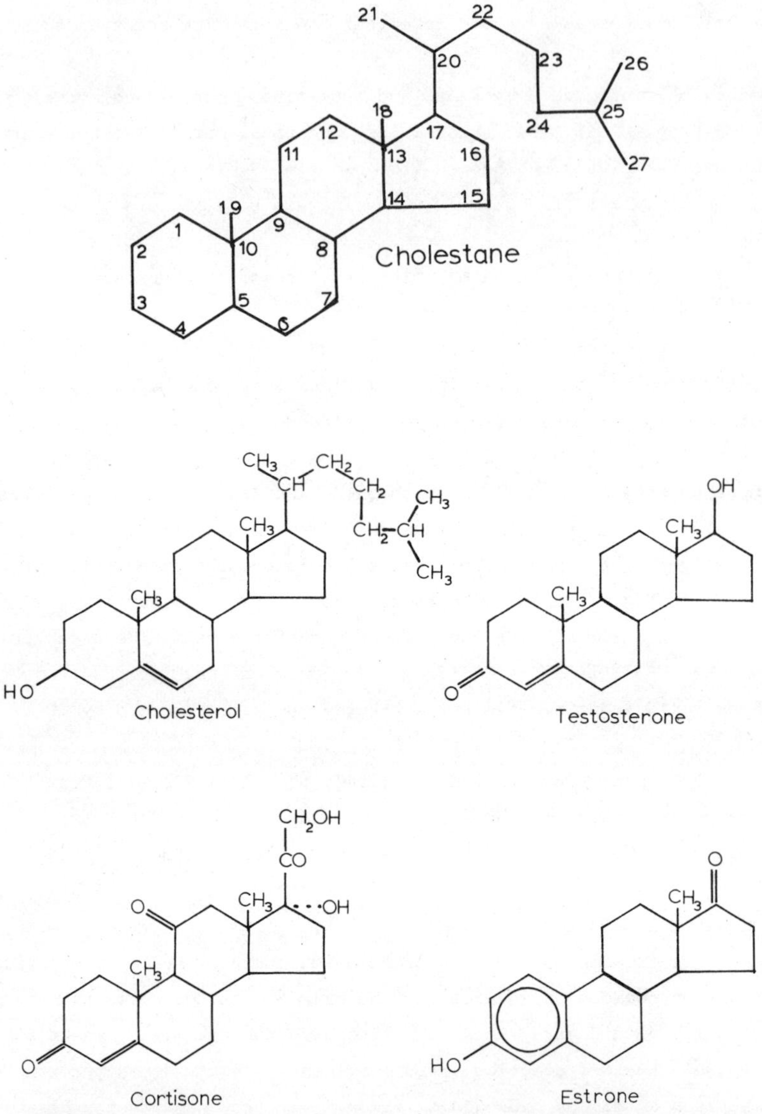

FIGURE 12.10 The steroid numbering system, and four illustrative steroids. Cholestane is the reference hydrocarbon for steroid numbering. Δ in a name indicates a double bond. Testosterone and cortisone have Δ^4 structures, for instance. α and β indicate stereoisomeric placement of hydrogen above or below the equatorial plane of the structure.

"From the point of view of decreasing column adsorption effects and gaining thermal stability, the conversion of all hydroxy groups to TMS ether groups is attractive" [15]. "The silylating mixture TSIM-BSA-TMCS [3:3:2, vol/vol, in pyridine] at 60° C., will yield fully silylated steroids..." [197]. Alternatively, "fully silylated MO-TMS and TMS derivatives of human steroid metabolites" may be prepared "by reaction with N-trimethylsilylimidazole [TSIM], alone or in combination with bis(trimethylsilyl)acetamide [BSA] and with or without trimethylchlorosilane [TMCS] as a catalyst. At or above 150°C, the rate of catalyzed and non-catalyzed silylation reactions with these reagents are similar" [198].

(Among silylating reagents, TSIM is atypical. TSIM "reacts particularly well with the sterically hindered 17α-hydroxyl group but not so easily with non-hindered hydroxyl groups" [17]. However, TSIM is not particularly powerful, comparatively: Neither TSIM nor TSIM-MSTFA can silylate the 11β-hydroxyl group that is, however, successfully silylated by any of the following: BSA-TMCS, BSTFA-TMCS, MSTFA-TMCS, or trimethylsilyldiethylamine (TMSDEA)-TMCS (24 hr, room temperature) [17].)

Compared to trimethylsilyl ethers, the tert-butyldimethylsilyl ethers present drastically simpler mass spectra that are based at M-57. They are also more stable toward hydrolysis [199].

Silylation nicely counters adsorption on the column, but the matter of improving the thermal stability of the steroids remains: "From a quantitative point of view, the double type derivatives (e.g., methoxime-TMS) appeared the most satisfactory... [for] thermally unstable steroid structures, primarily the dihydroxy side chain" [200]. Two other approaches to derivatizing and stabilizing the dihydroxy acetone side chain are base-catalyzed silylation [200] or reaction with a boronic acid to form a cyclic boronate [201, 202].

Although silylating in the presence of a nucleophilic agent such as piperidine "offers a procedure to form TMS-enol-TMS from hydroxyl keto steroids," the approach cannot be used without attention and insight. "Hindered hydroxyl groups (11β, $17_{\alpha-tert}$, e.g., in cortol) are not silylated." Application "to complex biological mixtures, such as a total neutral urinary extract, would yield some undesirable enol-TMS by-products, particularly from the 17-keto steroids" [200].

The boronic acids--$RB(OH)_2$--can form "cyclic derivatives with a wide range of functional groupings," including 1,2 and 1,3 diols, 1,2 diamines, and β- and γ-hydroxyamines. Only "extremely mild and simple experimental conditions [are] required...[and] in many instances the reaction is rapid and essentially complete" [201] The derivatives show good gas chromatographic properties and, given the exclusion of moisture, are stable. Also, compounds can be further derivatized in the presence of the boronate protecting group. In one illustrative example, 3α-, 7α, 21-trihydroxy-5β-pregnane-20-one (THS) was first reacted with n-butyl boronic acid (room temperature, 15 min) to form the THS 17,21-n-butyl boronate. In succession, the 20-MO and the 3-TMS derivatives were then formed, culminating in THS 17,21-n-butyl boronate, 20-O-methyloxime, 3-trimethylsilyl ether. The fully derivatized product showed no hydroxyl or carbonyl infrared absorption bands [201].

Thermionic Flame Ionization Detection

Vogt et al. [203] have suggested sensitizing hydroxyl steroids to the phosphorus-sensitive thermionic flame ionization detector. As an example, they reacted estrone with dimethylamine-dimethyl phosphine (DIMAP) to produce the dimethylphosphine ester (2 hr, 90°C). The volatile DIMAP (b.p., 99°C), used in 100-fold excess with acetonitrile solvent, is removed after the reaction by vacuum evaporation. The ester is thermally stable but sensitive to hydrolysis. The detection limit found was 200 pg.

Electron Capture and Mass Spectrometric Detection

Wotiz and Clark [204] and Cummins [37] have reviewed electron-capturing derivatives.

The heptafluorobutyrates, introduced in 1963 by Clark and Wotiz [205], generally allow a detection limit of about 100 pg. The sensitivity to detection by electron capture endowed by HFB derivatization can vary with the laboratory by a factor of 10 [204] and with the structure of the molecule derivatized by a factor of at least 100 [37, 206].

Dehennin and Scholler [207], realizing that "unsaturation of the parent steroid is...a prerequisite for high [EC] sensitivities," increased the sensitivity of HFB-derivatized steroidal secondary alcohols by a factor of about 100 by introducing "an exocyclic conjugated heptafluorobutanoyl group." (HFBA is added to a solution of the steroid in acetone and TMCS; the mixture is let stand an hour at room temperature; and is dried by evaporation. The derivative is taken up in n-hexane.)

HFB derivatization of testosterone, the extremely sensitive and reliable detection of which is important, leads to two products. These are stable if the base pyridine is used as solvent, but they are not stable in benzene or acetone [208].

To increase the EC sensitivity of derivatized testosterone and of keto steroids generally, Koshy, Kaiser, and VanDerSlik prepared stable oximes with O-(2,3,4,5,6-pentafluorobenzyl)hydroxylamine hydrochloride (30 min, 65°C). The resultant "PFB oxime gave much higher [EC] sensitivities than the heptafluorobutyrates...the highest sensitivity ever reported for keto steroids...as little as 5 picograms [of testosterone] could be detected" [209].

The EC-sensitizing PFB moiety has also been used by Morgan and Poole [210] to form pentafluorophenyldimethylsilyl (flophemesyl) ethers. However, because the "strongly electronegative pentafluorophenyl group alters reactivity at the silicon atom," the resultant silylation conditions require that flophemesyl chloride be used as catalyst.

With the flophemsyl system, reactivity can be made to vary widely. Hydroxy groups can, if desired, be derivatized in the presence of unprotected ketones. A "10:1 mixture of flophemesyldiethylamine and flophemesyl chloride in pyridine...gives ...complete conversion of cholesterol to the silyl ether without affecting 5α-cholestane-3-one, even after 6 hours at 80°" [210]. On the other hand, increasing the catalyst proportion from 10:1 to 1:1 increases silylation potential enough to cause extensive enol ether formation and require methoxime protection of ketones.

Singly flophemesyl-derivatized steroids can be detected at the nanogram level. With two or more such groups, picograms can be detected.

The principal halomethyldimethylsilyl ethers are the chloro- (CMDMS) and the iodomethyldimethylsilyl (IMDMS). Introduced in 1966 [211] as derivatives that remain stable during thin layer and gas chromatographic separations and that enhance sensitivity to electron capture detection, the CMDMS ethers may eventually find their most important use in gas chromatography-mass spectrometry. "The mass spectra of most CMDMS ethers...show relatively intense peaks at high mass, particularly the $(M - CH_2Cl)^+$ peak. Because of this, these derivatives are ideally suited to the quantitative determination of hydroxylated steroids at low levels" [212].

Among the halomethyldimethylsilyl ethers, "electron capture response increases according to I > Cl > F... Some of the [IMDMS] ethers have been found to be more sensitive than their corresponding heptafluorobutyric esters" [213]. In an application to the measurement of plasma testosterone [214], the bromomethyldimethylsilyl ether was first formed, then converted quantitatively to the IMDMS ether by incubation for 30 min at 37°C with sodium iodide-saturated acetone. The limiting detection sensitivity was about 25 pg.

12.12 MISCELLANEOUS

A. Carbamates and Ureas

Cochrane [215] and Dorough and Thorstenson [216] have reviewed the derivatization of insecticide and herbicide carbamates and ureas.

Carbamates and ureas can be hydrolyzed to amines and, given the structure, phenols. Either the original, usually thermally unstable compounds or, more often though less desirably, the hydrolyzates, can be acylated or alkylated.

Using trifluoroacetic anhydride (TFAA), Evans [36] converted monoureas and symmetrically substituted ureas to their trifluoroacetates (2 to 3 min, room temperature). The products are solid, stable for weeks if protected from moisture, and yield symmetrical peaks.

Saunders and Vanatta [217] found that TFAA would trifluoroacetylate the secondary nitrogen of a potential herbicide (used by them as a model urea) to yield a thermally stable product with good gas chromatographic properties. Given the same approach with three -1,1-dimethylurea herbicides, fenuron (3-phenyl-) and monuron [3-(p-chlorophenyl)-] yielded stable derivatives, but not diuron [3-(3,4-dichlorophenyl)-].

However, they found that all three (fenuron, monuron, and diuron) could be methylated to stable and satisfactory trimethyl derivatives (methyl iodide and potassium tert-butoxide in tetrahydrofuran, 2 to 16 hr, 52°C)[217]. Tanaka and Wien [218] used flash-heater methylation, injecting a mixture of trimethylanilinium hydroxide (TMAH)(in methanol) and the herbicide (in acetone) into a 220°C injection port. Under their conditions, a TMAH/urea mole ratio of 10 ensures reliable monomethylation, but only that: The other three nitrogen positions must already have alkyl or aryl substitutions.

The hydrolyzed phenol or amine moieties are derivatized and sensitized for hypersensitive detectors in ways and with relative efficiencies already described in this chapter for such substances. For examples, in one widely used approach [219-223], carbamate phenols have been converted to their 2,4-dinitrophenyl ethers for ultimate detection by electron capture.

B. Purines, Pyrimidines, Nucleosides, and Nucleotides

Purines and pyrimidines receive attention because they proceed from the hydrolysis of nucleotides, as do nucleosides. Nucleosides are purine- or pyrimidine-pentose combinations. Purines and pyrimidines are derivatized quite generally by trimethylsilylation. This is done preferably with BSTFA (BSA may not silylate completely [12]).

For instance, Lakings and Gehrke [224] hydrolyzed ribonucleic acid (RNA) and deoxyribonucleic acid (DNA) (20 µg amounts) in 100 µl trifluoroacetic acid:formic acid (1:1, vol/vol) at 200°C for 1.5 hr (RNA) or 2.0 hr (DNA). After the reagents had then evaporated under N_2, the bases were silylated by BSTFA-CH_3CN for 15 min at 150°C. These silylation conditions, which are adequate, are not critical [225]. Eliminating oxygen by N_2 purge before the silylation is important to prevent oxygen incorporation during silylation [226]. A 50-fold BSTFA excess, used for 45 min at 150°C, gives 95% or better conversion with the model pyrimidines uracil and thymine [227].

Most nucleotides (30 µg in 30 µl pyridine) can be successfully trimethylsilylated by BSTFA (30 µl)-TMCS (1 µl)(30 min, either room temperature or 150°C); adenosine 3',5'-cyclic monophosphate required 3 hr at 100°C to form the tri-TMS derivative [228]

C. Indoles and Imidazoles

Ehrsson [229], using 3-methylindole as a model indole, obtained the N-trifluoroacetyl derivative after 20 min reaction at room temperature. The reaction of 100 mg of the indole was carried out in 5 ml of a 0.9 M trimethylamine solution in hexane to which 0.5 ml of TFAA was added. The product was extracted with water and dried under vacuum. The derivative is sensitive to detection by electron capture.

Begg and Grimett [230] report that most imidazoles can be quantitatively and almost instantaneously converted at room temperature to the 1-acetyl imidazoles by reaction with acetic anhydride in dry tetrahydrofuran. The derivatives are stable for several days and can be directly injected.

D. Epoxides

If pyridine is used as solvent, epoxides react quantitatively and specifically with TMCS (30 min, 70°C) to form Cl-TMS derivatives [231]. (Trace water catalyzes the reaction.)

E. Phosphoric and Phosphonic Acids

With diazomethane, Daemen and Dankelman [232] converted mono- and dialkyl (C_8-C_{16}) phosphoric acids and their salts into the corresponding methyl esters.

Harvey and Horning [233] satisfactorily trimethylsilylated a number of alkyl- and aminoalkylphosphonates (BSTFA:TMCS (2:1, vol/vol)). However, for the 1-aminoalkylphosphonates, conversion by CS_2 of the amino group into the isothiocyanate was found necessary subsequent to the silylation.

F. Thiols

Lofberg [234] silylated three series of thiols with BSTFA (overnight, room temperature), although as late as 1974 others [235] felt obliged to form the thioalkylbenzoates as derivatives. Donike [236] has applied MBTFA to thioglycol (mercaptoethanol) to form the O,S-bis(TFA) derivative.

G. Sulfonamides

Vandenheuvel and Gruber [237] quantitatively converted primary sulfonamides to the N-dimethylaminomethylene derivatives by reaction with dimethylformamide dimethylacetal (10 min, 75 to 80°C). The reaction mixture was injected. With electron capture, 25 ppb (wt/wt) of 3-bromo-5-cyanobenzene sulfonamide in sheep could be detected (the dosage was 15 mg/kg).

ACKNOWLEDGMENT

Charles Feit, collaborator in the preparation of this chapter, was responsible for the selection of papers from some of which the chapter was written, and for constructive criticism of the chapter itself.

REFERENCES

1. M. G. Horning, A. M. Moss, and E. C. Horning, *Biochem. Biophys. Acta 148*, 597 (1967).
2. J. A. Perry and C. A. Feit, in *GLC and HPLC Determination of Therapeutic Agents*, Vol. 1 (K. Tsuji and W. Morozowich, editors), Dekker, New York, 1978, Chapter 4, pp. 137-208.
3. J. C. Cavagnol and W. R. Betker, in *The Practice of Gas Chromatography* (L. S. Ettre and J. A. Zlatkis, editors), Interscience, New York, 1967, pp. 71-128.
4. T. Nanbara and J. Goto, *Jap.-Analyst 23*, 704 (1974).
5. J. Drozd, *J. Chromatog. 113*, 303 (1975).
6. S. Ahuja, *J. Pharm. Sci. 65*, 163 (1976).
7. K. Blau and G. King, *Handbook of Derivatives for Chromatography*, Heyden, London, 1977.

8. A. E. Pierce, *Silylation of Organic Compounds*, Pierce Chemical Co., Rockford, Ill, 61105, 1968.
9. C. Grunwald and R. G. Lackard, *J. Chromatog.* 52, 491 (1970).
10. A. L. Barta and C. A. Osmond, *J. Agr. Food Chem.* 21, 316 (1973).
11. C. M. Williams, *Anal. Biochem.* 11, 224 (1965).
12. I. A. Muni, C. H. Altschuller, and J. C. Neicheril, *Anal. Biochem.* 50, 354 (1972).
13. J. E. Mrochek and W. T. Rainey, *Anal. Biochem.* 57, 173 (1974).
14. M. B. Perry, *Can. J. Biochem.* 42, 451 (1964).
15. E. M. Chambaz and E. C. Horning, *Anal. Biochem.* 30, 7 (1969).
16. N. E. Hoffman and K. A. Peteranetz, *Anal. Lett.* 5, 589 (1972).
17. H. Gleispach, *J. Chromatog.* 91, 407 (1974).
18. G. D. Brittain, in *Recent Advances in Gas Chromatography* (I. I. Domsky and J. A. Perry, editors), Dekker, New York, 1971, pp. 215-222.
19. G. D. Brittain, J. E. Sullivan, and L. R. Schewe, in *Recent Advances in Gas Chromatography* (I. I. Domsky and J. A. Perry, editors), Dekker, New York, 1971, pp. 223-229.
20. H. Morita, *J. Chromatog.* 101, 189 (1974).
21. D. J. Harvey, D. B. Johnson, and M. G. Horning, *Anal. Lett.* 5, 745 (1972).
22. E. Änggård and G. Sedvall, *Anal. Chem.* 41, 1250 (1969).
23. F. Karoum, F. Cattabeni, E. Costa, C. R. J. Ruthven, and M. Sandler, *Anal. Biochem.* 47, 550 (1972).
24. N. K. McCallum and R. J. Armstrong, *J. Chromatog.* 78, 303 (1973).
25. H. Ehrsson, T. Walle, and H. Brötell, *Acta Pharm. Suec.* 8, 319 (1971).
26. F. K. Kawahara, *Anal. Chem.* 40, 1009 (1968).
27. H. Brötell, H. Ehrsson, and O. Gyllenhaal, *J. Chromatog.* 78, 293 (1973).
28. C. W. Gehrke, K. C. Kuo, R. W. Zumwalt, and T. P. Waalkes, in *Polyamines in Normal and Neoplastic Growth* (D. H. Russell, editor), Raven, New York, 1973, pp. 343-353.
29. J. A. F. Wickramasinghe, W. Morozowich, W. E. Hamlin, and S. R. Shaw, *J. Pharm. Sci.* 62, 1428 (1973).
30. E. Watson, S. Wilk, and J. Roboz, *Anal. Biochem.* 59, 441 (1974).
31. A. C. Moffat and E. C. Horning, *Anal. Lett.* 3, 205 (1970).
32. C. Moffat, E. C. Horning, S. B. Matin, and M. Rowland, *J. Chromatog.* 66, 255 (1972).
33. T. Walle and H. Ehrsson, *Acta Pharm. Suec.* 7, 389 (1970).
34. H. Ehrsson and B. Mellstrom, *Acta Pharm. Suec.* 9, 107 (1972).
35. R. H. Greeley, *Clin. Chem.* 20, 192 (1974).
36. R. T. Evans, *J. Chromatog.* 88, 398 (1974).
37. L. M. Cummins, in *Recent Advances in Gas Chromatography* (I. I. Domsky and J. A. Perry, editors), Dekker, New York, 1971, pp. 313-340.
38. D. O. Edlund and J. R. Anfinsen, *J. Assoc. Off. Anal. Chem.* 53, 289 (1970).
39. E. R. Atkinson and S. I. Calouche, *Anal. Chem.* 43, 460 (1971).
40. D. J. Harvey and W. D. M. Patton, *J. Chromatog.* 109, 73 (1975).
41. M. P. Rabinowitz, P. Reisberg, and J. I. Bodin, *J. Pharm. Sci.* 63, 1601 (1974).
42. W. H. Porter, *Anal. Biochem.* 63, 2743 (1975).

43. J. J. Myher and A. Kuksis, *J. Chromatog. Sci. 13*, 138 (1975).
44. C. Braestrup, *Anal. Biochem. 55*, 420 (1973).
45. R. V. Smith and A. W. Stocklinski, *Anal. Chem. 47*, 1321 (1975).
46. M. P. Heenan and N. K. McCallum, *J. Chromatog. Sci. 12*, 89 (1974).
47. N. K. McCallum and R. J. Armstrong, *J. Chromatog. 78*, 303 (1973).
48. J. N. Seiber, D. G. Crosby, H. Fouda, and C. J. Soderquist, *J. Chromatog. 73*, 89 (1972).
49. M. Margosis, *J. Chromatog. Sci. 12*, 549 (1974).
50. K. Tsuji and J. H. Robertson, *Anal. Chem. 43*, 818 (1971).
51. D. H. Calam, *J. Chromatog. Sci. 12*, 613 (1974).
52. L. W. Brown, and P. B. Bowman, *J. Chromatog. Sci. 12*, 373 (1974).
53. P. T. Russell, A. J. Eberle, and H. C. Ching, *Clin. Chem. 21*, 657 (1975).
54. F. Vane and M. G. Horning, *Anal. Lett. 2*, 357 (1969).
55. C. Pace-Asciak and L. S. Wolfe, *J. Chromatog. 56*, 129 (1971).
56. D. S. Erley, *Anal. Chem. 29*, 1564 (1957).
57. S. Nicosia and G. Galli, *Anal. Biochem. 61*, 192 (1974).
58. R. W. Kelly, *Anal. Chem. 45*, 2079 (1973).
59. B. Samuelsson, M. Hamberg, and C. C. Sweeley, *Anal. Biochem. 38*, 301 (1970).
60. L. Baczynskyj, D. J. Duchamp, J. F. Zieserl, and U. Axen, *Anal. Chem. 45*, 479 (1973).
61. L. J. Papa and L. P. Turner, *J. Chromatog. Sci. 10*, 744 (1972).
62. H. Kaalio, R. R. Linko, and J. Kaitaranta, *J. Chromatog. 65*, 355 (1972).
63. H. M. Fales and T. Luukkainen, *Anal. Chem. 37*, 955 (1965).
64. W. C. Butts, *J. Chromatog. Sci. 8*, 474 (1970).
65. J. W. Vogh, *Anal. Chem. 43*, 1618 (1971).
66. J. Korolczuk, M. Daniewski, and Z. Mielniczuk, *J. Chromatog. 88*, 177 (1974).
67. O. Mlejnek, *J. Chromatog. 70*, 59 (1972).
68. M. J. Levitt, *Anal. Chem. 45*, 618 (1973).
69. R. H. Greeley, *J. Chromatog. 88*, 229 (1974).
70. J. C. West, *Anal. Chem. 47*, 1708 (1975).
71. L. Giuffrida, *J. Assoc. Off. Anal. Chem. 47*, 293 (1964).
72. P. Schulz and R. Vilceanu, *J. Chromatog. 111*, 105 (1975).
73. R. V. Smith and S. L. Tsai, *J. Chromatog. 61*, 29 (1971).
74. A. J. Sheppard and J. L. Iverson, *J. Chromatog. Sci. 13*, 448 (1975).
75. Association of Official Analytical Chemists, *Official Methods of Analysis*, 10th ed., Washington, D. C., 1965, Sec. 26.052, p. 429.
76. A. J. Sheppard and L. A. Ford. *J. Assoc. Off. Anal. Chem. 46*, 947 (1963).
77. D. Firestone, *J. Assoc. Off. Anal. Chem. 52*, 254 (1969).
78. C. J. F. Bottcher, F. P. Woodford, E. Boelsma-Van Haute, and C. M. VanGent, *Rec. Trav. Chim. 78*, 794 (1959).
79. S. W. Christopherson and R. L. Glass, *J. Dairy Sci. 52*, 1289 (1969).
80. J. L. Iverson, 87th Annual Meeting, Association of Official Analytical Chemists, Washington, D. C., Oct. 9-12, 1973, Paper No. 203.

81. E. A. Moscatelli, *Lipids 7*, 268 (1972).
82. J. MacGee and K. G. Allen, *J. Chromatog. 100*, 35 (1974).
83. J. P. Thenot, E. C. Horning, M. Stafford, and M. G. Horning, *Anal. Lett. 5*, 217 (1972).
84. H. Ko and M. E. Royer, *J. Chromatog. 88*, 253 (1974).
85. U. Hintze, H. Röper, and G. Gercken, *J. Chromatog. 87*, 481 (1973).
86. H.-P. Klemm, U. Hintze, and G. Gercken, *J. Chromatog. 75*, 19 (1973).
87. C. C. Alley, J. B. Brooks, and G. Choudberry, *Anal. Chem. 48*, 487 (1976).
88. J. B. Brooks, C. C. Alley, and J. A. Liddle, *Anal. Chem. 46*. 1930 (1974).
89. J. M. Tedder, *Chem. Revs. 55*, 787 (1955).
90. T. J. Sprinkle, A. H. Porter, M. Greer, and C. M. Williams, *Clin. Chim. Acta 25*, 409 (1969).
91. N. E. Hoffman and K. M. Gooding, *Anal. Biochem. 31*, 471 (1969).
92. Y. H. Loo, L. Scotto, and M. G. Horning, Personal communication, 1975.
93. R. A. Chalmers and R. W. E. Watts, *Analyst 97*, 958 (1972).
94. S. Wilk, E. Watson, and S. D. Glick, *Eur. J. Pharmacol. 30*, 117 (1975).
95. E. Watson, B. Travis, and S. Wilk, *Life Sci. 15*, 2167 (1975).
96. J. Eagles and M. E. Knowles, *Anal. Chem. 43*, 1697 (1971).
97. J. S. Parsons, *J. Chromatog. Sci. 11*, 659 (1973).
98. R. H. DeWolfe (editor), *Carboxylic Ortho Acid Derivatives*, Vol. 14: *Organic Chemistry*, Academic, New York, 1970.
99. J. P. Thenot and E. C. Horning, *Anal. Lett. 5*, 519 (1972).
100. M. W. Scoggins, *J. Chromatog. Sci. 13*, 146 (1975).
101. H. V. Street, *J. Chromatog. 73*, 73 (1972).
102. S. F. Sisenwine, J. A. Knowles, and H. W. Ruelius, *Anal. Lett. 2*, 315 (1969).
103. D. D. Clarke, S. Wilk, and S. E. Gitlow, *J. Gas Chromatog. 4*, 310 (1966).
104. S. B. Matin and M. Rowland, *J. Pharm. Sci. 61*, 1235 (1972).
105. T. Walle, *J. Chromatog. 111*, 133 (1975).
106. P. H. Degen and W. Riess, *J. Chromatog. 85*, 53 (1973).
107. P. Hartvig and J. Vessman, *Anal. Lett. 7*, 223 (1974).
108. J. Vessman, P. Hartvig, and M. Molander, *Anal. Lett. 6*, 699 (1973).
109. P. Hartvig and J. Vessman, *Acta Pharm. Suec. 11*, 115 (1974).
110. P. Hartvig and J. Vessman, *J. Chromatog. Sci. 12*, 722 (1974).
111. M. Donike, *J. Chromatog. 103*, 91 (1975).
112. M. Donike, *Chromatographia 7*, 651 (1974).
113. G. Schwedt and H. H. Bussemas, *J. Chromatog. 106*, 440 (1975).
114. H. G. Lovelady and L. L. Foster, *J. Chromatog. 108*, 43 (1975).
115. J. Willner, H. F. LeFevre, and E. Costa, *J. Neurochem. 23*, 857 (1974).
116. E. Gelpi, E. Peralta, and J. Segura, *J. Chromatog. Sci. 12*, 701 (1974).
117. D. J. Edwards and K. Blau, *Anal. Biochem. 45*, 387 (1972).
118. J.-C. Lhuguenot and B. F. Maume, *J. Chromatog. Sci. 12*, 411 (1974).
119. J. B. Brooks, C. C. Alley, and R. Jones, *Anal. Chem. 44*, 1881 (1972).
120. K.-A. Karlsson, *Biochim. Biophys. Res. Commun. 39*, 847 (1970).

121. P. Erdtmansky and T. J. Roehl, *Anal. Chem. 47*, 750 (1975).
122. F. E. Kaiser, C. W. Gehrke, R. W. Zumwalt, and K. C. Kuo, *J. Chromatog. 94*, 113 (1974).
123. C. W. Gehrke, H. Nakamoto, and R. W. Zumwalt, *J. Chromatog. 45*, 24 (1969).
124. C. W. Gehrke and K. Leimer, *J. Chromatog. 57*, 219 (1971).
125. J. F. Klebe, H. Finkbeiner, and D. M. White, *J. Amer. Chem. Soc. 88*, 3390 (1966).
126. D. L. Stalling, C. W. Gehrke, and C. D. Ruyle, *Biochem. Biophys. Res. Commun. 31*, 616 (1968).
127. R. W. Zumwalt, Masters thesis, University of Missouri, Columbia, Missouri, June, 1968.
128. E. D. Smith and K. L. Shewbart, *J. Chromatog. Sci. 7*, 704 (1969).
129. P. W. Albro and L. Fishbein, *J. Chromatog. 55*, 297 (1971).
130. J. P. Hardy and S. L. Kerrin, *Anal. Chem. 44*, 1497 (1972).
131. J. Rosmus and Z. Deyl, *J. Chromatog. 70*, 221 (1972).
132. W. H. Lamkin, J. W. Weatherford, N. S. Jones, T. Pan, and D. N. Ward, *Anal. Biochem. 58*, 422 (1974).
133. P. Husek, *J. Chromatog. 91*, 475 (1974).
134. P. Husek, *J. Chromatog. 91*, 483 (1974).
135. J. Jönsson, J. Eyem, and J. Sjöquist, *Anal. Biochem. 51*, 204 (1973).
136. C. W. Moss and M. A. Lambert, *Anal. Biochem. 59*, 259 (1974).
137. C. W. Gehrke and H. Takeda, *J. Chromatog. 76*, 63 (1973).
138. D. J. Casagrande, *J. Chromatog. 49*, 537 (1970).
139. F. Raulin, P. Shapshak, and B. N. Khare, *J. Chromatog. 73*, 35 (1972).
140. P. Cancalon and J. D. Klingman, *J. Chromatog. Sci. 12*, 349 (1974).
141. R. W. Zumwalt, K. Kuo, and C. W. Gehrke, *J. Chromatog. 57*, 193 (1971).
142. I. M. Moodie, *J. Chromatog. 99*, 495 (1974).
143. N. N. Nihei, M. C. Gershengorn, T. Mitsuma, L. R. Stringham, A. Gordy, B. Kuchmy, and C. S. Hollander, *Anal. Biochem. 43*, 433 (1971).
144. S. Wilk and M. Orlowski, *Fed. Eur. Biochem. Soc. 33*, 157 (1973).
145. S. Wilk and M. Orlowski, *Anal. Biochem.*, in press.
146. E. Brochmann-Hanssen and T. O. Oke, *J. Pharm. Sci. 58*, 370 (1969).
147. E. W. Robb and J. J. Westbrook, *Anal. Chem. 35*, 1944 (1963).
148. J. MacGee, *Int. Cong. Ser. No. 286*, 111 (1972).
149. E. B. Solow, J. Metaxas, and T. R. Summers, *J. Chromatog. Sci. 12*, 256 (1974).
150. J. Pecci and T. J. Giovanniello, *J. Chromatog. 109*, 163 (1975).
151. W. D. Hooper, D. K. Dubetz, M. J. Eadie, and J. H. Tyler, *J. Chromatog. 110*, 206 (1975).
152. M. Kowblansky, B. Scheinthal, G. D. Cravello, and L. Chafetz, *J. Chromatog. 76*, 467 (1973).
153. J. Barrett, *Clin. Chem. Newsletter 3*, 16 (1971).
154. G. Kananen, R. Osiewicz, and I. Sunshine, *J. Chromatog. Sci. 10*, 283 (1972).
155. H. V. Street, *Clin. Chim. Acta 34*, 357 (1971).
156. W. Dünges and E. Bergheim-Irps, *Anal. Lett. 6*, 185 (1973).
157. R. Osiewicz, V. Aggarwal, R. M. Young, and I. Sunshine, *J. Chromatog. 88*, 157 (1974).

158. W. Dünges, *Anal. Chem. 45*, 963 (1973).

159. V. S. Venturella, V. M. Gualario, and R. E. Lang, *J. Pharm. Sci. 62*, 662 (1973).

160. C. F. Johnson, W. A. Dechtiaruk, and H. M. Solomon, *Clin. Chem. 21*, 144 (1975).

161. M. G. Horning, K. Lertratanangkoon, J. Nowlin, W. G. Stillwell, R. N. Stillwell, T. E. Zion, P. Kellaway, and R. M. Hill, *J. Chromatog. Sci 12*, 630 (1974).

162. H. L. Rice and G. R. Pettit, *J. Amer. Chem. Soc. 76*, 302 (1954).

163. R. Gerstl and K. Ranfft, *Z. Anal. Chim. 258*, 110 (1972).

164. C. Pantarotto, A. Bossi, G. Belvedere, A. Martini, M. G. Donelli, and A. Frigerio, *J. Pharm. Sci. 63*, 1554 (1974).

165. J. H. Sloneker, in *Biomedical Applications of Gas Chromatography*, Vol. 2 (H. A. Szymanski, editor), Plenum, New York, 1968, pp. 87-135.

166. G. S. Dutton, in *Advances in Carbohydrate Chemistry and Biochemistry*, Vol. 28, (R. S. Tipson and D. Horton, editors), Academic, New York, 1973, pp. 11-160.

167. C. C. Sweeley, R. Bentley, M. Makita, and W. W. Wells, *J. Am. Chem. Soc. 85*, 2497 (1963).

168. S. H. Langer, R. A. Friedel, I. Wender, and A. G. Sharkey, *Anal. Chem. 30*, 1353 (1958).

169. A. H. Weiss and H. Tambawala, *J. Chromatog. Sci. 10*, 120 (1972).

170. R. Bentley and N. Botlock, *Anal. Biochem. 20*, 312 (1967).

171. K. M. Brobst and C. E. Lott, Jr., *Cereal Chem. 43*, 35 (1966).

172. J. Casals-Stenzel, H-P. Buscher, and R. Schauer, *Anal. Biochem. 65*, 507 (1975).

173. O. A. Mamer and S. S. Tjoa, *Clin. Chem. 19*, 58 (1973).

174. I. M. Morrison and M. B. Perry, *Can. J. Biochem. 44*, 1115 (1966).

175. R. E. Hurst, *Carbohydrate Res. 30*, 143 (1973).

176. H. Jacin, J. M. Slanski, and R. J. Moshy, *J. Chromatog. 37*, 103 (1968).

177. L. J. Griggs, A. Post, E. R. White, J. A. Finkelstein, W. E. Moeckel, K. G. Holden, J. E. Zarembo, and J. E. Weisbach, *Anal. Biochem. 43*, 369 (1971).

178. R. A. Laine and C. C. Sweeley, *Anal. Biochem. 43*, 533 (1971).

179. J. Szafranek, C. D. Pfaffenberger, and E. C. Horning, *J. Chromatog. 88*, 149 (1974).

180. H. Ehrsson, T. Walle, and S. Wikström, *J. Chromatog. 101*, 206 (1974).

181. M. Zinbo and W. R. Sherman, *J. Amer. Chem. Soc. 92*, 2105 (1970).

182. D. J. Harvey and M. G. Horning, *J. Chromatog. 76*, 51 (1973).

183. D. A. Craven and C. W. Gehrke, *J. Chromatog. 37*, 414 (1968).

184. S. W. Gunner, J. K. N. Jones, and M. B. Perry, *Chem. and Ind. 1961*, 155 (1961).

185. M. B. Perry and A. C. Webb, *Can. J. Biochem. 46*, 1163 (1968).

186. F. Polvani, M. Surace, and M. Luisi (editors), *Gas Chromatographic Determination of Hormonal Steroids*, Academic, New York, 1967.

187. E. M. Chambaz, G. Maume, B. Maume, and E. C. Horning, *Anal. Lett. 1*, 749 (1968).

188. W. L. Gardiner and E. C. Horning, *Biochim. Biophys. Acta 115*, 524 (1966).

189. K. M. McErlane, *J. Chromatog. Sci. 12*, 97 (1974).

190. M. Delaforge, B. F. Maume, P. Bournot, M. Prost, and P. Badieu, *J. Chromatog. Sci. 12*, 545 (1974).

191. M. Axelson, G. Schumacher, and J. Sjovall, *J. Chromatog. Sci. 12*, 535 (1974).

192. E. C. Horning and M. G. Horning, in *Recent Advances in Gas Chromatography* (I. Domsky and J. A. Perry, editors), Dekker, New York, 1971, pp. 341-376.

193. M. G. Horning, A. M. Moss, and E. C. Horning, *Anal. Biochem.* 22, 284 (1968).
194. T. A. Baillie, C. J. W. Brooks, and E. C. Horning, *Anal. Lett.* 5, 351 (1972).
195. P. G. Devaux, M. G. Horning, and E. C. Horning, *Anal. Lett.* 4, 151 (1971).
196. J. P. Thenot and E. C. Horning, *Anal. Lett.* 4, 683 (1971).
197. E. M. Chambaz and E. C. Horning, *Anal. Lett.* 1, 201 (1967).
198. N. Sakauchi and E. C. Horning, *Anal. Lett.* 4, 41 (1971).
199. R. W. Kelly and P. L. Taylor, *Anal. Chem.* 48, 465 (1976).
200. E. M. Chambaz, G. Defaye, and C. Madani, *Anal. Chem.* 45, 1090 (1973).
201. G. M. Anthony, C. J. W. Brooks, I. MacLean, and I. Sangster, *J. Chromatog. Sci.* 7, 623 (1969).
202. C. J. W. Brooks and I. MacLean, *J. Chromatog. Sci.* 9, 18 (1971).
203. W. Vogt, I. Fischer, and M. Knedel, *Z. Anal. Chem.* 267, 28 (1973).
204. H. H. Wotiz and S. J. Clark, in *Methods of Biochemical Analysis,* Vol. 18 (D. Glick, editor), Interscience, New York, 1970, pp. 340-372.
205. S. J. Clark and H. H. Wotiz, *Steroids 2,* 535 (1963).
206. L. Dehennin, A. Reiffsteck, and R. Scholler, *J. Chromatog. Sci.* 10, 224 (1972).
207. L. Dehennin and R. Scholler, *J. Chromatog.* 111, 238 (1975).
208. P. G. Devaux and E. C. Horning, *Anal. Lett.* 2, 637 (1969).
209. K. T. Koshy, D. C. Kaiser, and A. L. VanDerSlik, *J. Chromatog. Sci.* 13, 97 (1975).
210. E. D. Morgan and C. F. Poole, *J. Chromatog.* 104, 351 (1975).
211. B. S. Thomas, C. Eaborn, and D. R. M. Walton, *Chem. Commun.* 1966, 408 (1966).
212. J. R. Chapman and E. Bailey, *Anal. Chem.* 45, 1636 (1973).
213. D. Exley and A. Dutton, *Steroids 14,* 575 (1969).
214. B. S. Thomas, *J. Chromatog.* 56, 37 (1971).
215. W. P. Cochrane, *J. Chromatog. Sci.* 13, 246 (1975).
216. H. W. Dorough and J. H. Thorstenson, *J. Chromatog. Sci.* 13, 212 (1975).
217. D. G. Saunders and L. E. Vanatta, *Anal. Chem.* 46, 1319 (1974).
218. F. S. Tanaka and R. G. Wien, *J. Chromatog.* 87, 85 (1973).
219. I. C. Cohen and B. B. Wheals, *J. Chromatog.* 43, 233 (1969).
220. I. C. Cohen, J. Norcup, J. H. A. Ruzicka, and B. B. Wheals, *J. Chromatog.* 49, 215 (1970).
221. J. H. Caro, D. E. Glotfelty, H. P. Freeman, and A. Taylor, *J. Assoc. Off. Anal. Chem.* 56, 1319 (1973).
222. J. H. Caro, H. P. Freeman, D. E. Glotfelty, B. C. Turner, and W. M. Edwards, *J. Agr. Food Chem.* 21, 1010 (1973).
223. B. C. Turner and J. H. Caro, *J. Environ. Qual.* 2, 245 (1973).
224. D. B. Lakings and C. W. Gehrke, *Clin. Chem.* 18, 810 (1972).
225. S. Y. Chang, D. B. Lakings, R. W. Zumwalt, C. W. Gehrke, and T. P. Waalkes, *J. Lab. Clin. Med.* 83, 816 (1974).
226. D. L. Von Minden, R. N. Stillwell, W. A. Koenig, K. J. Lyman, and J. A. McCloskey, *Anal. Biochem.* 50, 110 (1972).
227. V. Rehak and V. Pacakova, *Anal. Biochem.* 61, 294 (1974).
228. A. M. Lawson, R. N. Stillwell, M. M. Tacker, K. Tsuboyama, and J. A. McCloskey, *J. Amer. Chem. Soc.* 93, 1014 (1971).

229. H. Ehrsson, *Acta Pharm Suec.* *9*, 419 (1972).
230. C. G. Begg and M. R. Grimett, *J. Chromatog.* *73*, 238 (1972).
231. D. J. Harvey, D. B. Johnson, and M. G. Horning, *Anal. Lett.* *5*, 745 (1972).
232. J. M. H. Daemen and W. Dankelman, *J. Chromatog.* *78*, 281 (1973).
233. D. J. Harvey and M. G. Horning, *J. Chromatog.* *79*, 65 (1973).
234. R. T. Lofberg, *Anal. Lett.* *4*, 77 (1971).
235. J. Korolczuk, M. Daniewski, and Z. Mielniczuk, *J. Chromatog.* *100*, 165 (1974).
236. M. Donike, *J. Chromatog.* *78*, 273 (1973).
237. W. J. A. Vandenheuvel and V. F. Gruber, *J. Chromatog.* *112*, 513 (1975).

Chapter 13

QUALITATIVE ANALYSIS

The three principal means of qualitative analysis associated with gas chromatography are, in order of increasing preference, gas chromatography itself, infrared spectroscopy, and gas chromatography-mass spectrometry (GC-MS). The treatments of these three are emphasized in this chapter.

By their elegance and power, certain other methods also demand description. For those with no infrared or mass spectrometers, we discuss several of these other means.

Except for a special section on GC-MS, our organization reflects sample flow and handling.

A well-prepared book on qualitative analysis, published in 1970, is *Identification Techniques in Gas Chromatography*, by D. A. Leathard and B. C. Shurlock [1]. Their treatment occasionally complements ours in emphasis and point of view. Leathard has also prepared a more recent review, published in 1975 [2].

13.1 IDENTIFICATION OF ELUTED COMPONENTS

Components can be identified either on elution or by inference from sample treatment before application of gas chromatography. We consider first the identification of eluted components.

Eluted components can be identified either during elution, which we discuss first, or after collection from the carrier gas, which we take up later.

During Elution

Components are identified during elution by means of either standards or special detectors.

By Means of Standards

We distinguish the direct from the indirect use of standards. By *the direct use of a standard* we mean injecting both the standard and the unknown into the same column in either immediate sequence or deliberate admixture. By *the indirect use of a standard* we mean identifying an unknown component solely by matching two retention data: Numerically matching (datum one) the retention of an unknown by a present column with (datum two) the retention of a standard by a comparable column, presumably at some other time or place. A comparable column would be one having the same stationary phase and used at the same temperature as the one at hand.

In gas chromatography, qualitative analysis by the direct use of standards is best (and most often) done when the sample being inspected has only one or two components to be identified among a relatively few in all. This case occurs frequently. Therefore, the direct use of standards is the most used form of qualitative analysis in gas chromatography.

The indirect use of standards depends on retention data of high precision. It is inherently unsuited to most qualitative analyses that involve gas chromatography. The unknown must be measured as precisely as the standard. Moreover, the requisite degree of precision depends on the sample being analyzed and is much higher than is either obtainable or generally realized. These difficulties are considered presently

The direct use of standards. In direct experimentation, a standard can be injected to identify an unknown. If the retentions match, the identity is apparent (Fig. 13.1) but not conclusive. A peak often contains more than one component.

Establishing the true number of components in a mixture may require elaborate proof. For instance, two columns of different polarity might be used to see if any peaks split or show broadening or asymmetry, thus indicating more than one component [3]. Or the peaks may be collected individually and each separated further, either on a column of different polarity or by a completely different method such as thin layer chromatography. Applying approaches such as these obviously is practical only for simple mixtures. (Just consider trying either of these approaches with the chromatogram shown in Fig. 10.9.)

The injection of a standard to establish an identity on the grounds of a retention match is the simplest method of qualitative analysis in gas chromatography. It is immediate. The method requires only that the retention times do not shift between sample and standard chromatograms. If only a few peaks are involved, this is possible.

A complex sample, however, can require not only a long time to run but also temperature programming. Such a sample might be, for example, a flavor or a petrole fraction. Then, reproducing retentions precisely enough may be impossible. See, fo instance, the retention variation afflicting the chromatograms shown in Fig. 10.5.

In such a case, some of the possibly relevant standard may be added to a portic of the sample and the resultant mixture injected (Fig. 13.2). This way, experimenta conditions need not be so strictly reproduced. The gas chromatograms of the sample-plus-standard will show one peak emphasized with respect to the sample chromatogram, and thus identified. However, this method tends to require n + 1 runs for n unknowr plenty of sample, and a little of each component to be added. Even if all the

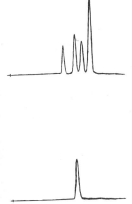

FIGURE 13.1 Peaks can be identified by separate injection of each component.

possible components are commercially available, acquiring and running the standards can take months (or years). Often, standards are not commercially available, but must be synthesized. And the problem of establishing the true number of components remains, too. Finally, some compounds may be essentially new, requiring not standards but structure elucidation.

A variation of this approach is to use as standards not single compounds but mixtures of composition known and relevant to the composition being investigated. For instance, a group of components that comprise part of a homologous series may allow several components to be identified at one time. Again, the gas chromatogram of any complex mixture, the peaks of which have been identified, constitutes a standard-in-being for the elucidation of similar mixtures. For instance, the gas chromatogram of blackberry serves to identify the constituents of raspberry, a "happy coincidence... of extreme value" [4].

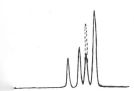

FIGURE 13.2 Adding a possible component to the mixture and reinjecting identifies a peak by the increase in peak height. This method does not require the identical column conditions between first and subsequent injections that the method shown in Fig. 13.1 does.

The indirect use of standards. Retentions can be expressed in several ways: As specific retention volumes; or, as either absolute or relative retentions expressed as either times or volumes. Specific retention volumes are never calculated for or used in qualitative analysis. They are difficult to calculate and the precision of them is no more than routine. Absolute retention times and volumes are never used for retention tables because they reflect all possible operating variables. In summary, only relative retention data are precise enough to be potentially useful for qualitative analysis.

The term *relative retention* usually implies a simple ratio of adjusted retention times. Although such a ratio helps overcome variations in parameters such as flow rate, it leaves considerable room for improvement in precision [5]. The most reproducible expression of a "relative" retention is the Kovats Retention Index I [6-11]. (The Kovats Retention Index has come up previously, in the discussion of stationary phase characterization, Sec. 5.5C.)

In the indirect use of a standard, one might measure an isothermal Kovats Retention Index either for use in a table of such indices or for comparison of the index of a component to be identified with an index already in such a table. In either case, one should observe several requirements [9, 10]: The sample should be so minute that the sample size does not affect the retention times observed; throughout the measurements, the column temperature should be accurately and precisely known and controlled, and carrier gas flow should be precisely constant; the stationary phase should be exactly definable, preferably a given pure chemical rather than a trade material that is not exactly reproducible; any peak to be measured should not overlap another, because overlapping affects retention time; and the support should be inert. The degree of difficulty in meeting a given requirement in this list increases in the order of its citation. Meeting any of the last four requirements with normal equipment and a mixture of any complexity is next to impossible.

In the Kovats system, all retention times are referred to those of the n-paraffin homologous series. To find for substance X the Kovats Retention Index I_X, the adjusted retention time t'_X is related to the nearest, bracketting adjusted retention times t'_{n-C_z} and $t'_{n-C_{z+1}}$ of the two relevant n-paraffins, as follows:

$$I_X = 100 \frac{\log t'_X - \log t'_{n-C_z}}{\log t'_{n-C_{z+1}} - \log t'_{n-C_z}} + 100 z$$

For example, if for some given stationary phase and column temperature the adjusted retention times of, say, toluene (t'_X), n-heptane (t'_{n-C_z}), and n-octane ($t'_{n-C_{z+1}}$) were 16.5, 13.7, and 29.2, respectively, then for toluene under these conditions

$$I = 100 \frac{\log 16.5 - \log 13.7}{\log 29.2 - \log 13.7} + (100)(7) = 724.6$$

Thus without excessive calculation the index indicates the location of the given peak on the scale defined by the n-paraffins for that stationary phase at that temperature. Here, toluene (724.6) lies between n-heptane (700) and n-octane (800). (Note that by the properties of logarithms, the numerator and denominator are both logarithms of ratios. The gas compressibility factor j, if used, would just factor out from the two elements of each adjusted time and then cancel out from the ratio. The gas compressibility factor therefore need not be applied in calculations of the isothermal Kovats Retention Index.)

The usual error in the Kovats Retention Index for isothermal gas chromatography, which is what we have just described, is about 2 index units [12]. This uncertainty, already too large, becomes even larger with temperature programming.

Guiochon [13] found that the temperature-programmed indices [Eq. (11.30)] can be made to agree reasonably well with the isothermal, but only with several precautions: Low initial column temperature, the presence in the chromatographed mixture of the bracketting n-paraffins, and exactly reproducible experimental conditions. Even given these precautions, Guiochon concluded that an identification from such an index would be "only an indication or, better, a presumption to be confirmed [because] all the sources of error and imprecision in isothermal chromatography play an equal role in temperature programmed" chromatography.

Nevertheless, under highly restricted, well-controlled conditions, the Kovats index or some analog of it can be used for qualitative analysis. In such a case, the index error can be reduced from 2 units to perhaps 0.2 unit [10, 14]. In the most favorable cases the error can be even less [15, 16].

Reliance for identification on a gas chromatographic index similar to the Kovats index was described by Miwa [17]. This was used in a long-term analysis of mixtures of fatty acid methyl esters and fatty alcohols. Again, Caesar, with the help of computer monitoring and calculation, successfully identified by means of the Kovats Retention Indices the hydrocarbons routinely eluted from a 100-m squalane open tubular column [15]. Both cases are unusually favorable to this approach.

Molecular randomness introduces still another difficulty. Klein and Tyler [18] pointed out that the probability of simultaneous elution increases exponentially with a linear increase in the number of components with similar retention indices, that is, in the "density" of similar components in a sample. (See Fig. 13.3). As a result, the resolution and retention time precision required for identification by index matching quickly becomes impossibly high. For instance, if 10 peaks from a given mixture fall randomly between the peaks of two successive n-paraffins, then to identify each of these peaks with 99.5% certainty requires an error in retention time no greater than 0.001%. This precision is clearly unachievable [19].

It can be argued that certain types of components are not likely to occur together, and that therefore simultaneous elution may not be so probable as a list of similar indices implies. One should consider, goes the argument, the sample origin. Such an approach, however, converts what should be an identification into a

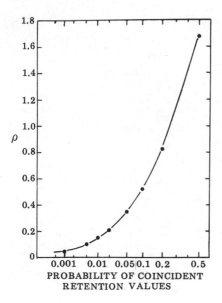

FIGURE 13.3 The exponential increase in the probability of simultaneous elution of similar components quickly limits the possibility of identification by matching retention indices. The horizontal axis--note the logarithmic scale--shows the probability; the vertical axis, scaled linearly, shows the "density" of components with similar retention indices. (Reprinted with permission from P. D. Klein and S. A. Tyler, Ref. 18. Copyright by the American Chemical Society.)

supposition. Rather, one desires an unequivocal instrumental reading that does *not* require using information about the background of the sample. Methods such as infrared spectroscopy and GC-MS *do* supply such readings. They are treated presently in some detail.

Complementary columns. Although the sure identification of some given isolated substance by using retention data alone may be next to impossible with just one column, it should become much more nearly feasible with two Kovats indices measured with two columns of differing selectivity [20]. Nevertheless, the goal proves elusive.

In 1976, Möller reported an interlaboratory study on the feasibility of identifying drugs by using the retention indices determined on two complementary columns. Standardization of the column temperature was found absolutely necessary, as might be expected. But even given such standardization, the precision of the retention indices from OV-1 and OV-17 was still found "not sufficient to identify unknown substances" [21].

An early approach called for four columns of complementary selectivities to be used at the same time and in parallel [22]. The four peaks of a given, already isolated substance emerging from such an arrangement and passing through the same detector produce a highly characteristic pattern. However, the library of patterns ma with this approach requires a completely reproducible set of columns that are absolutely stable with respect to the stationary phase and the mutual distribution of th

13.1 Identification of Eluted Components

carrier gas. Such sets cannot be reproduced with reasonable effort. Although usable within a given laboratory, the method has not become popular.

Special Detectors

Selective detectors. Selective detectors increase the certainty of identification. Two of these are the electron capture and the thermionic, both reviewed in Chap. 9. The electron capture detector, as was pointed out, is insensitive to most organic substances but highly sensitive to a few. Thus it can be used to confirm the identification of insecticides or drugs that may have been isolated by thin layer chromatography. Similarly for phosphorus-containing organic compounds with the thermionic detector, which can be "...highly selective in response to phosphorus in an organic compound... The relative sensitivity of the detector can be adjusted to measure only phosphorus" [23].

Identifying detectors. In particular, the GC-MS. Identifying detectors receive the column effluent more or less directly. If the gas chromatographic separation is complete, then such a detector can in general provide data sufficient for the unequivocal identification of a given substance as it is being eluted from the column. The pyrolyzer-pyrolyzate gas chromatograph is one such detector system. We describe it briefly later in this chapter in Sec. 13.3B, Pyrolysis of Pure Compounds. The rapid-scanning infrared spectrometer [24], especially the Fourier transform interferometric type [25-27], is another such system. But the identifying detector most widely used and best adapted to the purpose is the GC-dedicated mass spectrometer. We now provide here a short account of the GC-MS, referring primarily to a review by Merritt [28]. The GC-MS is described in greater detail in its own section, Sec. 13.4.

In the mass spectrometer, molecules are bombarded by electrons. In consequence, ions that are fragments of the bombarded molecules are produced. After the fragments are electrically or electromagnetically separated by mass with a precision characteristic of the given mass spectrometer, they are detected and the detector output is displayed or recorded. The fragment patterns are in general highly characteristic of the molecules bombarded (Fig. 13.4) [29].

With many modern mass spectrometers, even sharp GC peaks can be sampled many times across the peak, so that the GC column need not supply completely separated peaks to the mass spectrometer (Fig. 13.5) [30].

The resulting mass spectra are highly detailed, specific, systematic, reproducible, and referable to the mass spectra accumulated from over 30 years of mass spectrometry. The interpretation of mass spectra is well developed and ostensibly suited for completely automatic handling by computers (a matter also treated in greater detail in Sec. 13.4). We note here, however: Computer interpretation of mass spectra remains only a goal at this writing, not an accomplishment.

The modern GC and MS are nicely matched in almost every respect, including sample size, sensitivity, range of sensitivity, and effective internal diameter--but not operating pressure. The GC operates above atmospheric pressure with a given

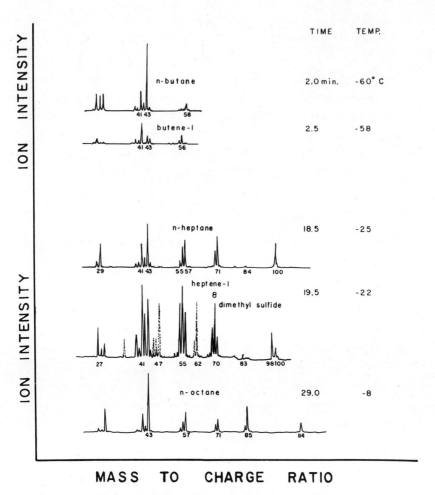

FIGURE 13.4 The masses and relative abundances of fragments produced by electron bombardment of molecules are highly characteristic of the parent molecules. (Reprinted with permission from C. Merritt, Jr., et al., Ref. 29. Copyright by the American Chemical Society.)

constituent highly diluted in the carrier gas. The MS requires an operating pressure that is lower by a factor of 10^4 to 10^6 and, preferably, a decreased ratio of carrier to constituent.

The decrease in pressure and the enrichment of the constituent is accomplished by a GC-MS interface. The GC-MS interface may effect sample-to-helium enrichment by allowing the lighter and faster-diffusing helium to escape through a porous glass tu (F, in Fig. 13.6a)[31] or to move away from a jet of the two ($d_1 d_2$ and $d_3 d_4$ in Fig. 13.6b), or by presenting a selectively permeable membrane (M_1 and M_2 in Fig. 13.6c) as an entrance to the MS for the sample molecules while retaining the lower MS pressure.

Sec. 13.1 Identification of Eluted Components   303

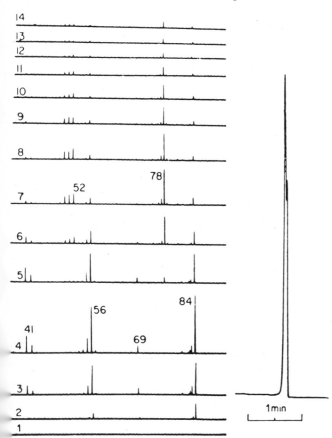

FIGURE 13.5 A mass spectrum can be obtained very quickly, so that a gas chromatographic peak can be sampled repeatedly. These mass spectra were obtained from successive half-sec scans of the cyclohexane-benzene peak shown. (Reproduced from A. E. Banner, R. M. Elliott, and W. Kelly, Ref. 30, with permission of Associated Electrical Industries Limited.)

Alternatively, particularly if the ion source is differentially pumped, a simple restrictor may bridge the pressure difference, admitting the total but minute effluent from an open tubular column into the mass spectrometer.

The modern GC-MS has only to be seen in operation to be admired. There is just no comparison in information per time or in potential capability between this elegant device and the alternative methods that are about to be described, however well developed and ultimately effective they may be.

The mass spectrometer has supplanted the infrared spectrophotometer as an identifying detector because the requirements and capabilities of the mass spectrometer so nicely match, complement, and supplement those of the gas chromatograph. In contrast, the noninterferometric infrared spectrophotometer acting as a detector of

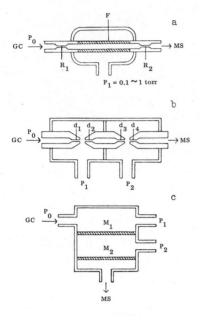

FIGURE 13.6 The GC-MS interface decreases GC outlet pressure to MS inlet level and increases the constituent/carrier gas ratio. The designs diagrammed are known as the Watson-Biemann (a), the Ryhage (b), and the Llewellyn (c). (Reproduced with permission of Marcel Dekker, Inc., from *Flavor Research*, by R. Teranishi, I. Hornstein, P. Issenberg, and E. L. Wick, Ref. 31.)

vapors in gas chromatographic carrier gas requires more sample, higher concentrations and more time. These particular handicaps are largely overcome by the Fourier transform interferometric infrared spectrophotometer, but there are other difficulties.

The infrared spectra of vapors differ appreciably from those of liquids. These differences increase with increasing polarity of the material [32]. Allowing for these differences is not prohibitive, but does make interpretation less clear cut. This further limits the applicability of the approach with respect to GC-MS.

Another handicap is the much greater, irreducible complexity of infrared spectra. In the next section we describe infrared spectroscopy briefly, to show why it is such a powerful adjunct to GC in qualitative analysis.

B. After Collection from the Carrier Gas

Derivative Formation

Components can be collected by functional group. The effluent is bubbled through test tube-held solutions of the class reagents normally used for derivative formation in classical organic analysis. If the effluent is divided into n parallel streams, can be tested simultaneously by classes. The derivatives so formed can be identified by normal methods [33] or, if they can be vaporized without decomposition, by gas chromatography [34], in which case the derivatives need not be recrystallized.

Trapping and Identification: Infrared Spectroscopy

Usually, if a component is to be identified, it is first trapped as it comes out of the column. Subsequently, it is identified. We shall describe some of the methods for doing both.

We cannot overemphasize that in any of these methods the tube that carries the column effluent from the column exit to the trapping device must be kept uniformly hot. The whole tube must be kept at the maximum temperature the column will reach during the separation. Condensation can occur instantaneously at any "cold" spot, partly or wholly removing the constituent to be trapped from the carrier gas and certainly vitiating the separation. The exit tube usually is made of stainless steel which, though a metal, is not a good conductor. It can have such cold spots. Therefore the tube must be heated evenly along its entire length. Merely a loose helix of heating tape will not do.

Infrared spectroscopy (IR) is the most usual and powerful means for identifying trapped, condensed GC fractions. IR is the tool of choice for the identification or structure analysis of organic substances. This has been true since about 1945, when IR instrumentation became commercially available. Since then, many tens of thousands of infrared spectra intended for reference use have been made, refined, correlated, standardized, and filed for retrieval. Enormous files of such correlated, retrievable infrared spectra are available from several sources.

Infrared spectroscopy. The infrared spectrum of a substance reflects the frequency of the vibrations of the atoms and atomic groups that comprise the substance, and also the distribution of energy among these vibrations [35-39]. As a result, no two pure compounds have the same infrared spectrum. Generally, spectra can be quickly and easily told apart. See, for instance, the three in Fig. 13.7 [39].

Yet if infrared spectra are mutually distinct, they are nevertheless also mutually similar. The infrared spectra of related substances can be related to each other. This has allowed the many aspects of infrared spectra to be correlated with the corresponding molecular structures. Hence to the sufficiently trained eye an infrared spectrum immediately reveals, in astonishing detail, a corresponding molecular structure.

For example, consider in Fig. 13.7 the sharp labeled peaks between 700 and 800 cm^{-1}. These all show the presence of hydrogens attached to an aromatic ring. Yet the frequencies and shapes of these peaks vary somewhat from A to B to C. These variations reflect corresponding variations in the associated molecular structures: ortho- *versus* para-substituted phenyl, for instance, in A and B, respectively; and four adjacent hydrogens on one of the anthraquinone rings, in C.

These peaks illustrate the fluid, expressive nature of infrared spectra. Not only does each peak inform, but so also does each variation of that peak from an implied norm. Infrared spectra are complex and rich. But they are not simple, and cannot be treated simply.

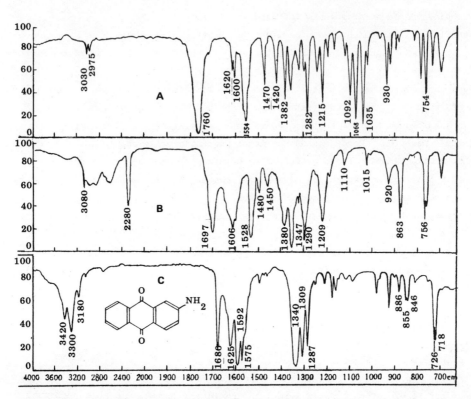

FIGURE 13.7 The infrared spectrum of a pure compound is singular. The infrared spectra of related compounds resemble each other, yet remain distinct. Therefore, they can be correlated with molecular structure, yet furnish clear choice among alternate possibilities at the end of a search. (A) 3-(nitromethyl) phthalide; (B) p-nitrophenylpropiolic acid; (C) 2-aminoanthraquinone. (Spectra reprinted from *Infrared Absorption Spectroscopy* by K. Nakanishi, Ref. 39, pp. 183, 189, and 206; with permission of the Holden-Day Publishing Company.)

With computers, infrared spectra can be reduced to numerical form and then reproduced. This does not affect the spectra themselves (computers can store data without distorting them), but neither does it clarify the relationships among them (computers do not help interpretation).

Identifying or interpreting an infrared spectrum still requires human judgment. The mind easily accommodates and uses the shifts in peak frequency and changes in peak shape, readily provides the necessary background information from organic and physical chemistry. A search program to encompass all this would be extremely difficult to write, and then too expensive in computer time to use.

Mass spectra, in contrast, are indeed expressible in digital form. More than one structure can be written for a given mass, but that mass is discrete and does not shift from one possible structure to another, as infrared peaks routinely do shift in frequency and thus reflect the molecular neighborhood of the vibrating entity. Further mass spectra can be condensed, at least for use as reference spectra, albeit with

Sec. 13.1 Identification of Eluted Components

loss of information. Infrared spectra cannot. This partial list of contrasts helps explain why the MS works so well as a GC detector--and why the IR does not.

However, for many of the same reasons that IR is not the identifying detector of choice for GC effluents, IR is very much the identifier of choice for trapped, condensed GC fractions. Thus given the spectrum of an unknown, a routine search of a spectrum collection quickly produces a few candidate spectra of known compounds. Selecting among them is no trouble. Because IR spectra are so complex, the identification of the unknown from among the final candidates becomes immediately and strikingly obvious.

IR is quite generally applied to almost all types of samples--all phases, gums, polymers, and foams--and to quantities down to micrograms, without unusual effort or technique. Thus, extending IR to GC-size samples was no great jump. Littlewood [32] reviewed the coupling of IR with GC. The review contains a table citing applications from fruit flavors and terpenes to impurities in ethylene and liquid oxygen.

Usually, the GC fraction is condensed. The condensate is then either transferred or, if formed directly on KBr, made into a KBr pellet [40, 41].

Trapping in capillaries. Sparagana [42] developed and described methods for obtaining good infrared spectra from microgram amounts of steroids. The paper cited is a veritable manual for this microtechnique. In it, a small part of the effluent is sent to a flame detector, but most to a collection vent. When the desired peak appears, a 9-in. glass capillary pipet is inserted into the vent. It is withdrawn when the peak is completed. Aerosol formation is usually avoided if the inner end of the capillary is at column temperature and the outer at room temperature. The peak constituent deposits within the capillary as a fine film.

The condensate can be washed out of the capillary with 50 µl of acetone onto a few milligrams of KBr held in a small hard mortar. The slurry is then dried, with mixing, under a heat lamp. (It is most important also to run blanks for bleed from the stationary phase and impurities in the acetone.) A KBr micropellet 0.5 mm in diameter can then be made, and an infrared spectrum obtained, preferably with a beam condenser. (For details on infrared technique and instrumentation, see such texts as White [38]; Alpert, Keiser, and Szymanski [35]; and Potts [36].)

The empty melting point capillary is an elegant vehicle for peak component trapping and transfer. (Frank Kabot originated this use of the melting point capillary.) This capillary is cheap, clean, ubiquitous, standard, effective. Its internal area is minimal and accessible. (More scientific-looking traps, on the other hand, spread out the condensate over elaborate and extended internal areas from which it cannot be retrieved except perhaps by vapor transfer under vacuum, thus demanding yet more equipment and time.) If necessary, the capillary can be cooled by a handkerchief holding some crushed dry ice.

For trapping, the empty melting point capillary is simply inserted at the appropriate time for about a quarter of its length into the hot effluent port, which should admit the capillary easily but snugly. The relatively slow cooling of the vapor as it passes through the tube apparently eliminates "fogging," that vapor-to-solid transformation that, by preventing condensate recovery, becomes the bane of preparative gas chromatography.

With a melting point capillary, enough material may be collected from only one peak to cause coalescence of the condensate within the capillary. When this happens, the condensate can be brought as a drop to the end of the capillary by either gravity or careful tapping. Otherwise, a microliter of solvent can be carefully injected into one end of the capillary. Then, careful tilting of the capillary causes the drop of solvent to move downward onto the condensed component. As the drop of solvent dissolves the component and thus forms a solution, its surface tension drops suddenly and the drop accelerates equally suddenly toward the lower end of the capillary. The effect is startling if not anticipated.

The drop of condensate or solution can be removed from the lower end into an infrared absorption microcell or onto either KBr or the surface of an internal reflection crystal, where any solvent evaporates. The microquantity of component is thus deposited both uncontaminated and nicely situated to yield an infrared spectrum. The whole of the manipulation just described takes perhaps 2 min.

Other trapping methods. As traps, Kubeczka [43] and Howlett and Welti [44] used 100 × 3 mm i.d. columns filled with the same column packing used in the gas chromatographic column proper. The peak component is trapped in this temporarily affixed column segment. The segment is then, in a different arrangement, carefully heated under a slow stream of carrier gas so that the trapped component is gently transferred under good control into a miniature cold trap [44] or other characterization device such as a hydrogenation microreactor [43].

Trapping can be done for batch mass spectrometric analysis. McFadden [45] reviewed the various means of introducing components to a mass spectrometer. As mentioned earlier, GC and MS are similar in sample-size requirement--10^{-7} to 10^{-10} g-- if the GC effluent is led directly into the MS. A much-used means of accomplishing essentially this introduction is the *direct probe*, onto which the GC peak is condensed and then introduced directly into the MS ionization chamber. But if instead the mass spectrometer is used as an independent unit with a standard batch inlet (a set of vacuum flasks permanently mounted on a manifold connected with the MS inlet), then it minimum requisite sample size is 10 to 10^4 times greater. In their use of the MS as an independent unit, Howlett and Welti [44] applied their GC column segment as a trap as a sample transfer means, and finally as a sample reservoir to be attached to the inlet of their mass spectrometer.

A third powerful means of collecting and identifying GC fractions is thin layer chromatography (TLC) coupled to GC in a way we shall describe. TLC is a type of liquid-solid chromatography in which the solid exists as a uniform, thin (0.25 mm) layer of adsorbent held on a relatively rigid support. Nicely matching GC, TLC also requires a microsample. Stahl [46] developed its wide use by showing its manifold capacity and devising the means for applying it (see also Truter [47, 48] and Stock and Rice [49]).

Sec. 13.1 Identification of Eluted Components

In TLC, a minimal-diameter spot of the substance to be chromatographed is deposited near the edge of the TLC plate. The plate is put into a closed chamber, with the edge dipping into solvent. The solvent moves by capillary action up the plate, carrying the substance partway with it. If necessary, the plate can be dried and redeveloped with a different solvent, which advances this time at right angles to the first direction of solvent advance and yields a different type of separation. TLC has been described as a "very powerful technique" that furnishes "quite amazing" resolution [47].

GC fractions can be collected directly and automatically on a TLC plate. The plate can then be developed.

(This powerful concept was repeatedly skirted before its elements were finally put together. In 1961, Mangold and Kammereck [50] used TLC to prepare fractions that were then individually prepared for GC. In 1962 Janak [51] used GC to prepare fractions, independently trapped, for TLC. Also in 1962, Casu and Cavalloti [52] came even closer to the concept: They continuously collected the GC effluent on a TLC plate being moved under the GC vent by direct mechanical linkage to the GC recorder. However, they then carried out class reactions on the plate.

(In 1962 and 1963 Nigam et al. [53, 54] deliberately coupled GC and TLC and suggested a name for the combination. However, they did not mention and apparently did not have the steady and thus automatic movement of the TLC plate that does not demand the attention and participation of the operator. So it seems to have been Kaiser [55] who drew together the pieces into the most practicable sequence: (1) GC separation; (2) division of the GC effluent, 10% to a monitoring detector and 90% to a heated delivery tube; (3) direct and automatic collection of the GC fractions issuing from the delivery tube by a synchronously moving TLC plate; and (4), development of the TLC plate. Kaiser was the first to place prime emphasis on the technique itself rather than on a given problem, the solution of which happened to be aided by the new technique.)

In GC-TLC coupling as practiced by Kaiser [55], perhaps 90% of the column effluent is delivered through a uniformly and controllably heated 1-mm i.d. tube to the outer end of the tube (Fig. 13.8). This outer end is outside the gas chromatograph and points down. A horizontal thin layer plate lies about 1 mm below the vent. Peak-component deposition onto and retention by the plate is 30% to 50% efficient: Adequate, if not quantitative. Deposition efficiency can be increased by directing from beneath the plate a stream of chilled air aimed upward at the point of deposition.

In GC-TLC [55], the holder of the thin layer plate is moved by a motor along rails. The peaks are thus deposited on the plate along a line. The motor can be driven continuously (Fig. 13.9) or, by a peak-actuated switch on the recorder, intermittently in short, sharp advances (Fig. 13.10). When the GC is over, the TLC plate is developed at right angles to the line of deposition.

Using GC-TLC, Curtius and Miller separated 12 steroids as trimethylsilyl (TMS) ethers without decomposition. Then, because TLC could not separate the TMS ethers,

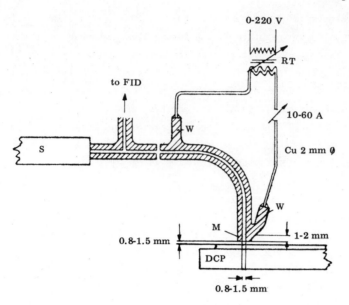

FIGURE 13.8 In this coupling of thin layer chromatography to gas chromatography, the column (S) effluent goes partly to a flame ionization detector (FID), but mostly through a heated tube down onto a thin layer plate (DCP). (The tube is heated by connections W to a variable transformer RT.) The mouth M of the tube is about 1 mm above the thin layer plate (DCP). The plate moves horizontally, either continuously or discontinuously, in peak-signalled steps. (Reprinted with permission of Springer-Verlag from R. Kaiser, in *Fresenius' Zeitschrift für Analytische Chemie*, Ref. 55.)

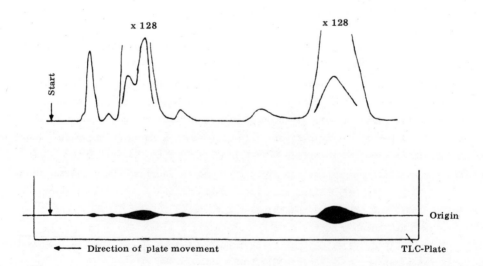

FIGURE 13.9 In coupled TLC-GC, the thin layer plate may move continuously under the delivery port. This allows no remixing and the densitometer tracing corresponds to the gas chromatogram, but each spot gets spread out. (Reprinted with permission of Springer-Verlag from R. Kaiser, in *Fresenius' Zeitschrift für Analytische Chemie*, Ref. 55.)

13.2 Identification by Inference from Sample Pretreatment

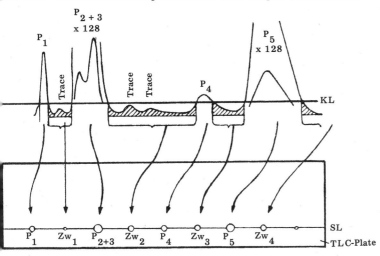

FIGURE 13.10 In coupled TLC-GC, the thin layer plate may move either continuously, as shown in Fig. 13.9, or discontinuously in peak-actuated steps (at level KL), as shown here. This concentrates the spots, but remixes those components originally present only in low concentrations. (Reprinted with permission of Springer-Verlag from R. Kaiser, in *Fresenius' Zeitschrift für Analytische Chemie*, Ref. 55.)

they recovered the free steroids on the plate by splitting with methanol-HCl. The free steroids were then separated by development of the TLC plate. The separated spots were made visible by distinctive color reactions [56].

Tumlinson et al. used only manual and intermittent TLC plate movement to examine alcohols [57] and carbonyl compounds [58] by GC-TLC. Their studies were aimed principally at establishing reproducibly usable conditions for direct GC-TLC with derivative formation on deposition. The derivatives were then separated during the TLC development. This type of approach requires either that the sample consist solely of components having the functional group corresponding to the class reagent to be used or that such components be first class-extracted from the original sample, as the carbonyls were [58].

13.2 IDENTIFICATION BY INFERENCE FROM SAMPLE PRETREATMENT

In contrast to the identification of components after separation by GC, a mixture can be chemically altered deliberately before or during gas chromatography. The resultant changes in the gas chromatogram cast light on the nature or composition of the sample and may enable the investigator to identify components. Chemical changes can also be made to assist structure elucidation.

A. Sample Pretreatment

   Syringe Reactions

   In his syringe reactions, Johann E. Hoff devised elegant means for extracting maximum information from microsamples with minimum equipment: "A technique for the gas chromatographic analysis of compounds in vapor mixtures...The vapor is brought into contact with classification reagents in the syringe used to inject the sample into the gas chromatograph...The technique may be applied to vapors at concentrations from 10^{-8} to 10^{-5} gram per ml of vapor diluent..." [59].

   Samples of the original mixture are repeatedly gas chromatographed, first straight, then once after each syringe reaction. The component of an altered or subtracted peak is thus characterized by functionality and by retention times from nonselective and selective stationary phases. The combination of these data should afford a positive identification of each component.

   Hoff's papers are replete with details on reagents, equipment, and procedures [59, 60]. For instance, a thin, 1-mm slice of sodium metal is put onto the top of a 10-ml syringe plunger, then exposed within the syringe for 3 min to humid air to make it more active. The sample vapor is then drawn into the syringe and held there for 3 min. As a result, all esters, ketones, aldehydes, and alcohols are removed completely, leaving only ethers and hydrocarbons. Fifteen such syringe reactions are described in detail [59]. Fredricks and Taylor extended the approach to compounds with boiling points well above 200°C [61].

   Component classes have also been removed by more usual means, before chromatography. Bassette and Whitnash, for instance, removed carbonyls and sulfides from organic mixtures by exposing them to aqueous sodium bisulfite and mercuric chloride [62].

   Another completely different type of identification by sample pretreatment is the pyrolysis of polymers.

   Polymer Pyrolysis

   Thorough reviews of polymer pyrolysis have been written by Levy [63] and by Stevens [64]. Shorter papers on polymer characterization have been presented by Zulaica and Guiochon [65, 66], and Tsuge and Takeuchi [67].

   Polymers can be conveniently heated so that they break down. The resulting fragments can be equally conveniently gas chromatographed. Well-developed and easily usable accessories for the pyrolysis of polymers are commercially widely available. Indeed, for the identification of a polymer or a mixture of polymers, nothing approaches polymer pyrolysis-gas chromatography for wide applicability, speed, ease of use (if the accessory is well-designed!), extension to microsamples, and wealth of detail leading to surety of identification.

   As simple a set of thermal fragments as possible is desired from polymer pyrolysis. The first fragments resulting from thermal breakdown come from primary reactions. In secondary reactions, these fragments can react mutually to form many more reaction products. A good pyrolyzer inhibits secondary reactions.

Sec. 13.2 Identification by Inference from Sample Pretreatment

In preparation for pyrolysis, a polymer is brought into maximal thermal contact with a heating element. Then, in the pyrolysis, the heating element is brought nearly instantaneously to some standard high temperature. Following closely, the polymer suddenly decomposes. The gaseous primary fragments explode into the surrounding cooler gas, cool at once while moving away from each other: The secondary reactions are minimized.

Ideally, the primary products comprise a minimally complex but maximally reproducible set. This set should characterize only the polymer. It should not reflect either the physical state of the polymer or the design of the pyrolyzer.

The flow of carrier gas through the pyrolyzer should be turbulent. Thus the pyrolyzate fragments should be swept away from the pyrolyzer into the column with full gas chromatographic injection efficiency [68].

At two-year intervals, the Pyrolysis-Gas Chromatography Subgroup of the Chromatography Discussion Group (London) has reported four cooperative studies in polymer pyrolysis [69-72]. These were designed "to rationalize and ultimately standardize the practice of pyrolysis-gas chromatography..." [72]. They found that good reproducibility demands certain elements of good practice in the pyrolysis, and bar graph presentation of the pyrograms.

The elements of good practice: A clean pyrolyzer; no cold spots in the pyrolyzer or the connections from the pyrolyzer to the column; a sample size of 100 µg or less; a solvent-free sample; a standard pyrolysis temperature (700°C) and duration (5-10 sec); a standard column used with standard operating parameters, *e.g.*, a column temperature of 100°C; and a minimum observed column efficiency associated with the pyrolysis, *e.g.*, 3,000 theoretical plates for the styrene peak *resulting from the pyrolysis of polystyrene* [71, 72].

Bar graph presentation of a pyrogram tends "to remove any artifact due to [the] optical illusion frequently met in visual assessment..." and "discriminates between the definitive and the trivial" [71]. The bars are scaled to the largest peak. "This procedure allows fingerprinting to be carried out easily and reliably and immediately detects the 'misfits' which must then be described as unsuitable pyrograms" [71].

By 1973, the Group had concluded that the practice of polymer pyrolysis had been improved and standardized enough to allow the start of a pyrolyzate library [71]. Some bar pyrograms from the 1973 study are shown in Fig. 13.11 [71].

In the experiment reported in 1975, the Group turned to "the problems associated with samples presenting particular handling difficulties...The sample consisted of a styrenated alkyd in the form of a base resin and a paint prepared from the resin, in the form of flakes of a dried and aged film. The base resin presented a minimum of handling difficulty, and was pyrolyzed [after deposition from] a dilute solution in acetone using 1-10 mm^3 of solution. The paint flake was somewhat more difficult since it was necessary to pyrolyze less than 100 µg as a solid" [72].

A further evolution in presentation can be seen in the normalization of the 1975 pyrogram coordinates (Fig. 13.12) [72]. As for the results, the relative retentions

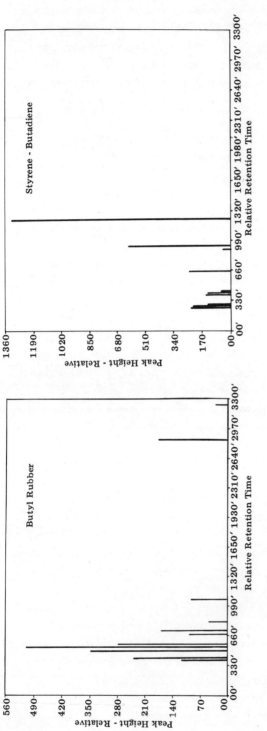
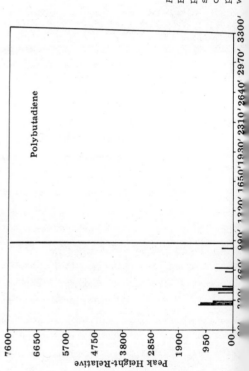

FIGURE 13.11 Pyrograms are best compared as bar charts. Bar heights are normalized with respect to the biggest peak. Retention is normalized with respect to the styrene peak from pyrolyzed polystyrene, flow rate or chart speed having been adjusted to have the styrene peak emerge 10 cm after the methane peak. (Reprinted with permission from N. B. Coupe, C. E. R. Jones, and

Sec. 13.2 Identification by Inference from Sample Pretreatment   315

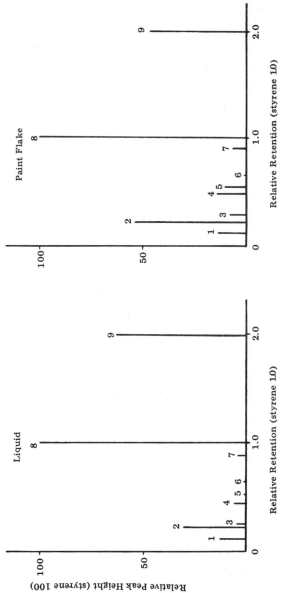

FIGURE 13.12 These are pyrograms of styrenated alkyds, one in the form of a base resin, the other a flake from a dried and aged film prepared from that resin. By 1975, pyrolysis conditions making such pyrograms essentially or exactly superimposable could be specified and applied in different laboratories with different types of pyrolyzers. (Reprinted with permission of T. A. Gough and C. E. R. Jones, Ref. 72, Pergamon Press, Ltd.)

are indistinguishable between the two pyrograms, and could hardly be improved upon. However, the relative peak heights are not independent of the original physical states of the polymer samples, although they are supposed to be. Nevertheless, it is clear that "provided the conditions of pyrolysis and chromatography are specified it is possible to achieve readily identifiable pyrograms from the same polymer in different matrices, and from one laboratory to another" [72].

B. Reaction Gas Chromatography Used for Structure Identification

Pyrolysis of Pure Compounds

The standard apparatus for single-molecule pyrolysis is a narrow gold tube connected to the head of a gas chromatographic column.

The pure sample to be pyrolyzed, or the gas chromatographic peak to be identified, is passed through the gold tube, which is held at 600°C (plus or minus 5°C). At this temperature, the pattern of molecular fragments produced varies only slightly with changes of up to 25°C in the temperature of the tube. The fragments produced are separated in the gas chromatographic column. The pyrolyzate chromatogram is highly characteristic of the molecule pyrolyzed.

Using pyrolysis to produce molecular fragments very similar in detail to those produced by low-voltage mass spectrometry was reported as early as 1968 [73]. Reviews of the approach have been written by Fanter et al. [74], Perry [75], and Walker and Wolfe [73].

Carbon Skeletal Determinations

Carbon skeletal determinations were invented and named by Morton Beroza. These determinations help establish the chemical structure of an injected pure substance. The determinations make clear the chemical "skeleton" of the unknown pure compound. The "flesh" of the skeleton is stripped away by catalytic hydrogenation, leaving only an easily discernible and identifiable carbon skeleton [76-79].

The sample is injected normally. Hydrogen, used as the carrier gas, sweeps the sample into a hot tube packed with a hydrogenation catalyst (palladium or platinum). In this tube, any multiple bonds in the sample become saturated. Hydrogen replaces any halogen, sulfur, oxygen, or nitrogen atoms. The resultant saturated hydrocarbon is swept into the column, identified by its retention time.

Replacement of oxygen or nitrogen may produce smaller skeletons. For example, although 3-pentanone yields n-pentane, ethyl propyl ether yields ethane and propane.

In situ Reactions

Other applications of reaction gas chromatography used for identification are catalytic deoxygenation [80], catalytic desulfurization [81], and chemical absorption [82, 83]. In catalytic deoxygenation, the sample is swept by hydrogen over 0.5% palladium-on-alumina pellets heated to about 250°C, then over the gas chromatographic column. In this, hydrocarbons are produced from all typical oxygen-containing classes. Catalytic desulfurization accomplishes the same purposes for compounds that contain sulfur. Chemical absorptive techniques, however, are subtractive. In reaction gas chromatography they involve an in-line absorber that does not increase peak broadening due to extracolumn dead space, that is, the absorber offers no larger an internal diameter than the column and negligible interaction with the unabsorbed components. Various mercury salts and sulfuric acid make up the absorbers used for determining hydrocarbon type [82]; boric acid is used as an alcohol subtracter [83].

13.3 MORE ON GAS CHROMATOGRAPHY-MASS SPECTROMETRY

In this section we outline the mass separation techniques most used in GC-MS, describe and compare electron impact ionization and chemiionization, cite an illustrative example in which both are used, and finally consider computer identification of mass spectra.

A. Mass Separation Techniques

Mass spectrometry began with the twentieth century. Papers describing the mass spectrometers of Dempster and Aston appeared in 1918 [84] and 1919 [85], respectively. So mass spectrometry was already mature when gas chromatography was yet an infant. (A chapter on the development and salient literature of mass spectrometry is included in the survey cited by McDowell [86].)

In all mass spectrometers, the molecules to be analyzed are first converted into one or more ions per molecule. This conversion takes place in the ionization source. The ions produced are then withdrawn from the source and accelerated into the *analyzer* by an electric field. The subsequent mass separation of the ions in the mass analyzer depends on (1) the mass-to-charge ratio of each ion and (2) the resultant motion of that ion. How the ionic motion is caused and used depends on the type of mass spectrometer.

The generally described types of mass spectrometers [87, 88] include those common to the GC-MS: The sectored magnetic field, the time of flight, and the quadrupole. Let us consider these briefly.

Direction-Focusing Sectored Magnetic Field

An ion moving normal to a magnetic field experiences a force normal to both its course and the magnetic field. As a result, the locus of an ion moving in a magnetic field tends to be circular. The radius of the circle varies as the square root of the mass-to-charge ratio of the ion.

Consider two ions of the same charge but different masses. If two such ions enter a magnetic field from the same direction and at the same velocity, they move along different, mass-dictated paths within the field and become separated according to mass (Fig. 13.13)[86].

With arrangement shown in Fig. 13.14 [86], the ions arrive at the exit slit one mass-to-charge ratio at a time. The usual ion has unit positive charge. Therefore,

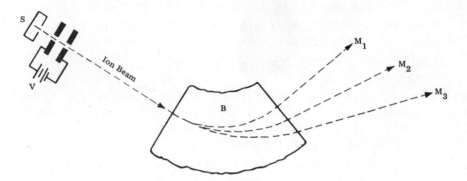

FIGURE 13.13 Ions of the same charge but different masses M move on paths of different radii on passing across a magnetic field. (Reprinted from L. Kerwin, in Ref. 86, p. 105, with permission of the author and the McGraw-Hill Book Company.)

steadily changing the magnetic field, and thus the ion deflection, sweeps a mass spectrum across the exit slit and the detector.

Although it is to be desired, ions of a given mass do not actually leave the source either at the same velocity or in the same direction. In simpler instruments such as are usually used in GC-MS, making the exit slit somewhat wider than the entrance slit tends to accommodate the velocity spread. The direction spread, if not too large, is cancelled to a degree by the symmetry of the second-field analyzer (Fig. 13.14). Such analyzers are therefore called first-order direction focusing [87] or, more simply and commonly, single focusing.

In mass spectrometry with sectored-field analyzers, resolution involves the mass dispersion produced at the exit slit and also the widths of the entrance and exit slits. The resolution often is described as the ratio of the peak mass to the peak width, with the width evaluated at some percentage of peak height. Whether that

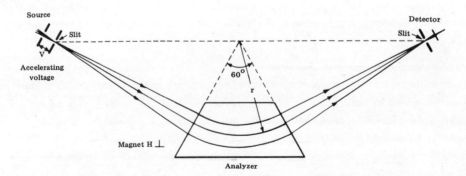

FIGURE 13.14 The sector magnetic analyzer exhibits direction focusing. (Reprinted from J. B. Farmer, in Ref. 86, p. 13, with permission of the author and the McGraw-Hill Book Company.)

percentage is taken as 1% [87] or 10% [88], the resolution of a GC-MS direction focusing mass spectrometer is generally less then 1,500.

Of the forms of mass spectrometer, the sectored magnetic field has the highest potential for resolution. The entrance and exit slits can be narrowed, for instance, in a given instrument. Double-focusing versions giving much higher resolution have been used for decades. In these, a velocity-focusing radial electric field precedes the direction-focusing sectored magnetic field [89-96]. However, high resolution and high scanning speed tend to be mutually exclusive with these mass spectrometers. Nevertheless, in 1975, with a double-focusing mass spectrometer furnishing a resolution of 10,000 and used in GC-MS combination, mass spectra ranging from mass 60 to mass 800 were obtained in 9.6 sec [97].

High scanning speed also decreases sensitivity with a slit-equipped sectored-field mass spectrometer. (Conversely, with a given instrument, increased sensitivity can be obtained by decreasing the speed of scan.)

Masses are normally scanned with sectored magnetic field spectrometers by steadily increasing the magnetic field. Mass scanning by changing the accelerating voltages in the ion gun ("voltage scanning") can be fast, but voltage scanning tends to give poorer results with increasing mass, just the opposite of what is desired. Hence the preference for magnetic scanning. Nevertheless, voltage scanning is intrinsically faster than magnetic, and becomes necessary if two ions are to be monitored by jumping back and forth from one to the other [98].

Fast scan potential and good sensitivity--comparable to that of a flame ionization detector--are the most important characteristics for a GC-MS mass spectrometer. Precisely because the GC column simplifies the sample before it is presented to the MS, relatively low resolution is usually quite acceptable. However, this combination of characteristics--fast scan combined with good sensitivity and low resolution--is not a natural combination for a sectored magnetic field mass spectrometer (although the modern sectored field MS is nevertheless adequate to the GC-MS task). But fast scan, good sensitivity, and low resolution *is* a natural combination for the time-of-flight and the quadrupole mass spectrometers, to which we now turn.

Time of Flight

The time-of-flight mass spectrometer, first reported in 1948 [99], differs thoroughly from the sectored magnetic field mass spectrometer. Although resolution-- not too important in GC-MS--was at first mediocre (though later improved [100]), the mass scan rate was always almost unbelievably rapid.

Because in the time-of-flight MS most of the ions produced reach the detector, whereas in the sectored magnetic field MS the slits cut off much of the ion beam, the sensitivities of the two are about the same [101], and adequate.

In short, the time-of-flight was the first mass spectrometer naturally suited to GC-MS.

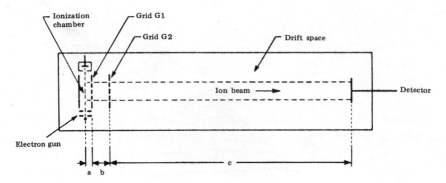

FIGURE 13.15 In a time-of-flight mass spectrometer, a tight bunch of ions is produced in an ionization chamber and accelerated out of it. As the ions move with mass-dependent velocities, they separate mutually. Once separated enough, they may be detected. (Reprinted from J. B. Farmer, Ref. 87, with permission of the author and the McGraw-Hill Book Company.)

In the time-of-flight mass spectrometer (Fig. 13.15)[87], a tight bunch of ions is briefly--ideally, instantaneously--produced in an ionization chamber. This bunch is then accelerated toward the mass analyzer. The ions then drift through a field-free tube--the mass analyzer--with uniform energy, but different velocities because they have different masses. The time of flight is longer for the heavier and therefore slower ions. If we reexamine these processes for the times involved, we are then able to understand how the scan rate can be so fast.

After having been produced within a few tenths of a microsecond, the ion bunch is accelerated for a few unit microseconds [87]. So propelled, the tight bunch of ions then moves along the meter-long drift tube for 1 to 30 μsec [102], while the lighter, faster ions move away from the heavier and slower ones. Thus, on a repetitive basis, a set of masses can arrive and be scanned once every 30 μsec. This corresponds to a repetition rate of about 33,000 times/sec. The usual rate is 20,000 scans--20,000 mass spectra--per second [102].

Quadrupole

The quadrupole mass spectrometer was first described in 1953 [103-107].

The quadrupole mass analyzer consists of four replicate parallel bars (Fig. 13.16) [88]. The bars of a given set may be either hyperbolic (Fig. 13.17)[86] or circular in cross section. The bars must be precisely parallel, uniform in cross section, smooth to the microinch.

The bar centers exist at the corners of a hypothetical square. (In one early example, this square was 8 cm on a side [108].) The center of that square is one center of symmetry for the quadrupole. That center, extended, is a straight line parallel to and equidistant from each bar. (The bars were 834 cm long in the example cited [108].)

Both a direct and a larger alternating voltage are simultaneously applied to the bars (in the example just cited, about 300 v d-c; and about 2,000 v a-c at about

Sec. 13.3 More on Gas Chromatography-Mass Spectrometry   321

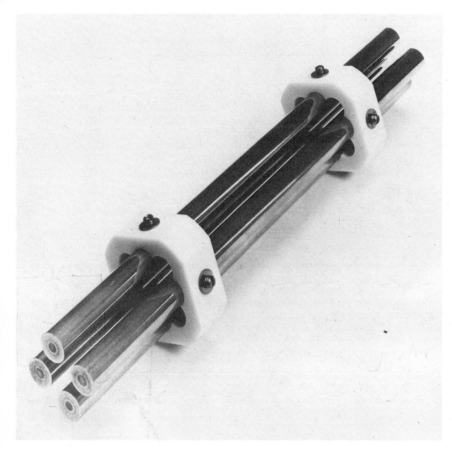

FIGURE 13.16 The mass-selective quadrupole assembly consists of four smooth parallel bars. (Photograph courtesy of finnigan Instruments, Inc.)

450 kc/sec [108]). Two bars mutually opposite across the quadrupole line-center of symmetry form a pair. The bars are connected, and exist at the same voltage. The pairs, however, differ in both the sign of the direct voltage and the phase of the alternating.

Ions are propelled out of the ion source along the quadrupole line-center of symmetry (at only a few tens of volts, in contrast to the thousands used with magnetic spectrometers). Any given ion, accelerated laterally by both the direct and alternating voltages, probably hits an electrode and is lost. Put more idealistically, only a small proportion of all possible ions oscillate stably enough to pass through the quadrupole (and be detected). This is as desired: The smaller the proportion of all possible ions--as opposed to the absolute number of them--the higher the resolution.

On the other hand, the quadrupole tends to focus those ions that *do* oscillate stably. The quadrupole can accept ions that have relatively wide initial variations

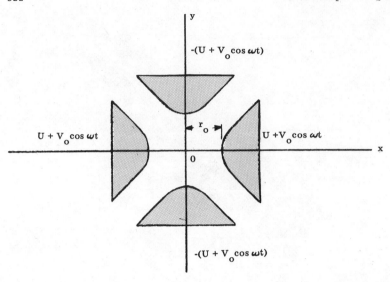

FIGURE 13.17 Opposite bars of the quadrupole are electrically connected. Thus the quadrupole functions as two electrical pairs. The pairs are driven by opposing direct (U) and alternating (V_o) voltages. In consequence, only ions of a particular mass-to-charge ratio can move for any distance between the bars without striking one. (Reprinted from J. B. Farmer, Ref. 87, with permission of the author and McGraw-Hill Book Company.)

in direction and energy [108]. Entrance and exit slit requirements therefore relax (no exit slit is necessary [108]), more ions are transmitted, sensitivity benefits.

The voltages applied to the quadrupole control its resolution. Resolution increases with the ratio of the direct voltage to the alternating voltage. (However, the direct voltage must never exceed the alternating, or all ions are lost.) These voltages are varied during each mass scan to provide uniform separation of one mass from the next throughout the spectrum. This is a resolution that is deliberately varied to achieve uniform peak width without regard to the mass at hand. Such dynamic variation of resolution during the mass scan is not possible with magnetic spectrometers, the resolution of which is set by mechanical parameters. Because low-mass resolution need not be so stringent as high-mass, low-mass sensitivity with quadrupoles may and does benefit. (On the other hand, for reference spectra *constant* sensitivity is important. Therefore quadrupoles must be "tuned" often, and intensity reference compounds referred to frequently.)

Linearly increasing the magnitude of the alternating voltage generates a linear mass spectrum. Therefore this is the usual method of mass scan. (Masses can also be scanned by alternating voltage frequency changes, but the resulting mass scan is not linear.)

With the quadrupole, monitoring two or more ions (multiple ion monitoring: MID is not difficult [109, 110]. In one example, each of eight ions ranging in mass from 45 to 140 was resampled 12 times/sec by automatically resetting a control voltage of

Sec. 13.3 More on Gas Chromatography-Mass Spectrometry   323

total range 0-10 volts [110]. The dwell time on each ion mass can be automatically adjusted so as to detect equal numbers of each ion, whatever the concentration of the parent substance in the sample. The resultant increase in precision is important in work involving stable isotopes, wherein the precision must be very high, about 0.1%. The ions scanned are under no mass restriction: They may be very different from each other in mass, as illustrated in the example just cited. (With magnetic spectrometers, on the other hand, MID by sharp, cyclic changes in the ion source is either disadvantageous or impossible: The ion source becomes detuned, the characteristically high ion source voltages may not be quickly switched, and the relative ion mass differences may not exceed about 40% [109]. Also, the magnetic field must be specially stabilized [111]. Nevertheless, attempts persist to make the magnetic spectrometer fully usable for MID [60, 112, 113].)

In sum, the quadrupole is the mass spectrometer of choice for most GC-MS instruments. Its characteristics closely match those needed for GC-MS, but in addition it brings a useful flexibility of its own--its dynamically variable resolution, for instance.

(In 1977, Hewlett-Packard (HP) and Varian each introduced GC-MS systems based on the quadrupole mass spectrometer, long the choice of the finnigan Corporation. The HP instrument had a mass range of 10 to 1,000; the Varian instrument, 2 to 1,200. The HP scan rate was 600 mass units/sec, with a 1.3-sec minimum cycle time. The Varian MS resolution was 3,000 at one-half peak height. For both, sensitivity was in picograms [114]. Nevertheless, LKB, a company of great experience in GC-MS, concurrently introduced a necessarily competitively performing GC-MS system based on a 90° sectored magnetic field spectrometer.)

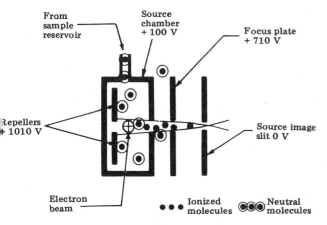

FIGURE 13.18 An electron impact (EI) ion source. The diagram shows the ions being accelerated by 1,000 volts into the mass spectrometer. (Reprinted with permission from E. M. Chait, Ref. 115. Copyright by the American Chemical Society.)

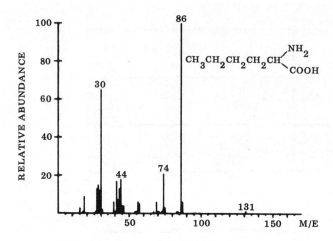

NORLEUCINE, M = 131, E.I.

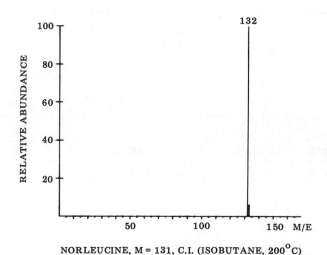

NORLEUCINE, M = 131, C.I. (ISOBUTANE, 200°C)

FIGURE 13.19 Compared to EI mass spectra, CI spectra are very simple. EI spectra are highly detailed, but tend to lack the molecular ion. CI spectra often consist only of the molecular ion--here, M + 1. (Reprinted with permission from H. M. Fales et al., Ref. 121. Copyright by the American Chemical Society.)

B. Electron and Chemical Ionization

Electron impact ionization (EI) "is the most commonly used and highly developed ionization method" [115] (see Fig. 13.18). In it, electrons emerge from a hot filament and are accelerated into the ionization source at a definite energy--usually 70 electron volts (eV). In the ionization source, these electrons ionize the sample molecules. Organic molecules so bombarded break into detailed, often individual patterns of ion fragments (Fig. 13.4)[29]. Mass spectral libraries consist mostly of electron impact spectra, which are the ultimate basis of identification in GC-MS.

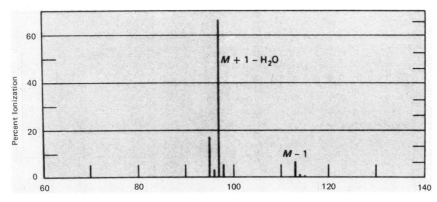

FIGURE 13.20 With methane as the reactant gas, the CI mass spectrum of 2-methylcyclohexanol consists principally of the $(M + 1 - H_2O)$ ion. (Reprinted with permission from B. Munson, Ref. 117. Copyright by the American Chemical Society.)

Nevertheless, there are several untoward features of EI mass spectra. Mainly, electron impact at 70 eV breaks up the molecules too thoroughly. Thus EI mass spectra have too many ion fragments, yet often lack the most useful one for qualitative

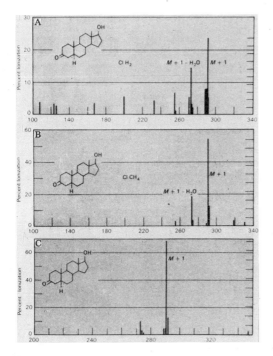

FIGURE 13.21 The different mass spectra resulting from different reactant gases make chemiionization a tool for structural elucidation. These are the CI mass spectra of 5α-dihydrotestosterone that result from use of hydrogen (A), methane (B), and isobutane (C) as reactant gases. (Reprinted with permission from Munson, Ref. 117. Copyright by the American Chemical Society.)

analysis, the parent mass. In 1966, as though in response to these failings, certainly in recognition of them, chemical ionization was invented [116, 117].

In chemical ionization (CI)[115, 117, 118], the ion fragments of the sample molecules are produced by reaction with ions from a reactant gas. The reactant gas, usually methane, is present in the ionization source in great (1,000-fold) excess and at high relative pressure (1 torr *versus* the 10^{-4} torr used for EI)(1 torr = 1 mmHg). Molecules of the reactant gas are ionized by high-energy electrons (50 to 500 eV). By proton transfer, the reactant ions in turn gently ionize the sample molecules [117, 119]. Among the few resulting fragments, two or three predominate. In these, if methane or isobutane has been the reactant gas, the original carbon skeleton is generally intact [119].

CI mass spectra quickly became "one of the most important analytical tools for use in the identification, structure elucidation, and quantitation of organic compounds" [120]. As we have seen, CI mass spectra are relatively simple (Fig. 13.19) [121], and usually provide ions that show minimal fragmentation. Such ions are either molecular, or quasimolecular such as $M-H_2O$ (Fig. 13.20)[117]. Also, CI mass spectra show different structural information with different reagent gases (Fig. 13.21)[120, 122]. Finally, with respect to EI mass spectra, CI mass spectra are at least as sensitive [123] and may be 10 to 100 times more sensitive [120, 124].

With respect to GC-MS instrumentation, CI and EI mass spectrometry differ only in the nature and handling of the ion source. As we have described, for CI a reactant gas must exist in the ionization source at a pressure 10,000 times higher than normal for EI. For CI the source must therefore be made as nearly gas-tight as possible, and also pumped separately from the mass spectrometer and unusually effectively, because the source-to-detector path of the ion within the mass spectrometer still requires its own high vacuum. But once this more flexible accommodation of the source has been accomplished, the same gas chromatograph and mass spectrometer can be used for both CI and EI spectra [125]. In consequence, one requirement of a modern GC-MS has come to be an easy CI/EI interchangeability. Indeed in 1977 one instrument offered simultaneous CI/EI on the same sample [126].

We can now reconsider qualitative analysis by GC-MS.

C. Qualitative Analysis by GC-MS

In this section, we first distinguish low resolution (LR) GC and MS from high resolution (HR). We then consider in some detail a highly informative example of LRGC-LRMS. Following this, we consider the application of computers to GC-MS qualitative analysis.

For GC, high resolution begins with a certain minimum number of theoretical plates--say, 25,000. For MS, high resolution begins at perhaps 10,000 [97].

With present techniques (1978), HRGC excludes sample loads greater than perhaps 10 μg per component, yet many applications require much more. Similarly, HRMS cannot at present accommodate scan speeds faster than perhaps 1 mass decade--as, 80 to 800--in 5 sec [97]. Yet many analyses, especially those from HRGC, require higher scan rates. Also, high resolution makes sense only if applied to EI spectra, yet CI spectra, which are highly simplified, are very useful. In short, in GC-MS high resolution entails restrictions that exclude wide utility.

The bulk of GC-MS is done with LRGC-LRMS: Short packed columns; high-speed, low-resolution MS; and, frequently, CI excitation. This implies, correctly, that most qualitative GC-MS analysis treats samples of well-known origin and relatively few components. Dealing with such a case, a paper by Szczepanik, Hachey, and Klein [119] illustrates how the power of the LRGC-LRMS combination can far transcend that of either technique alone.

An Exemplary Study

This paper [119] constitutes a veritable text for this subject. In a highly pertinent manner it reviews for us the nature and uses of electron impact ionization and chemical ionization. It shows how to perform GC-MS qualitative analysis *without* a computer. And it illustrates how relative retention times can complement mass spectra--a topic of growing interest [127, 128].

The paper is entitled, "Characterization of Bile Acid Methyl-Ester Acetate Derivatives Using Gas-Liquid Mass Spectrometry" [119]. On one injection, the 183-cm packed column used for GC in this work can fully separate only a few of the 54 highly similar derivatives considered. Many of the derivatives are either isomers or stereoisomers. They cannot be mutually distinguished by either EI or CI mass spectrometry alone. However, the derivatives do have different retention times and do break up or ionize differently under either electron impact or chemical ionization, respectively. Such retention time and ionization differences can be systematically exploited, particularly with mixtures of such restricted qualitative composition as these.

For the electron-impact mass spectrometry of these derivatives, Fig. 13.22 [119] shows a sample of already known ion structures for some seven sets of mass-to-charge (m/e) ratios arbitrarily selected for illustration here. In the larger display from which Fig. 13.22 is taken, 67 such correlations are arranged by increasing m/e values, *i.e.*, as the mass spectra are presented to the gas chromatographer-mass spectrometrist. The authors comment:

> The identification of bile acid methyl ester acetates without using library search techniques or computer-matching routines can be simplified by the use of [this display]...and is perhaps quicker than the use of the above methods alone. One need only look up the mass ions in the unknown spectrum and compile a set of ion structures that are consistent for the unknown compound. The epimeric configuration or position of the functional groups may not always be apparent from the intensity of the mass ions and their representative structures. However, at this point, the identification of the unknown compound can usually be narrowed to two or three compounds whose mass spectra and relative retention times can easily be compared with that of the unknown compound to establish the identification.

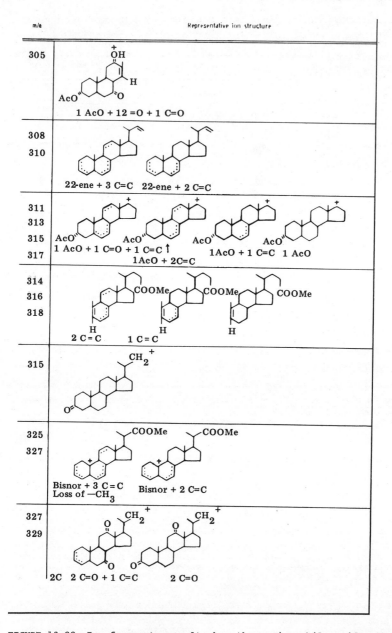

FIGURE 13.22 Ion fragments result when the various bile acid methyl ester acetates are subjected to electron impact. The structures of these ion fragments are known and can indicate which acetate was being bombarded. This figure samples the m/e structure correlations. (Reprinted with permission of P. A. Szczepanik, D. L. Hachey, and P. D. Klein, Ref. 119, and of the *Journal of Lipid Research*.)

As the description of chemical ionization has suggested, mass spectra of ions produced by chemical ionization are very simple. The first 15 of the 52 spectra from our "text" [119] are listed in Table 13.1.

TABLE 13.1 An Illustrative Sample of the Chemical Ionization Characteristics of Bile Acid Methyl Ester Acetates When Isobutane Is Used as the Reagent Gas

No.	Functional groups	Mol. wt.	MH+	MH+ − 60	MH+ − (2×60)	MH+ − (3×60)
				Percent of total ionization		
1	$C_{24}5\beta3\alpha OAc$	432	0.0	71.8		
2	$C_{24}5\beta3\alpha OAc\Delta11$	430	11.8	63.4		
3	$C_{24}5\beta12\alpha OAc\Delta3$	430	0.7	72.0		
4	$C_{24}5\beta3\alpha OAc6\alpha OAc$	490	0.0	11.9	63.0	
5	$C_{24}5\beta3\alpha OAc6\beta OAc$	490	0.0	10.0	62.8	
6	$C_{24}5\beta3\alpha OAc7\alpha OAc$	490	0.0	6.5	66.8	
7	$C_{24}5\beta3\alpha OAc7\beta OAc$	490	0.0	10.4	65.4	
8	$C_{24}5\beta3\alpha OAc12\alpha OAc$	490	0.0	25.7	43.8	
9	$C_{24}5\beta3\alpha OAc12\beta OAc$	490	0.0	1.6	61.5	
10	$C_{24}5\beta3\alpha OAc7\alpha OAc\Delta11$	488	1.2	21.5	45.4	
11	$C_{24}5\beta3\alpha OAc12\alpha OAc\Delta8(14)$	488	0.5	18.5	5.0	
12	$C_{24}5\beta3\alpha OAc6\alpha OAc7\alpha OAc$	548	0.8	7.8	25.4	24.0
13	$C_{24}5\beta3\alpha OAc6\beta OAc7\alpha OAc$	548	7.0	7.9	24.1	22.3
14	$C_{24}5\beta3\alpha OAc6\beta OAc7\beta OAc$	548	0.5	9.0	15.9	16.8
15	$C_{24}5\beta3\alpha OAc7\alpha OAc12\alpha OAc$	548	0.0	0.0	58.5	12.3

Source: Reprinted with permission of P. A. Szczepanik, D. L. Hachey, and P. D. Klein, Ref. 119, and of the *Journal of Lipid Research*.

[These] highly simplified spectra resulting from chemical ionization processes... have two further advantages in the analysis of bile acid mixtures. The first advantage is that the yield of ions at a mass characteristic of that bile acid is significantly increased over that obtained with electron impact ionization... Chemical ionization...results in an ion yield for a given mass fragment that is between one and two orders of magnitude larger, solely because of the uniformity and low energy with which ion formation takes place. There is an additional factor of one to two orders of magnitude in sensitivity, which results from the overall rates at which ions are formed and extracted in the chemical ionization source as opposed to the electron impact source...

The second advantage of the simplified ionization spectra is that there are virtually no ions at any masses other than those listed in [the table sampled by Table 13.1] and their isotopic satellites. This is particularly important in using the mass spectrometer as a selective ion monitoring device to detect the presence of a particular bile acid in a mixture where a major constituent may have a similar retention time. In electron impact mass spectrometry, the generation of ions at each and every mass number increases with the concentration of bile acid molecules entering the ion source. Hence the background against which the minor constituent must be detected is increased by the presence of high concentrations of a neighboring bile acid and sensitivity is diminished...

The GLC data and chemical ionization spectra...can be applied to the analysis of complex mixtures of bile acids...Fragmentation of the bile acid molecule in the chemical ionization process is generally limited to the removal of acetate groups. Under these circumstances, all bile acids of the same category--*i.e.*,

TABLE 13.2 Selective Ion Monitoring-Relative Retention Time Coordinates for Bile Acid Methyl Ester Acetates by Gas-Liquid Chromatography-Chemical Ionization Mass Spectrometry

m/e	Relative Retention Time on 0.5% SP-525									
369[a]	0.86 (11)	1.26 (10)	1.62 (15)	1.65 (16)	1.83 (35)	2.26 (13)	2.36 (12)	3.10 (14)		
371	0.31 (3)	0.70 (2)	0.93 (9)	1.00 (8)	1.14 (33)	1.30 (6)	1.45 (34)	1.69 (5)	1.82 (4)	1.93 (7)
373		0.71 (1)								
385	1.82 (26)	2.11 (29)	2.40 (27)							
387	1.14 (23)	1.62 (22)	1.62 (37)	1.95 (24)						
389	0.75 (21)	0.93 (36)								
401	0.925 (19)	3.62 (31)	4.20 (32)							
403	2.42 (38)	2.80 (30)								
415	0.915 (20)									
429[a]	0.86 (11)	1.26 (10)	1.62 (15)	1.65 (16)	1.83 (35)	2.26 (13)	2.36 (12)	3.10 (14)		
431	1.00 (8)	1.14 (33)	1.45 (34)	1.69 (5)	1.82 (4)	1.93 (7)				
445	2.11 (29)	2,40 (27)	3.12 (28)							
447	2.01 (25)									
461	3.62 (31)	4.20 (32)								
489[a]	2.26 (13)	2.36 (12)	3.10 (14)							
505	3.12 (28)									
549	2.26 (13)									

[a] In addition to the compounds listed, cholate epimers may occur at the relative retention times predicted in Table 3 of Ref. 119.
Source: Reprinted with permission of P. A. Szczepanik, D. L. Hachey, and P. D. Klei Ref. 119, and of the *Journal of Lipid Research*.

with the same number of hydroxyl groups and the same carbon skeleton--will give rise to an ion at the same mass, regardless of hydroxyl position or epimer configuration. Usually formation of this ion will be limited to these compounds a the corresponding unsaturated bile acids. Accordingly, if one uses the gas chromatograph-mass spectrometer in a selective ion monitoring mode, all members of the same class can be monitored at one mass ion but, unlike selected ion monitoring with electron impact ionization, no other classes will be detected. The specificity of the intersection between ion mass and retention time offers a great enhancement over conventional GC or GC-EI-MS identification, since the CI mass ion and the relative retention time combine to form a unique index for each bile acid [119].

Table 13.2 "lists the selective mass ion and relative retention time coordinates for the bile acids and related compounds in our collection. The numerical reference for each compound is given in parentheses below its relative retention time coordinate" [119].

Computer Identification of Mass Spectra

Let us now consider the identification of unknowns by GC-MS and computers.

Using a computer to identify GC-MS mass spectra was obviously a called-for development. By 1971, Hertz, Hites, and Biemann were writing that a GC-MS "is capable of producing 400 mass spectra within a half-hour gas chromatogram [and therefore presents] an acute identification problem..." [129]. The goal of the GC-MS-computer search system is "...to go through a run from sample injection to computer identification with no operator intervention..." [130]. However, as we shall see, that goal is not easily attained. Abramson, 1975: "My experience with library searching of mass spectral data has been unsatisfactory...conventional searches provide equivocal answers" [131].

In 1975, Abramson proposed the terms *forward search* and *reverse search* [131]. At that time the forward search was the conventional one (see, for instance, Refs. 132-135).

Forward searches characteristically seek only to match the unknown spectrum with one of the spectra in the library, differ mutually only in the method of matching, apply a given method inflexibly, and cannot be trusted: "The goals of [forward] search methods are to rank the five to ten compounds in the library which most resemble the unknown. The correct answer may appear on the list, often at the top, when it is contained in the library. Incorrect compounds will *sometimes* head the list when the correct compound *is* in the library and *always* when the correct compound is *not* in the library" (italics mine)[131].

In a forward search, the purity of the unknown spectrum is taken for granted. In contrast, a reverse search handles the unknown spectrum as a multicomponent determination that includes identification of the components [131, 136]. (Ideally, there is only one component.) Also, the reverse search is flexible rather than fixed [131].

Part of the difficulty in computer searching of mass spectral libraries stems from spectrum condensation. Spectrum condensation attempts to reconcile the great number of organic compounds and of mass spectral peaks per organic compound with the finite size and speed of computers. Probably the most influential algorithm for spectrum condensation is that of Hertz, Hites, and Biemann [129]. (An algorithm is a specific way of handling a problem in a finite number of steps [137].) They tell why it was needed: "The most general comparison [of the mass spectrum in question with a

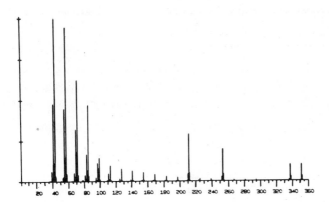

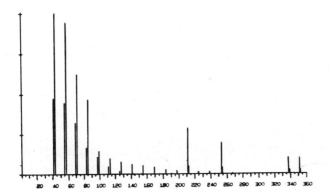

FIGURE 13.23 The complete EI mass spectrum is so detailed that it must be condensed for time-efficient computer searches of libraries of mass spectra. The top figure shows the complete 10-ethyl-10-n-propyl-docosane EI spectrum diagram. The bottom shows the condensation that retains only the two most prominent peaks from each 14-mass unit span. (Reprinted with permission from H. S. Hertz, R. A. Hites, and K. Biemann, Ref. 129. Copyright by the American Chemical Society.)

collection of all known spectra] would involve the use of all peaks. However, a mass spectrum could consist of several hundred peaks and a collection of reference spectra must consist of several thousand spectra, if it is to be useful. Thus storage limitations as well as the speed of the comparison dictate that the spectra must be condensed in some standardized manner" [129].

In their algorithm, "the spectrum is abbreviated by selecting the two largest peaks in each 14 mass unit interval throughout the spectrum, fourteen being the mass of a methylene group." The intervals 6 to 19, 20 to 33, 34 to 47, etc., are used rather than 1 to 14, 15 to 28, etc, in order not to split common peak clusters. The resultant algorithm "selects the *interpretively* significant peaks in a standardized manner" [129]. Figure 13.23 shows an example of a complete mass spectrum and of its condensation according to this algorithm [129]. By 1973, this algorithm was in use at the Environmental Protection Agency Data Bank [138], the National Institutes of Health [139], and the Massachusetts Institute of Technology [129].

TABLE 13.3 The Forward and the Reverse Search: A Summary of the Differences Between Them

Forward search	Reverse search
Data basis of search is unknown spectrum	Data basis of search is library spectrum
Arbitrary intensities are selected from unknown	Only intensities corresponding to library compound are selected
Positive or negative deviations are approximately equal in weight	Sign and number of the deviations are diagnostic
Spectra are not adjusted for fit	Automatic renormalization
Relatively sensitive to interference or mixtures	Relatively insensitive to interference or mixtures
Qualitative data only	Qualitative and quantitative data simultaneously
Identifies complete unknowns from a large library	Identifies preselected compounds from a limited library
Ranked library compound suggestions as output	Yes/No responses for each library compound
Substantial operator interaction and judgment required	Automatic operation
Library size limited by peripheral storage	Library size limited by core storage
Search algorithms fixed	Search algorithms flexible

Source: Reprinted with permission from F. P. Abramson, Ref. 131. Copyright by the American Chemical Society.

The limited capability of computers *seems* to require the condensation of mass spectra. But that very condensation then radically compromises computer searching. Abramson again, on the purity of unknown spectra: "The presence of significant levels of interference may artificially suppress the relative intensity of relevant masses and produce a bad fit. Even more importantly, when data are compressed (e.g., saving only the two largest peaks in a 14-amu region), interference of any nature may cause relevant masses to be excluded. To eliminate interferences, the operator must first detect such admixed spectra, then identify some other spectrum to subtract from the first to remove this interference, and, finally, determine how much of this second spectrum from the first. In addition, the operator must decide which, if any, of the multiple suggestions reported by most forward search methods is the correct answer. These human interventions make the automation of the identification process difficult" [131].

In other words, the reference spectra in the library may be condensed without loss of significant information, but the unknown spectrum should not be condensed. In the unknown spectrum, each mass must be considered potentially relevant. A reverse search can then determine relevance and, eventually, component identity or identities.

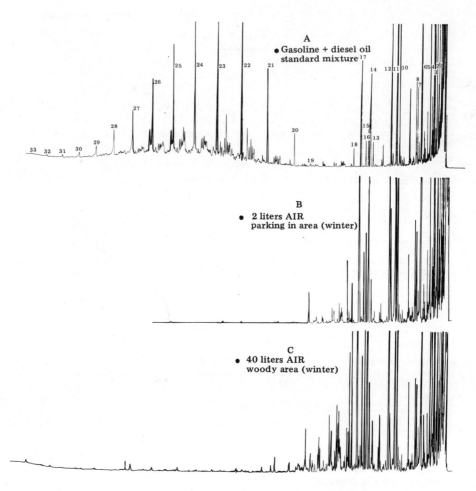

FIGURE 13.24 A qualitative analysis of organic micropollutants in air. This study was reported in 1976. The gas chromatographic separation was made with a glass capillary column, 60 m × 0.3 mm, temperature programmed. Such complex mixtures of traces demand highest resolution and sensitivity, and computer handling of the data. (Reproduced with permission of the Elsevier Scientific Publishing Company from B. Versino et al., Ref. 140, in the Journal of Chromatography.)

Another basic difference between reverse and forward searches is the assumption regarding the minimum useful library size. As we have seen, a forward search tends to require spectrum condensation because it is assumed that all available spectra are to be searched ("a collection of reference spectra must consist of several thousand spectra, if it is to be useful" [Hertz, Hites, and Biemann, Ref. 129]). In a reverse search, on the other hand, it is assumed that only limited groups of spectra will be searched: "...a 100-compound library seems sufficient...If a category needs more than 100 compounds, sequential searches may be called up on two or more libraries containing the compounds in question and the separate reports combined. Each library to be tailored to the problem at hand and therefore specific" [Abramson, Ref. 131].

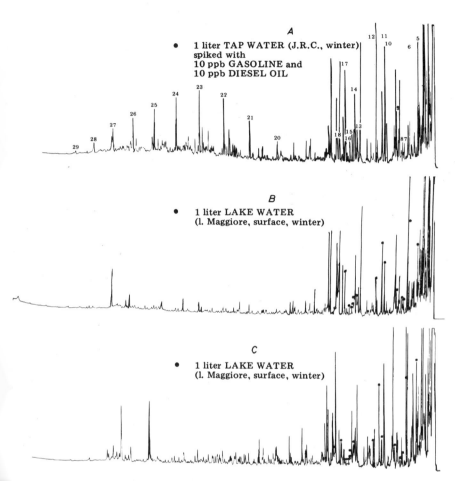

FIGURE 13.25 A qualitative analysis of organic micropollutants in water, from the same study sampled in Fig. 13.24. (Reproduced with permission of the Elsevier Scientific Publishing Company from B. Versino et al., Ref. 140, in the *Journal of Chromatography*.)

Abramson's summary of the differences between the forward and the reverse search is presented in Table 13.3 [131].

The reverse search has become firmly established in GC-MS. We consider an illustrative application reported in 1976 by Versino et al. [140]. They conducted a qualitative analysis of organic micropollutants in air and water. Their instrumental approach, HRGC-LRMS: glass open tubular column GC (Figs. 13.24, 13.25) [140], quadrupole MS. A two-step procedure cut the cost of the computer search. First, known pollutants were identified by reverse search [131]. Next, "the mass spectra of [the] GC peaks not identified by the reverse search" were checked by the Clerc algorithm for structural identification of organic compounds from spectroscopic data [128]. This algorithm finds similarly structured compounds "even if the spectrum of an unknown compound itself is not available in the library" [140]. "Unknown" compounds identified by the second search were added to the reverse search library.

The reverse search has been elaborated in Probability Based Matching (PBM)[141, 142]. In PBM, peaks are weighted "according to their 'uniqueness'"[142] and applied by reverse search. Application of PBM to a data base of 23879 spectra showed "that weighting of the mass as well as the abundance values improves retrieval performance, and confirms the value of 'reverse searching' for the spectra of mixtures" [131, 141].

In a later refinement, "spectrum subtraction [is used to complement] reverse search [in] the identification of [minor] mixture components" [143]. In this, the reference spectrum of the major component is subtracted, in proportion to its apparent concentration in the unknown, from the spectrum of the unknown. The "resulting residual spectrum" can be used to identify "a component whose concentration in the original spectrum was below detection limits" [143].

REFERENCES

1. D. A. Leathard and B. C. Shurlock, *Identification Techniques in Gas Chromatography*, Wiley, New York, 1970.
2. D. A. Leathard, in *Advances in Chromatography*, Vol. 13 (J. C. Giddings, E. Grushka, R. A. Keller, and J. Cazes, editors), Dekker, New York, 1975, pp. 265-304.
3. R. C. Crippen and C. E. Smith, *J. Gas Chromatog. 3*, 37 (1965).
4. J. J. Broderick, *Amer. Perf. Cosmetics 80*, 39 (1965).
5. J. H. Dhont, *Nature (Lond.) 198*, 990 (1963).
6. E. Kovats, *Helv. Chim. Acta 41*, 1915 (1958).
7. E. Kovats, *Helv. Chim. Acta 46*, 2705 (1963).
8. E. Kovats, in *Advances in Chromatography*, Vol. 1 (J. C. Giddings and R. A. Keller, editors), Dekker, New York, 1965, pp. 229-247.
9. L. S. Ettre, *Anal. Chem. 36*, 31A (1964).
10. L. S. Ettre, *Chromatographia 6*, 489 (1973).
11. J. K. Haken, in *Advances in Chromatography*, Vol. 14 (J. C. Giddings, E. Grushka, J. Cazes, and P. R. Brown, editors), Dekker, New York, 1976, pp. 367-407.
12. E. Kovats, in *Advances in Chromatography*, Vol. 1 (J. C. Giddings and R. A. Keller, editors), Dekker, New York, 1965, p. 232.
13. G. Guiochon, *Bull. Soc. Chim. France 1965*, 3420 (1965).
14. G. Schomburg, in *Advances in Chromatography*, Vol. 8 (J. C. Giddings and R. A. Keller, editors), Dekker, New York, 1968, pp. 211-246.
15. F. Caesar, *Chromatographia 5*, 173 (1972).
16. A. Sabatier, M. Goedert, and G. Guiochon, *Chromatographia 7*, 560 (1974).
17. T. K. Miwa, *J. Amer. Oil Chem. Soc. 40*, 309 (1963).
18. P. D. Klein and S. A. Tyler, *Anal. Chem. 37*, 1280 (1965).
19. G. Schomburg, H. Husmann, and F. Weeke, *Chromatographia 10*, 580 (1977).

20. D. A. Leathard and B. C. Shurlock, *Identification Techniques in Gas Chromatography*, Wiley, New York, 1970, pp. 57-63.
21. M. R. Möller, *Chromatographia 9*, 311 (1976).
22. R. Goulden, E. S. Goodwin, and L. Davies, *Analyst 88*, 941 (1963).
23. L. Giuffrida, *J. Assoc. Off. Agr. Chem. 17*, 293 (1964).
24. A. M. Bertz and H. D. Ruhl, *Anal. Chem. 36*, 1892 (1964).
25. M. J. D. Low, in *Gas Effluent Analysis* (W. Lodding, editor), Dekker, New York, 1968, p. 155.
26. A. W. Mantz and H. K. Morita, *Appl. Spec. 30*, 587 (1976).
27. A. W. Mantz, *Ind. Res. 19*, 90 (1977).
28. C. Merritt, Jr., in *Applied Spectroscopy Reviews*, Vol. 3 (E. G. Brame, Jr., editor), Dekker, New York, 1970, pp. 263-325.
29. C. Merritt, Jr., J. T. Walsh, D. A. Forss, P. Angelini, and S. M. Swift, *Anal. Chem. 36*, 1502 (1964), and Figure 16.
30. A. E. Banner, R. M. Elliott, and W. Kelly, in *Gas Chromatography 1964* (A. Goldup, editor), Elsevier, Amsterdam, 1965, pp. 180-189.
31. R. Teranishi, I. Hornstein, P. Issenberg, and E. L. Wick, *Flavor Research*, Dekker, New York, 1971, and Figure 6.1.
32. B. Littlewood, *Chromatographia 1*, 223 (1968).
33. J. T. Walsh and C. Merritt, Jr., *Anal. Chem. 32*, 1378 (1960).
34. R. J. Soukup, R. J. Scarpellino, and E. Danielczik, *Anal. Chem. 36*, 2255 (1964).
35. N. L. Alpert, W. E. Keiser, and H. A. Szymanski, *IR. Theory and Practice of Infrared Spectroscopy*, 2nd. ed., Plenum, New York, 1970.
36. W. J. Potts, Jr., *Chemical Infrared Spectroscopy*, Wiley, New York, 1963.
37. N. B. Colthup, L. H. Daly, and S. E. Wiberley, *Introduction to Infrared and Raman Spectroscopy*, Academic, New York, 1964.
38. R. G. White, *Handbook of Industrial Infrared Analysis*, Plenum, New York, 1964.
39. K. Nakanishi, *Infrared Absorption Spectroscopy*, Holden-Day, San Francisco, California, 1962.
40. L. Giuffrida, *J. Assoc. Off. Agr. Chem. 48*, 354 (1965).
41. J. Grasselli and M. K. Snavely, *Appl. Spec. 16*, 190 (1962).
42. M. Sparagana, *Steroids 8*, 219 (1966).
43. K. H. Kubeczka, *J. Chromatog. 31*, 319 (1967).
44. M. D. D. Howlett and D. Welti, *Analyst 91*, 291 (1965).
45. W. H. McFadden, *Sep. Sci. 1*, 723 (1966).
46. E. Stahl, *Thin Layer Chromatography*, 2nd ed. Springer-Verlag, New York, 1969.
47. E. V. Truter, *Thin Film Chromatography*, 2nd ed., Wiley, New York, 1971.
48. E. V. Truter, in *Advances in Chromatography*, Vol. 1 (J. C. Giddings and R. A. Keller, editors), Dekker, New York, 1965, pp. 113-152.
49. R. Stock and C. B. F. Rice, *Chromatographic Methods*, 2nd ed., Chapman and Hall, London, 1967.
50. H. R. Mangold and R. Kammereck, *Chem. and Ind. 1961*, 27 (1961).
51. J. Janak, *Nature (Lond.) 195*, 696 (1962).
52. B. Casu and L. Cavalloti, *Anal. Chem. 34*, 1514 (1962).
53. C. Nigam, M. Saharabudhe, T. W. M. Davis, J. C. Bartlet, and L. Levi, *Perfum. Essen. Oil Record 53*, 614 (1962).

54. I. C. Nigam, M. Saharabudhe, and L. Levi, *Can. J. Chem. 41*, 1535 (1963).
55. R. Kaiser, *Zeit. f. Anal. Chem. 205*, 284 (1964), and Figures 2 to 4.
56. H. Curtius and M. Miller, *J. Chromatog. 32*, 222 (1968).
57. J. P. Minyard, J. H. Tumlinson, A. C. Thompson, and P. A. Hedin, *J. Chromatog. 29*, 88 (1967).
58. J. H. Tumlinson, J. P. Minyard, P. A. Hedin, and A. C. Thompson, *J. Chromatog. 29*, 80 (1967).
59. J. E. Hoff and E. D. Feit, *Anal. Chem. 36*, 1002 (1964).
60. J. E. Hoff and E. D. Feit, *Anal. Chem. 35*, 1298 (1963).
61. K. M. Fredericks and R. Taylor, *Anal. Chem. 38*, 1961 (1966).
62. R. Bassette and C. H. Whitnash, *Anal. Chem. 32*, 1098 (1960).
63. R. L. Levy, in *Chromatographic Reviews*, Vol. 8 (M. Lederer, editor), Elsevier, Amsterdam, 1966, pp. 48-90.
64. M. P. Stevens, *Characterization and Analysis of Polymers by Gas Chromatography*, Dekker, New York, 1969.
65. J. Zulaica and G. Guiochon, *Bull. Soc. Chim. France 1966*, 1343 (1966).
66. J. Zulaica and G. Guiochon, *Bull. Soc. Chim. France 1966*, 1351 (1966).
67. S. Tsuge and T. Takeuchi, *Anal. Chem. 49*, 348 (1977).
68. M. Dimbat, in *Gas Chromatography 1970* (R. Stock, editor), The Institute of Petroleum, London, 1971, p. 413.
69. N. B. Coupe, C. E. R. Jones, and S. G. Perry, *J. Chromatog. 47*, 291 (1970).
70. C. E. R. Jones, S. G. Perry, and N. B. Coupe, in *Gas Chromatography 1970* (R. Stock, editor), The Institute of Petroleum, London, 1971, pp. 399-406.
71. N. B. Coupe, C. E. R. Jones, and P. B. Stockwell, *Chromatographia 6*, 483 (1973) and Figures 1, 2, and 3.
72. T. A. Gough and C. E. R. Jones, *Chromatographia 8*, 696 (1975)., and Figure 1.
73. J. Q. Walker and C. J. Wolfe, *Anal. Chem. 40*, 711 (1968).
74. D. L. Fanter, J. Q. Walker, and C. J. Wolf, *Anal. Chem. 40*, 2168 (1968).
75. S. G. Perry, *J. Chromatog. Sci. 7*, 193 (1969).
76. M. Beroza, *Nature (London) 196*, 768 (1962).
77. M. Beroza, *Anal. Chem. 34*, 1801 (1962).
78. M. Beroza and R. Sarmiento, *Anal. Chem. 35*, 1353 (1963).
79. T. Okamoto and T. Onaka, *Chem. Pharm. Bull. 11*, 1086 (1963).
80. C. J. Thompson, H. J. Coleman, C. C. Ward, R. L. Hopkins, and H. T. Rall, *Anal. Chem. 32*, 1762 (1960).
81. C. J. Thompson, H. J. Coleman, C. C. Ward, and H. T. Rall, *Anal. Chem. 32*, 424 (1960).
82. W. B. Innes, W. E. Bambrick, and A. J. Andreatch, *Anal. Chem. 35*, 1198 (1963).
83. F. W. Hefendehl, *Naturwiss. 51*, 138 (1964).
84. A. Dempster, *Phys. Rev. 11*, 316 (1918).
85. F. W. Aston, *Phil. Mag. 38*, 709 (1919).
86. C. A. McDowell (editor), *Mass Spectrometry*, McGraw-Hill, New York, 1963.
87. J. B. Farmer, in *Mass Spectrometry* (C. A. McDowell, editor), McGraw-Hill, New York, 1963, pp. 7-44.
88. W. H. McFadden, *Techniques of Combined Gas Chromatography/Mass Spectrometry*, Wiley-Interscience, New York, 1973.

1. *Ibid.*, pp. 41-43.
2. C. A. McDowell (editor), *Mass Spectrometry*, McGraw-Hill, New York, 1963, p. 18.
3. J. Mattauch and R. Herzog, *Z. Physik 89*, 786 (1934).
4. J. Mattauch, *Phys. Rev. 50*, 617 (1936).
5. E. G. Johnson and A. O. Nier, *Phys. Rev. 91*, 10 (1953).
6. A. O. Nier and T. R. Roberts, *Phys. Rev. 81*, 507 (1951).
7. A. O. Nier, *Rev. Sci. Instr. 31*, 1127 (1960).
8. A. L. Burlingame, B. J. Kimble, and P. J. Derrick, *Anal. Chem. 48*, 373R (1976).
9. B. J. Kimble, F. C. Walls, R. W. Olsen, and A. L. Burlingame, *23rd Annual Conference on Mass Spectrometry and Allied Topics,* Houston, Texas, p. 503.
10. W. H. McFadden, *Techniques of Combined Gas Chromatography/Mass Spectrometry,* Wiley-Interscience, New York, 1973, pp. 38, 247, 252, 264.
11. A. E. Cameron and D. E. Eggers, *Rev. Sci. Instr. 19*, 605 (1948).
12. W. C. Wiley and I. H. McLaren, *Rev. Sci. Instr. 26*, 1150 (1955).
13. W. H. McFadden, *Techniques of Combined Gas Chromatography/Mass Spectrometry,* Wiley-Interscience, New York, 1973, p. 34.
14. *Ibid.*, p. 47.
15. W. Paul and N. Steinwedel, *Z. Naturforsch 8A*, 448 (1953).
16. W. Paul and M. Raether, *Z. Physik 140*, 262 (1955).
17. J. B. Farmer, in *Mass Spectrometry* (C. A. McDowell, editor), McGraw-Hill, New York, 1963, pp. 34-38.
18. E. J. Bonelli, M. S. Story, and J. B. Knight, *Dynamic Mass Spectrometry 2*, 177 (1971).
19. W. H. McFadden, *Techniques of Combined Gas Chromatography/Mass Spectrometry,* Wiley-Interscience, New York, 1973, pp. 51-58.
20. J. B. Farmer, in *Mass Spectrometry* (C. A. McDowell, editor), McGraw-Hill, New York, 1963, p. 37.
21. R. M. Caprioli, W. F. Fies, and M. S. Story, *Anal. Chem. 46*, 453A (1974).
22. D. J. Jenden and R. W. Silverman, *J. Chromatog. Sci. 11*, 601 (1973).
23. P. D. Klein, J. R. Haumann, and W. J. Eisler, *Anal. Chem. 44*, 490 (1972).
24. F. Artigas, E. Gelpi, M. Prudencio, J. A. Alonso, and J. Baillart, *Anal. Chem. 49*, 543 (1977).
25. N. D. Young, J. F. Holland, J. N. Gerber, and C. C. Sweeley, *Anal. Chem. 47*, 2373 (1975).
26. Staff Report, "Instrumentation '77," *Chem. Eng. News 55*, 30 (1977).
27. E. M. Chait, *Anal. Chem. 44* (3), 77A (1972), and Figures 2 and 3.
28. M. S. B. Munson and F. H. Field, *J. Amer. Chem. Soc. 88*, 1621 (1966).
29. B. Munson, *Anal. Chem. 43* (13), 28A (1971), and Figures 2 and 4.
30. F. Hatch and B. Munson, *Anal. Chem. 49*, 169 (1977).
31. P. A. Szczepanik, D. L. Hachey, and P. D. Klein, *J. Lipid Res. 17*, 314 (1976).
32. D. F. Hunt, C. N. McEwen, and T. M. Harvey, *Anal. Chem. 47*, 1730 (1975).
33. H. M. Fales, G. W. A. Milne, H. U. Winkler, H. D. Beckey, J. N. Damico, and R. Barron, *Anal. Chem. 47*, 207 (1975), and Figures 2-4.
34. D. F. Hunt and J. F. Ryan, III, *Anal. Chem. 44*, 1306 (1972).
35. D. Beggs, M. L. Vestal, H. M. Fales, and G. W. A. Milne, *Rev. Sci. Instr. 42*, 1578 (1971).

124. H. M. Fales, Y. Nagai, G. W. A. Milne, H. B. Brewer, Jr., and J. J. Pisand, *Anal. Biochem. 43*, 288 (1971).

125. R. Ryhage, *Anal. Chem. 48*, 1829 (1976).

126. Staff Report, "Instrumentation '77," *Chem. Eng. News 55*, 31 (1977).

127. B. E. Blaisdell, *Anal. Chem. 49*, 180 (1977).

128. P. R. Naegeli and J. T. Clerc, *Anal. Chem. 46*, 739A (1974).

129. H. S. Hertz, R. A. Hites, and K. Biemann, *Anal. Chem. 43*, 681 (1971).

130. W. H. McFadden, *Techniques in Combined Gas Chromatography-Mass Spectrometry* Wiley-Interscience, New York, 1973, p. 311.

131. F. P. Abramson, *Anal. Chem. 47*, 45 (1975).

132. E. R. Malinowski and M. McCue, *Anal. Chem. 49*, 284 (1977).

133. T. F. Lam, C. L. Wilkins, T. R. Brunner, L. J. Soltzberg, and S. L. Kaberlin *Anal. Chem. 48*, 1768 (1976).

134. G. Van Marlen and A. Djikstra, *Anal. Chem. 48*, 595 (1976).

135. S. L. Grotch, *Anal. Chem. 46*, 526 (1974).

136. T. L. Isenhour, *Anal. Chem. 45*, 2153 (1973).

137. R. N. Schmidt and W. E. Meyers, *Introduction to Computer Science and Data Processing*, 2nd ed., Dryden Press, Hinsdale, Ill., 1970, p. 43.

138. J. R. Hoyland and M. B. Neher, *Implementation of a Computer-Based Information System for Mass Spectral Identification of Pesticides*, Batelle Columbus Laboratories, Columbus, Ohio, 1972.

139. S. R. Heller, *Anal. Chem. 44*, 1951 (1972).

140. B. Versino, H. Knöppel, M. DeGroot, A. Peil, J. Poelman, H. Schauenberg, H. Vissers, and F. Geiss, *J. Chromatog. 122*, 373 (1976).

141. F. W. McLafferty, R. H. Hertel, and R. D. Villwock, *Org. Mass Spectrom. 9*, (1974).

142. G. M. Pesyna, R. Venkataragnavan, H. E. Dayringer, and F. W. McLafferty, *Anal. Chem. 48*, 1362 (1976).

143. B. L. Atwater, R. Venkataragnavan, and F. W. McLafferty, *Anal. Chem. 51*, 19 (1979).

Chapter 14

QUANTITATIVE ANALYSIS

14.1   INTRODUCTION

Quantitative analyses can be gleaned from gas chromatographic separations. Such analyses can satisfy most needs for precision and accuracy.

In this chapter, the tools, practices, and results of quantitative analysis by gas chromatography are discussed in the following order: Quantitation without an integrator; area percent; response factors; sample size *versus* detector range; trace detection; potentiometric recorders; and nondigital integrators. The application of digital technology to gas chromatography is then taken up in enough detail to lend insight. Finally, some critical studies of precision and accuracy in gas chromatographic analysis are examined.

14.2   QUANTITATION WITHOUT AN INTEGRATOR

. Peak Height

Ideally, the peak height of a component varies directly and linearly with the amount of that component injected.

Peak height is the distance, measured normal to the retention axis, from the peak apex to the extrapolated line from which the peak departs.

Basing a quantitative analysis on manual peak height measurement cannot be criticized on grounds of precision [1-3].

If the mixtures to be analyzed resemble each other and yield not too many peaks of interest, if internal standards are used, and if a mixture of known and similar composition is analyzed once every five or ten unknowns in order to track the analysis, then quantitative analyses based on peak height can be convenient, precise, and accurate. For examples, see Podmore [4]; Menini [5]; Cawley, Musser, and Tretbar [6]; and Charrier *et al*. [7]. Also, see the studies of Grant and Clarke [2] and of nik [3].

However, the height of a peak reflects every possible experimental variable--sample size, injection technique, injection temperature, carrier flow rate, stationary phase identity and loading, column length and diameter--even the design of the instrument. This is awkward. A well-designed quantitative method reflects only the concentrations determined, not the expected variations of experimental parameters. In general, for desirable analytic stability, peak areas must be measured.

B. Peak Area

For an isothermal separation, the most precise and easily used method of determining peak area without an integrator is to multiply the peak height by the peak retention time measured from injection [2, 8, 9]. The precision of this method is equivalent to that of peak height measurement which, as we have just pointed out, can be very good.

If the separation is not isothermal, then peak area can be determined as peak height times peak width, measured at either one-half peak height or peak base.

In practice, peak areas are usually determined with integrators, which we discuss presently. First, we discuss peak area relationships. Within these, first we take up simply area percent, which approximates weight percent with the workhorse detectors; and then we consider response factors, which are more precise.

14.3 AREA PERCENT

By about 1960, it had become clear that peak areas based on thermal conductivity detectors (TCD) are roughly proportional to weight concentration, as opposed to mole concentration [10-13]. Soon thereafter, it was also found that peak areas based on flame ionization detectors (FID) are roughly proportional to the weight of unoxidized carbon in the components detected [14-16]. So as early as 1962 it had been established that with the workhorse detectors, the TCD and the FID, area percent is a usable guide to weight percent. This is why computer programs routinely present area percent, with more precise calculations usually offered only as options [17, 18].

In favorable cases such as mixtures comprising only methyl esters of fatty acids, area percent can indeed indicate weight percent within 3 to 4% of the amount present [19]. Nevertheless, the correspondence is unreliable. Good results--dependable and accurate, as well as precise--require calibration.

The next section, Response Factors, gives examples of the calculations associated with types of hypothetical calibrations and the determinations based on them.

14.4 RESPONSE FACTORS

A. Normalized Analyses

Weight Percent

To relate area percent to weight percent in a normalized analysis, we calibrate with a mixture having known concentrations of the components we wish to determine.

Sec. 14.4 Response Factors

We can then use the resultant response factors to determine the concentrations of these components in mixtures of them. Hypothetical measurements and related calculations for calibration and for analysis are shown in Tables 14.1 and 14.2, respectively.

TABLE 14.1 Response Factor Determination for Normalized Weight Percent

Calibration mixture; components	Known weight percent	Measured areas (counts)	Area percent	Response factor (weight percent/area percent)[a]
F	30.00	3,500	33.33[b]	0.9001[c]
G	45.00	4,000	38.10	1.181
H	25.00	3,000	28.57	0.8750

[a] To be used as area multiplier.
[b] 33.33 = 100 [3,500/(3,500 + 4,000 + 3,000)]
[c] 0.9001 = 30.00/33.33

TABLE 14.2 Normalized Analysis of Unknown for Weight Percent

"Unknown" mixture; components	Measured areas (counts)	Response factor (weight percent/area percent)	Weight product	Weight percent
F	4,000	0.9001[a]	3,600[b]	24.40[c]
G	5,000	1.181[a]	5,905	40.02
H	6,000	0.8750[a]	5,250	35.58

[a] Data from Table 14.1.
[b] 4,000 (area units) × 0.9001 (weight percent/area percent) = 3,600 (weight units)
[c] 24.40 wt% = 100[3,600 weight units/(3,600 + 5,905 + 5,250 total weight units)]

Mole Percent

If answers are to be in mole percent, in addition to the information required for answers in weight percent the molecular weights of the components must be known. Table 14.3 shows the conversion to mole percent of answers already in weight percent. Tables 14.4 and 14.5 show a hypothetical calibration and analysis, respectively, to give a normalized analysis directly in mole percent. The same hypothetical mixtures are employed.

TABLE 14.3 Conversion from Normalized Weight Percent to Normalized Mole Percent

Components	Weight percent	Molecular weight	Moles	Mole percent
F	24.40[a]	60.00	0.4067[b]	28.74
G	40.02[a]	80.00	0.5002[b]	35.34
H	35.58[a]	70.00	0.5083[b]	35.92

[a] Data from Table 14.2.
[b] Moles per 100 g mixture from Table 14.2: Moles/100 g mixture = $\frac{\text{grams}/100 \text{ g mixture}}{\text{grams/mole}}$

$$0.4067 = 24.40/60.00$$

TABLE 14.4 Response Factor Determination for Normalized Mole Percent

Calibration mixture components	Known weight percent	Gram-molecular weight	Moles	Mole percent	Measured areas (counts)	Area percent	Response factor (Mole %/area %)[d]
F	30.00	60.00	0.5000[a]	35.22[b]	3,500	33.33	1.057[c]
G	45.00	80.00	0.5625[a]	39.62	4,000	38.10	1.040
H	25.00	70.00	0.3571[a]	25.15	3,000	28.57	0.8803

[a] Moles per 100 g mixture.
[b] 35.22 = 100[0.5000/(0.5000 + 0.5625 + 0.3571)]
[c] 1.057 = 35.22/33.33
[d] To be used as area multiplier.

TABLE 14.5 Normalized Analysis of Unknown for Mole Percent

"Unknown" mixture components	Measured areas (counts)	Response factor (Mole %/area %)	Mole-proportional product	Mole percent
F	4,000	1.057[a]	4,228[b]	28.74
G	5,000	1.040[a]	5,200	35.35
H	6,000	0.8803[a]	5,282	35.91

[a] Data from Table 14.4.
[b] 4,228 = 4,000 × 1.057
[c] 28.74 = 100[4228/(4228 + 5200 + 5282)]

B. The Internal Standard

A good internal standard is one that is added early in sample processing. It then behaves as the components to be determined. Such a standard incomparably aids analytical accuracy, automatically compensating for errors that affect the components tracks. For example, a normal error in the quantity of sample injected then become

Sec. 14.4 Response Factors 345

TABLE 14.6 Determination of Weight Response Factors Using an Internal Standard

Calibration mixture components[a]	Known weight percent[c]	Measured areas (counts)	Weight ratio	Area ratio	X/H Response factor (weight ratio/area ratio)
F	30.00	3,500	1.200	1.167	1.028
G	45.00	4,000	1.800	1.333	1.350
H[b]	25.00	3,000			

[a] As in Tables 14.1 and 14.4.
[b] H taken as internal standard.
[c] In calculations such as these, the weight percents must be known but need not total 100%.

TABLE 14.7 A Weight Analysis of a Sample, Using an Internal Standard

Sample mixture components	Measured areas (counts)	H/Sample weight ratio	Area ratio	X/H Response factor (Weight ratio/area ratio)	Weight percent[c]
F	4,000		0.6667	1.028[b]	37.86[d,f]
G	5,000		0.8333	1.350[b]	62.13[e,g]
Internal standard					
H	6,000	0.5523[a]			

Data from Table 14.2: $0.5523 = 35.58/(100.00 - 35.58)$
Data from Table 14.6.
Weight percent $X = 100[\text{(area ratio)}_{(X/H)} \times \text{(wt. ratio/area ratio)}_{(X/H)} \times \text{(wt. ratio)}_{(H/S)}]$
  $37.86 = 100 \times 0.6667 \times 1.028 \times 0.5523$
  $62.13 = 100 \times 0.8333 \times 1.350 \times 0.5523$
From Table 14.2: In F, G binary, $37.86 = 100[24.40/(24.40 + 40.02)]$
From Table 14.2: In F, G binary, $62.13 = 100[40.02/(24.40 + 40.02)]$

practically undetectable. (A sample too complex for the addition of an internal standard can instead be run twice. On the second run, the internal standard can be an additional known proportion of an already present component [20, 21].)

Using the same figures as before, we indicate the use of an internal standard. We use component H as the internal standard, and calibrate for weight percent.

The weight response factor expresses the weight ratio of internal standard to component that would produce equal areas from each.

Table 14.6 shows the determination of the F/H and G/H weight response factors, considering H now as the internal standard. The corresponding illustrative analysis is shown in Table 14.7. Note that the weight percents of F and H and of G and H must be known, but as pairs do not and need not total 100%. Calibration mixtures may have any convenient number of components at any convenient concentration level.

14.5 SAMPLE SIZE *VERSUS* DETECTOR RANGE

We have just suggested that a response factor can be expressed as a constant. This assumes a linear detector response from the origin through the highest concentration in the analysis. However, the studies at the end of the chapter show that obtaining full accuracy and precision from gas chromatography requires abandoning most assumptions, including this one. For such performance, instruments must be calibrated individually and daily with mixtures of known composition. These mixtures closely bracket the component concentration ranges within the sample group.

Nevertheless, for now we shall continue to assume detector linearity. At the very least, this requires staying within the linear dynamic range of the detector. How much may we inject?

The maximum sample quantity acceptable to the linearly responding detector depends on the concentration at the peak maximum. A peak maximum represents a weight-per-volume concentration C_{max} that is a function of the corrected volume V_R^o and the number n of theoretical plates of the peak, along with the injected weight w_I of the solute that comprises the peak [22]:

$$C_{max} = \frac{\sqrt{n}}{V_R^o} \frac{w_I}{\sqrt{2\pi}} \qquad (14.1)$$

The maximum detector-acceptable weight-per-time concentration C_D can be related to the weight-per-volume peak-maximum concentration C_{max} by the temperature-corrected volume-per-time flow rate F_c:

$$C_D \frac{\text{Gram}}{\text{Second}} = F_c \frac{\text{milliliter}}{\text{second}} \; C_{max} \frac{\text{gram}}{\text{milliliter}}$$

Substituting C_D/F_c for C_{max}, and $F_c t_R^o$ for V_R^o, we can now express an acceptable though maximum injected weight w_D of a given component. This weight of component will appear in a peak of n theoretical plates after a corrected retention time t_R^o. The concentration of the component at the peak maximum will place it at the upper limit of the linear dynamic range of the detector:

$$\frac{C_D}{F_c} = \frac{\sqrt{n}}{V_R^o} w_D \sqrt{2\pi}$$

We solve for w_D:

$$w_D = \frac{1}{\sqrt{2\pi}} \frac{C_D}{\sqrt{n}} \frac{V_R^o}{F_c}$$

and

$$w_D = 0.4 \frac{C_D t_R^o}{\sqrt{n}} \qquad (14.2)$$

Here, C_D has the dimensions of grams per second; t_R^o, seconds; and w_D, grams.

Sec. 14.6 Trace Detection

The maximum concentration C_D permissibly deliverable to a detector without exceeding its linear dynamic range equals the sensitivity S of the detector multiplied by its linear dynamic range R_L. Because R_L is a pure number, S will have the dimensions of C_D, grams per second in Eq. (14.2). We write

$$w_D = 0.4 \frac{SR_L t_R^o}{\sqrt{n}} \qquad (14.3)$$

For the flame ionization detector, $C_D = SR_L = (10^{-12} \text{ g/sec})(10^7) = 10^{-5}$ g/sec.

With a flame ionization detector and thus a C_D of 10^{-5} g/sec, a column delivering 10^4 theoretical plates, and a solute emerging with a corrected retention time of 500 seconds (about 8 min), we find for w_D,

$$w_D = 0.4 \frac{C_D t_R^o}{\sqrt{n}} = 0.4 \frac{10^{-5} \text{ g/sec}}{10^2} \, 5 \times 10^2 \text{ sec} = 2 \times 10^{-5} \text{ g}$$

This is a *maximum* of about 2×10^{-2} µl of injected liquid per component, or about 4 µl of injected gas per component.

Thus, for instance, if a substance is to be injected for calibration, it should never be injected neat, but rather as perhaps 0.1 µl of a 1% solution.

Equation (14.3) applies only to isothermal separations. An equivalent expression for programmed temperature operation is not easily developed [23-25]. In any case, predictions such as those from Eq. (14.3) merely indicate limits. Whether the calibration is linear should always be tested [26].

14.6 TRACE DETECTION

Trace detection has continuing importance. A principal method in trace detection is gas chromatography. The discussion below is very brief and envisages the use of only normal equipment. A much more extended treatment is the book by H. Hachenberg, *Industrial Gas Chromatographic Trace Analysis* [27], which includes a review of special trace-enhancing equipment.

A. The Isothermal Column

For a given column, retention time, and detector sensitivity, what detection limit can be predicted? One necessary guide to finding that limit is current reported practice. Prediction from theory is another. By removing the linear dynamic range R_L from Eq. (14.3), we gain such a predictive device:

$$w_S = 0.4 \frac{S t_R^o}{\sqrt{n}} \qquad (14.4)$$

In Eq. (14.4), w_S is the minimum detectable weight in grams of a trace component injected into a column, emerging after t_R^o seconds as a peak of n theoretical plates, and detected by a detector of sensitivity S expressed in grams per second.

For instance, for the flame ionization detector, $S = 10^{-12}$ g/sec. If the peak emerges in 100 seconds and shows 1600 theoretical plates, it should be possible to detect $[(0.4)(10^{-12}$ g/sec$)(10^2)]/\sqrt{1600}$, or 10^{-12} g of the substance, 1 pg. (However, few chromatographs do not limit the inherent sensitivity of the flame ionization detector.)

B. Temperature Programming: Trace Enrichment

Trace detection is aided by concentrating the trace from a large sample. Injection into a cold column does this, where "cold" refers to the later, higher column temperature at which the trace component will be eluted. The resultant trace enrichment is V_o/V_a, where V_o and V_a are the isothermal retention volumes of the trace at the injection and elution temperatures, respectively [28, 29].

The theory and practice of this concept were studied by Hollingshead, Habgood, and Harris, and reported in a paper entitled "Sample Injection in Programmed Temperature Gas Chromatography and Its Significance for Trace Analysis" [29]. For trace analysis, they concluded, "...optimum sample injection conditions may be achieved by proper choice of initial temperatures. In other words the operator has independent control of the injection conditions and of the retention characteristics."

Also, they were able to introduce a 500-ml gas sample onto a deactivated alumina column at -60°C without finding any peak broadening for components heavier than propane: "...apart from differences in...[n]...between isothermal and programmed conditions, the apparent plate height for large samples is the same for...[programmed temperature gas chromatography]...and for isothermal elution at the initial temperature" [29].

The variation of the retention volume V with the absolute temperature is fairly well described by Eq. (14.5):

$$\log_{10} V = c \frac{1,000}{T} - \text{constant} \tag{14.5}$$

The trace enrichment factor V_o/V_a can therefore be evaluated from Eq. (14.6):

$$\log_{10} \left(\frac{V_o}{V_a}\right) = c \left(\frac{1,000}{T_o} - \frac{1,000}{T_a}\right) \tag{14.6}$$

The value of the slope c depends on the solute and the stationary phase. It can be read from plots such as those for SE-30 shown in Fig. 14.1 [30].

Some values of the slope c are given in Table 14.8. They can be seen to range from about 1.2 to about 2.0.

Some hypothetical values of c, T_o, and T_a, and therefore of the trace enrichment factor V_o/V_a, are shown in Table 14.9. It is obvious that injecting at low temperatures and eluting at much higher temperatures can increase trace concentrations by several orders of magnitude. Having a favorable stationary phase-solute combination is more effective in this increase than an extra 100° rise in elution temperature.

Sec. 14.6 Trace Detection

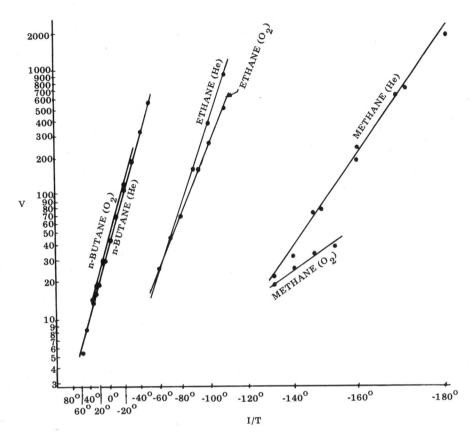

FIGURE 14.1 Over the temperature range shown here, from −180°C to about 80°C, SE-30 shows a linear relationship between log V and 1/T. The column is SE-30, 5% on Chromosorb W. (Reproduced from *Advances in Gas Chromatography 1965*, 1966, Ref. 30, with permission of A. G. Altenau and Preston Publications, Inc.)

TABLE 14.8 Sample Constants in Eq. (14.6)[a]

Stationary phase	Solute	Constant c
SE-30	n-Butane	1.32
Carbowax 20M[b]	n-Butane	1.5
DC 702[c,d]	n-Propyl alcohol	1.5
Tricresyl phosphate[d]	n-Propyl alcohol	2.0
Tricresyl phosphate[d]	i-Propyl alcohol	1.8

[a] $\log (V_o/V_a) = c[(1{,}000/T_o) - (1{,}000/T_a)]$
[b] Frozen [30].
[c] Liquid silicone.
[d] Data from Ref. 31.

TABLE 14.9 Possible Trace Enrichment Factors[a]

| Temperature, °C | | Trace enrichment factor[b] | |
Injection	Elution	c = 1.5	c = 2.0
−73	127	5,000	100,000
−73	227	30,000	1,000,000

[a] Evaluated from Eq. (14.6).
[b] V_0/V_a

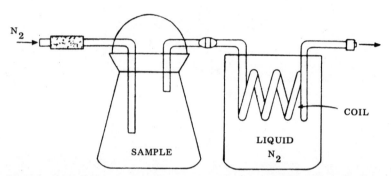

FIGURE 14.2 In trace enrichment by sample injection at low temperatures, the "head" of the column may be a separate cooled precolumn. The arrangement shown here avoids artifacts and loss of trace components. (Reprinted with permission of John Wiley and Sons, Inc., from Harris and Habgood, Ref. 25, *Programmed Temperature Gas Chromatography*, 1966. Original data published by I. Hornstein and P. F. Crowe in *Analytical Chemistry*, Ref. 32.)

The point of cooling the column for sample injection is to strip the trace from the sample and deposit it on a small number of theoretical plates at the head of the column. It follows that only these initial plates need be cooled, and that therefore they may be in the form of a separate coolable precolumn. After the trace has been stripped from the sample, flow may be stopped, the precolumn warmed, and the concentrated trace swept by recommenced flow onto the column for separation.

This was the approach used by Hornstein and Crowe and illustrated in Figs. 14.2 and 14.3 [32], and used by Zlatkis et al. and illustrated in Fig. 14.4 [33]. The technique of Zlatkis et al. involved "adsorption of [sample] volatiles on Tenax GC, heat desorption with helium, trapping on a [dry ice-] cooled precolumn, and chromatography on 100-m × 0.50-mm. i.d. nickel columns...After a trapping period of 20 min with dry ice as the coolant, the gas chromatographic separation was begun at room temperature after the container with the dry ice had been removed" [33].

Sec. 14.6 Trace Detection

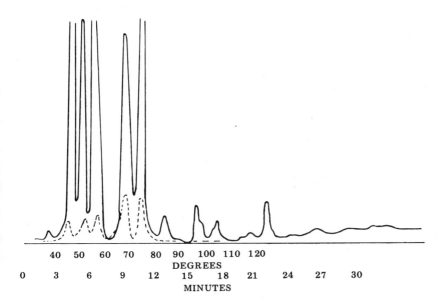

FIGURE 14.3 The sampling arrangement shown in Fig. 14.2 produced the solid curve, whereas conventional freeze-out and transfer yielded the dotted line. Both are chromatograms from heated lamb fat. (Reprinted with permission from I. Hornstein and P. F. Crowe, Ref. 32. Copyright by the American Chemical Society.)

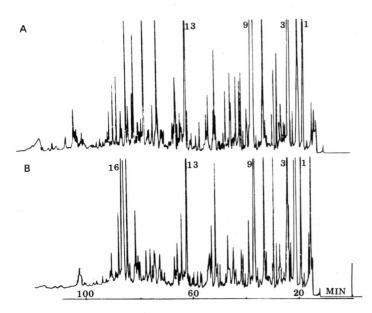

FIGURE 14.4 These chromatograms show the volatile components from 24-hr urines of a normal male (A) and a normal female (B). (Reprinted with permission from Zlatkis et al., Ref. 33. Copyright by the American Chemical Society.)

352   Chap. 14   Quantitative Analysis

14.7 POTENTIOMETRIC RECORDERS

Recorders can be galvanometric or potentiometric. Nearly exclusively, potentiometric recorders are the ones used in gas chromatography. As its name suggests, the potentiometric recorder records the measurement of a potential, which in GC is often changing.

A. The Concept of the Potentiometric Recorder

The potentiometric recorder measures potential by generating within itself a potential equal to the one being measured. In this way, the potential being measured does not become changed by the measurement itself.

The basic concept of the potentiometric recorder is shown in simplified diagram in Fig. 14.5.

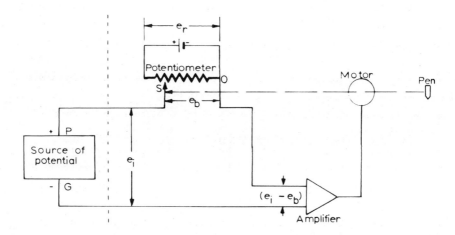

FIGURE 14.5 The potentiometric recorder in functional diagram.

In Fig. 14.5, the vertical dotted line indicates the physical boundary of the recorder. Two wires connect the source of potential to the recorder. We define the potential at G as zero. The wire at P is considered to be positive and to carry the nominal potential—the signal—being measured.

Within the recorder, the signal e_i is carried to the movable contact S of a potentiometer. The potentiometer has a resistor across which is imposed a potential e_r. Potential e_r comes from a source of constant reference potential that is a part of the potentiometer and that is shown in Fig. 14.5 as a battery. (In a battery symbol, the longer, thinner line is positive and the shorter, thicker line is negative.) Thus a potential e_b that is some part of e_r exists between the movable contact S and the end

Sec. 14.7 Potentiometric Recorders

O (for zero) of the slide-wire that is connected to the negative terminal of the battery.

We can now see by tracking the potential from point G that it rises to $+e_i$ at P, is carried unchanged to point S, and then is decreased from the potential at point S to a certain potential at point O. The potential difference between points G and O is therefore $e_i - e_b$. This difference is presented to the amplifier, which is shown as a triangle.

The amplifier has one salient feature: High gain. The gain is usually in the millions. If a detectable potential difference $e_i - e_b$ exists, it becomes greatly amplified at the output of the amplifier.

The output of the amplifier, symbolized as coming from the apex of the triangle, drives a motor, symbolized as a circle.

The motor is reversible. It is mechanically connected to the movable contact S of the recorder potentiometer. The motor is electrically and mechanically arranged to tend to make any potential $e_i - e_b$ vanish, whether that difference is positive or negative.

Consider the interaction of the amplifier, the motor, and the movable contact S, and consider the electrical effect of this interaction. If e_i happens at any moment to be larger than e_b, the amplified difference is applied to the motor, which moves to increase e_b and thus to make $e_i - e_b$ equal to zero.

When $e_i - e_b$ is equal to zero, then e_i is equal to e_b. The recorder is then said to be *in balance*.

Current flows from one point to another only if a potential difference exists between these points. If e_i is equal to e_b, and if these are the only potential differences in the input circuit to the amplifier, then no current will flow between points G and O nor, correspondingly, between points P and S. The conclusion: A potentiometric recorder at balance draws no current from the source being measured.

Also, Ohm's law states that $e = ir$: In response to a potential difference e, a current i will flow through a circuit of resistance r. Now if the current drawn by a measuring device is zero, then unless the source exhibits zero potential--not the case we are considering--the measuring device must be offering infinite resistance to the source of potential being measured. The corollary conclusion: A potentiometric recorder at balance offers an infinite resistance to the source. (This is usually referred to as an infinite impedance rather than as an infinite resistance.)

In sum, as we stated at the start, a potentiometric recorder tends to measure a potential without causing any change in the potential being measured.

We can now consider more exact terminology. The recorder amplifier drives the motor so as to reduce $e_i - e_b$ toward zero, but how close to zero? Just how large can or must $e_i - e_b$ become without activating the recorder motor, if the recorder was originally in balance? The answer introduces the concepts and terms *span* and *deadband*.

B. Terminology

Span

Span is that potential that causes pen deflection just equal to the full scale of the recorder. We shall here symbolize it as $e_{i\text{-span}}$.

A recorder with a span of 1 mV is referred to in the spoken language as a one-millivolt recorder.

Let a voltage e_i of 1 mV be presented to a recorder of 1 mV span, that is, to a 1-mV recorder. If the recorder was in balance at 0 mV e_i just before the 1-mV voltage was presented, then the recorder will immediately proceed to generate a 1-mV voltage within itself. In this, the moving contact S will be driven completely across the potentiometric resistor. The pen, reflecting this full-span movement of S, will be moved from one side of the recorder chart to the other. When the matching 1 mV has been generated, $e_i - e_b$ will again be zero and the recorder will again be in balance. No parts will be moving (except perhaps for pen jitter, which we shall be considering) and no current will be flowing from the e_i source.

In the earlier days of gas chromatography, usually only thermal conductivity detectors were available. By today's standards, these were quite insensitive. Therefore, to get at that time peaks as large as possible with the small TCD signals, only 1-mV recorders were considered acceptable. Now, however, TC detectors have become much more sensitive, and the sensitivities of ionization detectors and the associated amplifiers are not recorder-limited, so 5-mV and 10-mV recorders are often used.

Deadband

Deadband is a proportion of span. It is defined by the American Standards Association as "the range through which the measured quantity can vary without initiating action."

The specified deadband of the usual potentiometric recorder is 0.15% to 0.25% of span. This is about one-fifth of one division.

Deadband is *the* recorder-based source of error in gas chromatographic determinations. For instance, deadband crucially affects the ability of a recorder to show the top of a peak as the signal stops increasing, temporarily halts, and then begins to decrease. Deadband masks the ability of a recorder to show small peaks, or to show the true baseline before and after large peaks. But quantitative analysis *depends* on the measurements of these phenomena.

Do not use a recorder until you have tested it for deadband. How else do you know that the recorder is really working? The deadband should be no more than the width of the pen trace.

The width of the deadband can be estimated merely by shoving the recorder pen a little to one side and then a little to the other. The lateral distance between the two returns of the pen is the deadband.

Sec. 14.7 Potentiometric Recorders

Even better, manually turn a drive wheel or gear within the recorder so as to displace the recorder pen a few chart divisions. Keep the displacement small enough that momentum is not involved in the return. (Do not displace the pen an inch or two and then release the wheel, letting the mechanism snap back.)

When the pen is displaced in this test, a corresponding displacement of the contact S occurs. This changes e_b and causes a potential difference $e_i - e_b$ to appear at the input to the amplifier. As we have already described, a train of events then begins that should reduce $e_i - e_b$ to zero. Deadband measures any residual $e_i - e_b$. Within that deadband, the recorder is insensate.

If the deadband is appreciable, find the gain control of the amplifier and increase the gain. (The gain control is usually visible and labeled GAIN, but on some instruments it is only a screwdriver-slotted shaft available only through a hole in the amplifier housing.) When you have increased the gain too much, the recorder balancing mechanisms will usually break into oscillation. Reduce the gain until this oscillation just stops. Recheck the deadband. If the deadband has again become larger than a pen width, increase the gain again until the deadband is reduced to the acceptable pen width.

Ideally, the recorder pen will show a constant, minute, but observable tremor--*pen jitter*. A recorder trace that shows jitter can be read as precisely as one that does not, and much more confidently.

In contrast, the true signal masked by a large deadband can only be guessed at. However, a deadband-distorted recorder trace can usually be recognized by its peculiar terrace or staircase appearance.

Deadband is by no means a function only of gain. Interference rejection and input impedance also profoundly influence deadband.

Interference is any signal extraneous to the one of interest. If interference is not rejected, it will be amplified and tend to saturate the recorder amplifier. The amplifier will then be unable to afford due amplification to the signal proper. An enlarged deadband results.

To help reduce interference, ground the signal circuit at only one point. The directions of the GC manufacturer should indicate whether the signal circuit should be grounded by the operator or left floating, *i.e.*, ungrounded.

Interference is characterized as longitudinal and transverse. In longitudinal interference the signal leads rise and fall in potential *together* (whereas the signal consists of a potential difference *between* the leads). The usual minimum rejection of longitudinal interference in modern recorders is 60 decibels (db), *i.e.*, $10 \log_{10} 10^6$. It is often advertised as over 80 db. A 60-db rejection implies a rejection ratio of 10^6. With such rejection, a longitudinal interference equal to the span (100 chart divisions) would in theory move the pen 10^{-4} divisions, in actuality no distance at all.

Transverse interference, however, appears between the signal leads and thus is superimposed on the signal itself. Some of it can be removed by filters, for instance

60-cycle pickup, which has a frequency far higher than that of the GC signal. A floating shield around the input circuit, filter, and chopper (the chopper, situated at the input to the amplifier, converts the incoming direct current signal to a more easily handled alternating current signal), and another shield around the input transformer provide rather more effective protection than filtering. Such shields are built into good modern recorders.

Interference rejection depends not only on the internal construction of the recorder but also on the manner of connecting the signal cable to the recorder. The signal cable should be a two-wire cable inside a connecting shield that is stranded (made of fine braided wire). Good modern recorders offer three terminals to the signal cable. Two are for the signal leads; the third extends the just-described electrostatic guard shield of the recorder signal circuitry to the cable shield and thus to the output of the gas chromatograph. In this case, the jumper that may be connecting the negative signal lead to the guard terminal should be removed, particularly if the GC operating instructions call for a floating input.

The input impedance is that resistance of the signal source that requires the signal to be doubled to produce the same given recorder response. Modern recorders have input impedances that are satisfactorily high compared to the resistances of the usual gas chromatographic attenuator. Input impedances of 25,000 ohms--usually higher--are to be required of any recorder to be used in gas chromatography.

Dynamic Response

The response time of any system is that time required for the system to move 63% toward full response to a step change in signal (Fig. 14.6) (63% is 100 times $[1 - (1/e)]$) [34].

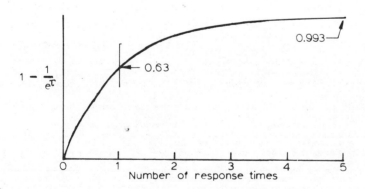

FIGURE 14.6 Correctly adjusted, the potentiometric recorder of Fig. 14.5 responds as shown to a step change in signal. In one response time, it moves $[1 - (1/e)]$ or 63% of the way toward its eventual full response. In five response times, it will have moved 99% of the way.

FIGURE 14.7 The potentiometric recorder diagrammed in Fig. 14.5 is a position servomechanism. Any position servomechanism inherently lags the signal, even when it is in correct adjustment. Here, the dotted line shows the signal, and the solid line the response. The response, lagging the signal both going up and coming down, distorts the peak.

A gas chromatographic peak should be recorded by a detector-recorder system having a response time not more than one-twentieth of the 4σ-duration of the peak at its base [35].

If a gas chromatographic peak were to be eluted 200 sec after injection and then be 2 sec wide at its base, that would indicate a separation efficiency of $16(200/2)^2$, or 160,000 theoretical plates. To record such a peak, a recorder should have a response time of 2/20 or 0.1 sec. This corresponds roughly to a balance time of 0.5 sec. For such a peak, the recorder would be just limiting the system. However, such performance on the part of the gas chromatograph is unusual, if it exists at all in commercial instruments. The typical gas chromatographic recorder of today, which can come to balance in less than a second when the span signal is suddenly and discontinuously imposed, is adequate for the great bulk of GC tasks.

Nevertheless, the potentiometric recorder inherently distorts a changing signal. The distortion increases as the rate of change increases, that is, as the peaks become sharper and closer. This inherent distortion comes about because the various elements of the recorder comprise an electromechanical system. The input $e_i - e_b$ to the amplifier must differ appreciably from zero before any action ensues. Even after the motor begins to be driven by the amplifier (e_b is already lagging e_i), each mechanical component contributes inertia and thus further retardation. Any device such as this--any position servomechanism--inherently lags the signal, and the lag increases as the signal changes faster (Fig. 14.7).

As peaks sharpen, the peak sides more and more represent not the signal but the maximum ability of the mechanism to get under way and to move. As peaks sharpen, the increasingly truncated peak apexes more and more represent not the signal but the maximum ability of the mechanism to turn around and to go the other way. Indeed, in the open tubular separations of highly complex materials that stretch the performance limits of gas chromatography, the pen track of a potentiometric recorder becomes only a qualitative guide to the course of the separation. Accurately recording maximally fast and sharp gas chromatographic peaks in great numbers is not impossible, but a recorder equal to the task would cost more than the gas chromatograph, handling the voluminous records it makes would be monumentally inconvenient, and measuring such records for quantitation would be unimaginably tedious. In such cases the detector output is directly monitored by a computer; the recorded gas chromatogram is used only as a guide to the quantitative results printed out by the computer.

14.8 INTEGRATORS

A. Analog

An analog device converts the magnitude of a continuous input into the magnitude of some sort of continuous output. An automobile speedometer is an analog device, converting wheel rotational velocity into speedometer needle rotational position.

A potentiometric recorder is an analog device. The input is a voltage that changes in a continuous fashion, as in representing the peaks of a gas chromatogram. The output is the corresponding continuous change in the rotational position of a shaft, displayed as the position of the recorder pen.

Analog integrators are usually dependent on the electronics and mechanisms of a potentiometric recorder, and are physically driven by the recorder. The input to such an analog integrator is the output from the recorder. Therefore the area measurements from such integrators reflect any deficiency in either the recorder or the operation of it. Two common deficiencies are recorder deadband and operator error.

Deadband-produced error in locating the baseline very effectively distorts an area measurement, because the area-per-division error is greatest at the baseline [36].

With samples of almost any complexity, operator error is almost inevitable. Using the attenuator, the operator must keep the whole of the large peaks on scale, yet also record the smaller peaks at high sensitivity. The resulting necessary reruns can easily cut the productivity of a laboratory in half.

Nevertheless, if the recorder deadband is kept to a pen width and the operator is not required to act like a machine for too long, a recorder-driven integrator is not only adequately precise but also very useful.

The basic ball-and-disc integrator (Fig. 14.8) of the Disc Instrument Company was introduced about 1961 [38]. It has proven dependable and precise--0.1% of full scale.

Within the Disc integrator is a disc that rotates at constant speed. A ball attached to the recorder pen is in contact with this disc. The ball is held at the center of the disc when the recorder pen is at the baseline of the chromatogram, at the perimeter of the disc when the recorder pen is at the upper edge of the scale. The ball thus does not rotate in the baseline case, but rotates at maximum speed in the full-scale case.

A cam-driven auxiliary pen mounted on the same recorder is caused--by a second ball driven by the first, a roller driven by the second ball, and the pen-driving,

Sec. 14.8 Integrators   359

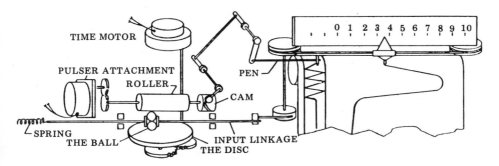

FIGURE 14.8 The principle of Disc integrator action is shown here in diagram.
An explanation of it may be found in the text. The Disc integrator is surprisingly
robust and highly precise. (Diagram courtesy of the Varian Instrument Division,
Walnut Creek, Calif., Ref. 37.)

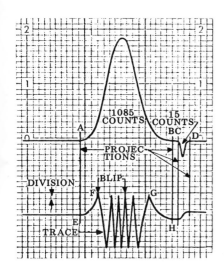

FIGURE 14.9 The integral of the peak (top) can be obtained from the Disc integrator
trace (bottom). Each division counts 10. Traverse EF:40. The 10 traverses from F
to G: 1,000. Traverse GH: 45. Total: 1085. (Diagram courtesy of the Varian Instrument Division, Walnut Creek, Calif., Ref. 37.)

pen-direction-reversing cam that is driven by the roller--to move back and forth over
a 10-division strip at a speed proportional to the speed of rotation of the first
ball. This speed is also proportional to the pen deflection.

The number of divisions crossed by the auxiliary pen is proportional to the area
above the baseline (Fig. 14.9).

Measuring areas with the auxiliary-pen Disc integrator just described becomes unprofitable if many peaks per time are to be integrated. If the laboratory produces
many quantitative analyses, the next step beyond a Disc integrator is a device to
print out a number representing the area integrated. However, here the buyer should
beware.

A number printed out by a recorder-driven integrator is as limited in meaning as any other output of a recorder-driven integrator. It is more convenient than manual division counting *if the device is reliable*. Such devices often depend on the correct and sometimes delicate operation of many moving parts. Recorder-driven printing integrators are not known for reliability. Before buying one, check with two or three long-time owners of identical units; do not assess reliability on the basis of a sales demonstration.

B. Analog Logic: The Digital Integrator

Compared to manual peak integration, an analog integrator such as the Disc integrator is a convenience. However, it requires continuous operator attention.

As just mentioned, in using either a recorder or a recorder-driven integrator for quantitative analysis, the chromatograph operator must keep the small peaks large (for good precision), and all peaks on scale. This requires constant, machine-like attention and action. Failure becomes inevitable, reruns routine, productivity poor. This serious problem was solved by the arrival in the late 1960s of the digital integrator [38-40]. Automatic in operation, it was interposed between the chromatograph and the recorder.

A digital integrator receives its signal directly from the gas chromatograph. The potentiometric recorder receives *its* signal from the digital integrator.

Peak measurement for quantitative analysis is now done by the digital integrator. The potentiometric recorder now does not affect quantitation, is used only for peak display, and indeed need not be used at all. Also, because the digital integrator operates automatically, the operator is not involved once he has injected the sample and activated the digital integrator.

Moreover, the digital integrator has a wide dynamic range (10^5 to 10^6): "This means that the operator can analyze samples varying in concentration between 50 ppm and 99+ percent without manual attenuation" [39].

A digital integrator has a slope detector that governs it. Because the slope detector is an analog device, the digital integrator is also known as an *analog-logic* integrator.

The operation of a digital integrator is enhanced by its memory, which may be short- or long-term: A buffer memory or magnetic tape, respectively. In either case, the analog signal from the gas chromatograph is first put into memory in analog form and only then played back to the slope detector. The rate of playback is thus divorced from the rate of signal arrival and can instead be optimized for slope detection.

The slope detector senses the rate of change of the signal presented to it. By analog techniques, both the magnitude and the direction of the signal change rate are measured. The analog output of the slope detector represents these measurements by its magnitude and sign.

Using criteria preset for it by the operator, the slope detector distinguishes signal from noise, i.e., real peaks from mere spikes in the baseline. It tells the integrator when to begin peak area measurement and when to halt. It allows the baseline drift corrector--another component of a digital integrator--either to hold the area integrator at zero in the absence of a peak or to permit peak area integration in the presence of a peak.

The "digital" aspect of a digital integrator lies primarily in that it counts pulses. The analog signal from the memory is converted to pulses--a digital signal-- by a voltage-to-frequency converter. During integration, these pulses are counted. If response factors are to be used, these are applied in analog fashion to the analog signal before it is sent to the voltage-to-frequency converter [39].

On the completion of the integration of a given peak area, that area is immediately printed out, usually in connection with the response time of the peak.

14.9 DIGITAL TECHNOLOGY APPLIED TO GAS CHROMATOGRAPHY

The "digital," analog-logic integrator constituted a great advance over its predecessor, the recorder-driven integrator. It released the operator from a machine-like and therefore inefficient involvement in the running of the sample. However, the printed output of the digital integrator was not fully usable as received. It still required judicious peak-by-peak appraisal, interpretation, and calculation. Being based on an inherently expensive and therefore economically inextensible analog logic, the "digital" integrator could not realistically be upgraded to yield a fully usable output.

For its output to be fully usable, a signal-handling device associated with a gas chromatograph should, at the choice of the operator, present anything from simple area or area percent to a complete qualitative and quantitative analysis. At least in principle, such an output requires the automatic accommodation of a variety of difficulties. These difficulties include baseline drift, solvent peaks that tail onto peaks of interest, poorly controlled carrier gas flow rates, the presence of quick fleeting peaks and broad shallow peaks in the same chromatogram, incompletely resolved ("fused") peaks of mixed orders of magnitude in size, and signal noise from diverse sources such as dirty electrodes and house circuit pickup.

Only devices based on digital technology could conceivably cope with these problems--which unfortunately are even worse than they seem [36, 41, 42]. In this section we briefly review the application of digital technology to gas chromatography (see also a short review by Gill [43] and a lengthy and detailed one by Leathard [44].) First we take up the handling of gas chromatographic signals. Then we review algorithms, which are the methods for solving the problems posed by the gas chromatogram. Next are treated the devices that apply this technology: The computing integrator, the small dedicated computer, and the large computer. Finally we consider the automation of gas chromatography.

A. Handling the Chromatograph Signal

Recorder Display

The chromatographer still may often wish to see the gas chromatogram rather than merely the printed analysis of the sample. To allow this optional display of the gas chromatogram by a potentiometric recorder, the full chromatograph signal is first presented to a buffer: An isolating, high-impedance, unit-gain amplifier/attenuator. This in turn presents the recorder with a signal replicate that can be manually attenuated without thereby inserting noise into the original signal [45].

Signal Transmission

Analog. Analog transmission systems can carry the signals from several monitored gas chromatographs to a central analog-to-digital converter (ADC). These instruments may be as far as 300 m from the converter [18, 45]. The signal from each instrument is then scanned briefly but frequently--say, 1 to 30 times/sec--for the ADC [18, 45].

Noise in the signal can be kept to a satisfactorily low level by proper shielding--at times requiring mutually isolated inner and outer shields--and careful single-point grounding of these shields. Should there be electrically separate parts of the gas chromatograph, these too must be brought to a common ground that is connected to one of the shields [46].

Pickup from power lines is usually the largest source of noise. This can be elegantly discriminated against by sampling at the power frequency (Fig. 14.10)[18]. Alternatively, an effective (90db) rejection filter sharply selective for the power frequency may be applied at the ADC [45].

The minimum noise level in an already quieted analog source and transmission system may be no lower than about 10 μV when reed relays are used in the scanning equipment (assuming, as is probable, that more than one instrument is being monitored) [45]. Guichard and Sicard reported receiving about 5 μV noise, even from a gas chromatograph 300 m away [47]. In one system, an arbitrarily chosen lowest level of interest of 25 μV was used [18]. This seems consistent with the average instrument noise in another system in which 141 analytical instruments are monitored [48].

Digital. In contrast to analog transmission, each chromatograph may be equipped with its own adjacent ADC [49-51]. Initially apparently more expensive, this is technically the better route. (The digital route can turn out to be the less expensive route, too [36, 49].) Digital data transmission is inherently superior to analog in that there is no limit to the distance of transmission, reliability against data distortion or loss is essentially complete, and a linear dynamic range of 10^6 is easily accommodated [49, 50].

Signal Sampling

The sampling rate should be continually adjusted to the current peak width. How the sampling is accomplished depends on whether the originally analog signal is one of many sent in analog form to a central analog-to-digital converter (*Analog*), or is converted to digital form at its source (*Digital*).

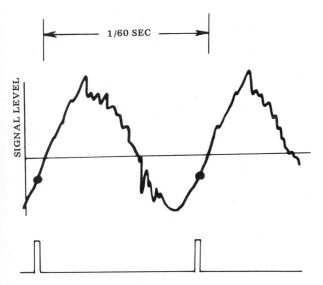

FIGURE 14.10 A neat way to discriminate against pickup from power lines is to sample (bottom trace) at the power frequency. (Reproduced with permission of Marcel Dekker-Inc., from D. Ford and K. Weihman, Ref. 18.)

Analog. Busch [49] and Baumann et al. [17] assume that the analog signal from a monitored gas chromatograph would be sampled 10 times/sec, and that the monitored instrument would be one of either 40 or 10, respectively, similarly and simultaneously monitored. The scan speed in another system is 30 points/sec (each presented to the ADC)[18]. In still another system, the scan rate is variable from 30 to 1 or fewer points/sec (of which only one-third to one-tenth are retained after data smoothing by a much-used routine [52]).

Digital. As already indicated, conversion of the analog signal to digital form at each chromatograph is technically preferable. "The conversion of GC signals into digital format can be done with...negligible...error. Digital data transmission is possible at failure rates of 1 bit a month or less" [49]. (A bit is one pulse in a group of pulses [53].) "Digital transmission [is] virtually immune to noise pickup and signal degradation" [51].

The basic technique for the conversion of gas chromatographic analog signals to digital form is integration [17, 49, 54]. The GC analog signal is integrated (without thereby using it as a source of current) until a fixed integral is attained. The attainment of one such fixed integral gives rise to one pulse.

At an analog signal level of 0.1 µV, such a fixed integral is formed one per second [54]. At an analog signal level 10^6 times higher, and with a voltage-to-frequency ADC, such fixed integrals would be attained at a rate of 10^6 per second, generating 10^6 pulses per second. This range of output would correspond to a linear dynamic range of 10^6 at the detector.

With other types of ADC, the output is less linear and less accurate, but is amenable to multiplexing, *i.e.*, to handling more than one ADC-equipped chromatograph with a single data processor [45, 49, 51].

The sampling rate is the result of an engineering decision that balances precision against cost and need. From their definitive study, Chesler and Cram state that "a peak which has a base width (at the 0.10% limits) of 10.0 seconds should be tracked by an ADC clock rate of 10 Hz. For very sharp peaks, such as those from capillary columns, conversion rates of up to 200 Hz may be required" [41]. On the other hand, Busch concludes more typically that for "1 sec wide Gaussian peaks (measured at half height) a band width from DC to 5 Hz will give an area and peak height distortion of less than 1%" [49]. One well-known dedicated computing integrator has a maximum sampling rate of 10 samples/sec [51].

B. Algorithms

An algorithm tells the data processor what to do with the digital data presented to it. ("An algorithm is a prescribed set of well-defined rules or processes for the solution of a problem in a finite number of steps" [55].) Algorithms for gas chromatography can be extraordinarily difficult to devise.

Speed and consistency apart, computers often do not perform so well as humans on complex, judgment-requiring tasks. The human brings far more to the task than is generally realized. Much of it cannot be well expressed in a computer's only source of judgment, an algorithm.

Consider, from among the many problems common to interpreting chromatograms, two easy ones. For one instance, interpreting a chromatogram may include recognizing and identifying peaks even though they have shifted with respect to a reference chromatogram. They may have shifted because the flow rate was lower than usual, or because the flow rate changed in the middle of the chromatogram. Or they may have "shifted" simply because the operator delayed in indicating the moment of sample injection. In another instance, a curving baseline may have to be drawn so as to measure the area of a peak above that baseline.

Such problems cause little trouble to a chromatographer because to them the chromatographer brings human capabilities. These include insight, a sense for what is probable, an aesthetic feeling for form, and a technical background. These various capabilities are applied with an easy and largely unconscious fluidity.

The algorithm, however, must express the application of these capabilties in explicit, elaborate detail. First, of course, the programmer must realize that the capabilities exist and analyze just what they are. They then must be translated into mathematical terms, and finally retranslated into the algorithm. As Fozard exclaimed "In trying to determine exactly what criteria the human operator uses in such circumstances [here, the apportioning of peak areas], one quickly realizes what a remarkable processor the human brain is. [While under] a given set of conditions various operators will use different methods, the computer, simple brain that it is, will only do what the programmer tells it" [56].

Sec. 14.9 Digital Technology Applied to Gas Chromatography

Nevertheless, computers are necessary and therefore algorithms, if imperfect, must be devised. First we consider several with respect to individual data-handling processes connected with digitizing gas chromatograms and with processing already digitized gas chromatograms. Finally, we review one high-precision algorithm for maximizing the information available from GC data.

If we are to do good gas chromatography with modern equipment, we must get a well-informed feeling for what algorithms can and cannot do.

Noise

Weimann has described a reasonably representative procedure for noise determination: "Nearly all algorithms are very dependent on...the noise-level...superimposed on the signal...[To determine the noise,] the average absolute difference between two consecutive points in the first 100 points is determined...[Of these differences,] only those are [averaged that] are less than four times the average absolute difference" [36].

Spikes

Spikes are differentiated from peaks by specifying a maximum spike width (usually 1 data point) that is less than a specified minimum peak width [36, 51], and a minimum spike height. Once identified, spikes are eliminated from data used to determine a baseline.

Data Smoothing

Decisions in data processing are based on smoothed rather than raw data. The choice of smoothing routine is therefore important. The digital data emanating directly from the ADC are smoothed either by least-squares fit [52] or group averaging (Fig. 14.11)[7, 17, 18, 51, 57].

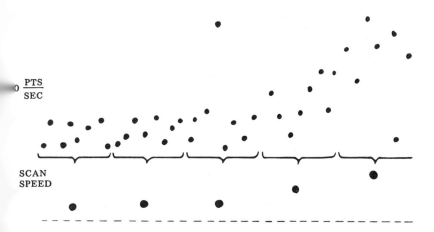

FIGURE 14.11 Groups of data points are averaged or otherwise smoothed to diminish the effects of noise. The number of points per group can be changed to reflect the current peak width. (Reproduced with permission of Marcel Dekker, Inc. from D. Ford and K. Weihman, Ref. 18.)

The duration of a single group is usually limited to no more than one-tenth the half-height peak width. That width may be either the current [18] or the narrowest expectable [51,54]. In the cited examples of the latter case, successive 0.1-sec integrals of the raw data are accumulated. It is these integrals that are then stored for further processing.

Baseline and Drift

Nominally, the lowest points of the chromatograms are taken as the baseline. (Close inspection, however, reveals that these lowest points are a function of sample size. Hence the unpalatable conclusion: "An acceptable analysis can [then] be obtained only if extreme care is taken with dual, matched columns to obtain a relative linear baseline which is independent of sample size and with injection procedures to ensure sample size reproducibility" [54]. These preconditions are general but they are not generally recognized. Unimposable, they are beyond the reach of any algorithm but, unless met, vitiate quantitation.)

"The largest source of error in the integration of a peak area is usually the uncertainty in the baseline determination," said Weimann, "and therefore we applied a thorough method for baseline detection" [36]. To be determined, the baseline must exist in usable segments near the peaks to be integrated. This is why the baseline "is not always easy to determine," even with a large, powerful computer [58]. Van Rijswick also makes this point, as we shall see at the end of the treatment of his algorithm. The lesson is presented repeatedly:

Computers yield good results only from good separations.

In Weimann's algorithm for baseline detection, 9 consecutive data points may be taken as 1 baseline point if the 8 consecutive differences among them do not exceed the noise level by more than some chosen factor, usually 4: "If the nine points are really on the baseline and if the noise exhibits Gaussian distribution then for [a factor of four] there is a probability of 97.3% that all eight absolute differences are smaller than [four times the noise level]" [36]. The absolute difference between the first and the ninth point must also not exceed the noise level by more than the chosen factor, or drift may be considered present.

Drift is defined in the Weimann algorithm as an upward rate of displacement from one baseline point to the next. (As just mentioned, 9 not-too-noisy data points comprise 1 baseline point.) The allowable drift, a value that is chosen with each run, is less than that taken as the beginning of a peak. Thus the larger the allowable drift, the less sensitive the detection of peaks [36].

Downward baseline displacement is allowable at any rate [36].

Finally, within a given set of consecutive baseline points, only the first and last are stored for later use in defining that baseline segment [36].

In the algorithm reported by Fok and Abrahamson [48], only data already smoothed by a least-squares routine [52] are used. The number of points per group specified this algorithm is related to the digital filter time constant. A 9-point group is suggested as typical. Free downward drift is allowed: "As long as the baseline keeps dropping, the baseline is reanchored at the lowest point" [48]. Otherwise, as long as the first and second (upward) derivatives are zero for the number of points specified in the routine, the signal is considered to be on the baseline.

Peak Detection and Definition

Peak start. Peak starts are detected when the first [36, 51] or the first and second [18, 48] derivatives of the data exceed assigned thresholds for assigned times.

In the visualization of a peak, as in the tracing of it by a recorder, the first derivative dy/dx calculated for point x shows the slope of the tracing at that point. The second derivative d^2y/dx^2 for that point shows the concavity of the tracing at that point.

In digital data processing, derivatives are calculated from the progression of successive group values. Each group has an assigned number of data points (say, five [36]). The value for a given group is found by applying to its data points one or another smoothing routine, usually that of Savitzky and Golay [52].

The succession of groups is obtained by forming a new group with each new data point. As the new data point is taken into the group, the oldest data point is dropped [18].

With digital data groups, then, the first derivative reflects the ongoing change in group values. The second derivative reflects the ongoing *rate* of change in group values; this is the same as the ongoing change in the values of the first derivative. (In actual algorithms, the second derivative is sometimes calculated separately from the first derivative [48].)

In one algorithm for the detection of peak starts, the first derivative--expressed as signal--must exceed twice the noise level--also expressed as signal--for three consecutive 5-point groups (a duration of 3 data points)[36]. In another algorithm, the slope must merely exceed a slope threshold preset by the operator [51]. In still other algorithms, both the first and the second derivatives must exceed preset threshold values for a preset number of consecutive times [18, 48].

To offset any increase in peak width during a run, the number of data points per group may be increased during that run [18, 51]. This gives more detailed scanning for the earlier, sharper peaks, and better slope sensitivity--better detection--for the later, broader peaks.

Peak maximum. At the peak maximum (Fig. 14.12), the slope decreases to zero after having been positive and before becoming negative. At the peak maximum, the downward concavity reaches a maximum.

In Weimann's algorithm, the peak maximum is considered found "when the product of two consecutive first derivatives is smaller than or equal to zero. The first of these two derivatives must be positive and the derivative of the next point after these two is not allowed to be positive"[36].

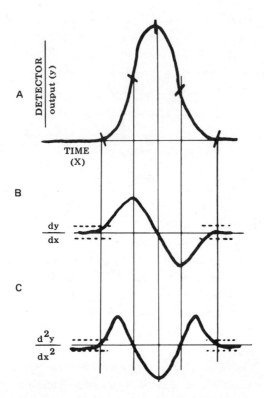

FIGURE 14.12 The first (B) and second (C) derivatives of the peak (A) are frequently used to calculate the beginning, top, and ending of a peak. (Reproduced with permission of Marcel Dekker, Inc., from D. Ford and K. Weihman, Ref. 18.)

Peak end. As illustrated in Fig. 14.12, the end of a peak is considered established when the first derivative, which had been negative, becomes nearly zero for a long enough time. "Nearly zero" may mean less than or equal to twice the noise level; "long enough" may mean lasting over 3 consecutive data points [36].

Peak Area

Once the perimeter of a peak--the start, the envelope, the end, and the bottom-- have been defined, integration "is carried out by the simple addition of y-coordinates" [36]. The start and end points define the baseline correction (Fig. 14.13) [

Area Apportionment

In any but the simplest gas chromatograms, resolution is incomplete. Peaks overlap. Other peaks may ride a solvent tail.

Fozard (among many others, for instance Maggs and Mead [59]) has described three cases: The resolved peak with baseline on either side (Fig. 14.14)[56], the pair of peaks that are not quite resolved (Fig. 14.15)[56], and the shoulder peak (Fig. 14. [56].

Sec. 14.9 Digital Technology Applied to Gas Chromatography   369

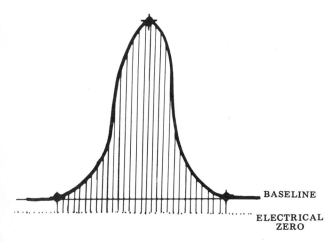

FIGURE 14.13 Once the beginning and end of a peak have been defined, the peak is integrated by the addition of its y-coordinates. The baseline correction is of the greatest importance in peak area determination. (Reproduced with permission of Marcel Dekker, Inc., from D. Ford and K. Weihman, Ref. 18.)

From the bottom of the valley between two not-quite-resolved-*but-nearly-equal* peaks, a perpendicular can be dropped to the interpolated baseline. Such a perpendicular nicely apportions the area between two such equal peaks (Fig. 14.15).

Unfortunately, the difference between the case of Fig. 14.15 and that of Fig. 14.16 is one merely of degree. As the size discrepancy between two peaks increases, one peak begins to ride the other, begins to become only a shoulder peak. A dropped perpendicular does not divide them equally (Fig. 14.17)[59].

Indeed a true shoulder peak requires not a dropped perpendicular but a separate baseline. The new baseline of a shoulder peak is the side of the peak it rides. (Note also, as we all know: Peak sides are curved, not straight tangents [54].)

An adequate area-apportioning algorithm must determine, must specify before the fact, when to abandon dropping perpendiculars and start "skimming." Skimming is using a tangent to the side of the larger peak as the new baseline for the shoulder peak. Each approach (dropping perpendiculars, skimming) involves errors that mount sharply with the peak height ratio [59]. These errors are always higher for skimming, at least theoretically.

Skimming is accepted not on grounds of theory but because it obviously yields less error--when it is applicable [59]. Hettinger, for instance, described an algorithm that calls for skimming if the larger peak is found to be skewed (Figs. 14.18, 14.19)[51]. To implement skimming, Fozard suggested looking either for a maximum rate of signal change before and after the smaller peak, or for a maximum area of the smaller peak as a function of the drawn-in tangent (Fig. 14.16)[56].

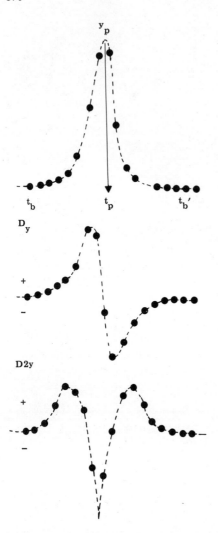

FIGURE 14.14 Figures 14.14, 14.15, and 14.16 are taken from the paper of Fozard *et al.*: "A typical resolved peak shown with the first and second derivatives of the curve. The points at which the computer samples the curve are shown by the large dots." (Reprinted with permission from A. Fozard, J. J. Frances, and A. J. Wyatt, Ref. 56, Pergamon Press, Ltd.)

Both Weimann [36] and Fozard [56] spelled out elaborate criteria to be incorporated within their area-apportioning algorithms. Both algorithms had been found highly usable. Nevertheless, Weimann commented, "Algorithms for peak-overlapping, shoulder detection etc. are very dependent on the shape of the peaks and therefore of the columns and compound class. Our experience suggests that a general solution is therefore hardly likely" [36].

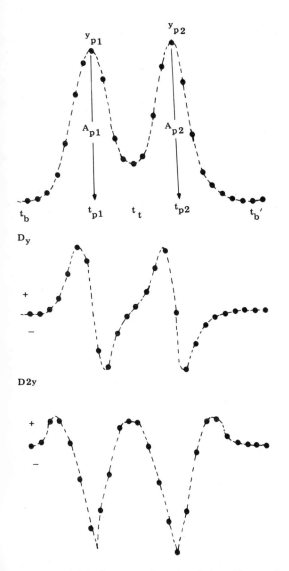

FIGURE 14.15 "Two overlapping but well resolved peaks. In this case area apportionment is made by dropping a perpendicular at the valley position." (Reprinted with permission from A. Fozard, J. J. Frances, and A. J. Wyatt, Ref. 56, Pergamon Press, Ltd.)

Presently, we discuss another approach to area apportionment: Curve fitting. But accurate curve fitting, we shall see, requires that all the peak shapes within a group of unresolved peaks be known and be mutually alike. Weimann's comment suggests that this ideal is not realizable in practice.

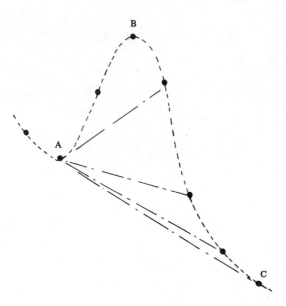

FIGURE 14.16 "Tangential skimming by computing the maximum skimmed area from the valley position. Point at which maximum skimmed area is obtained is C." (Reprinted with permission from A. Fozard, J. J. Frances, and A. J. Wyatt, Ref. 56, Pergamon Press, Ltd.)

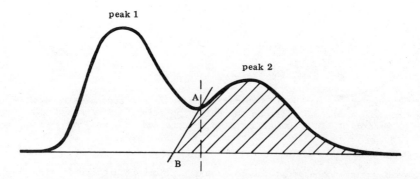

FIGURE 14.17 "The errors associated with perpendiculars and skimming are significant when the peak height ratios differ greatly from unity and the 'imposed cutting point' technique is found to be the better alternative. This method [which is] dependent on resolution and relative peak heights...is illustrated [here], where the line AB gives the optimum area distribution between the two peaks for the minimum error on both absolute and relative peak areas." The shaded area is allocated to peak 2, the unshaded to peak 1. (Reprinted with permission of the Institute of Petroleum from R. J. Maggs and A. S. Mead, Ref. 59.)

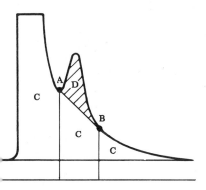

- PEAK START
- TANGENT CONDITION DETECTED
- CORRECTED MAJOR PEAK AREA
- CORRECTED TRACE PEAK AREA

FIGURE 14.18 "The tangential baseline correction is performed automatically and is brought into operation by either the skewed peak algorithm or the manual tangent action (described below). The skewed peak algorithm works by measuring the time of the peak downslope relative to the peak width parameter. If this downslope time is too long, the peak is determined to be skewed, and the logic for tangential correction is set. All peaks riding on the skewed peak tail will be split off." (Reproduced with permission of Preston Publications, Inc. from J. D. Hettinger, J. R. Hubbard, J. M. Gill, and L. A. Miller, Ref. 51, in the Journal of Chromatographic Science.)

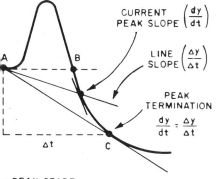

- PEAK START
- TANGENT TEST INITIATED
- TANGENT CONDITION DETECTED

FIGURE 14.19 In this algorithm (the same as in Fig. 14.18), the "zero slope...Point A is taken as the peak start...[Later,] when the analog signal on the trailing side of the trace peak drops below the trace peak start level, at point B, the firmware begins seeking a tangent condition. When the current slope of the backside of the peak (dy/dt) and the slope of the line through the peak start and the current level (Δy/Δt) are equal at point C, the tangent condition is met, and the end of the peak is detected." (Reproduced with permission of Preston Publications, Inc. from J. D. Hettinger, J. R. Hubbard, J. M. Gill, and L. A. Miller, Ref. 51, in the Journal of Chromatographic Science.)

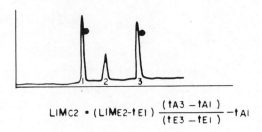

$$LIM_{C2} = (LIM_{E2} - t_{E1}) \frac{(t_{A3} - t_{A1})}{(t_{E3} - t_{E1})} - t_{A1}$$

where ✱ = reference peak
LIM = high or low time band limit
t = time in seconds
C = corrected
E = expected
A = actual

FIGURE 14.20 This shows the reasoning of an algorithm for the automatic identification of peak 2 with respect to reference peaks 1 and 3. (Reproduced with permis of Marcel Dekker, Inc., from D. Ford and K. Weihman, Ref. 18.)

Peak Identification

If response factors are to be used for quantitation, then they must be correc assigned. Such assignments first require accurate peak identification. The whole must be completely automatic, of course, or the worth of computer data handling is severely vitiated.

Retention times vary, particularly with the more complex samples. Therefore multiple reference peaks may often be used in routine analyses. These arise from major components that are known to occur reliably in each sample of a given type. The retention times of nearby peaks are then pragmatically related to the retentio times of the reference peaks:

> For example, if the first reference peak eluted 20 sec later than defined, th second reference peak would be expected 20 sec later also. [If] it instead w 10 seconds early, the third would be expected 10 seconds early, and so on...A the conclusion of a chromatogram as signaled by the operator or as defined in the computer control of the analysis, the number of reference peaks detected compared with the number defined. If one or more have been missed, no peak i dentification takes place and only a listing of retention times and area percentages for each peak is printed because no identification at all is better misidentification of peaks due to problems with reference peak retention time
>
> If all reference peaks have been detected, the remainder of the detected peak are identified. The time zones for each peak as defined by the chromatograph are modified according to the retention times of reference peaks on each side the zone as shown by the formula in [our Fig. 14.20]. Peaks which do not fal any defined time zone are called unknowns and are so reported.
>
> As peaks are identified, each can be treated as specified; sensitivity factc can be applied, peaks can be put in groups by type even if not adjacent in elution time, and specific baseline corrections can be made" [18].

Sec. 14.9 Digital Technology Applied to Gas Chromatography

High-Precision Off-Line Processing of Chromatograms

On-line (single-pass) algorithms call for logical decisions that would preferably involve a more considered view of a larger segment of the chromatogram. Off-line algorithms can take such a view. To gain depth in assessing what computers can and cannot do in gas chromatography, we now review an off-line algorithm of higher-than-normal adaptability and precision.

This algorithm was designed to extract from already digitized chromatograms "the maximum amount of information from each chromatogram:....no peaks neglected *a priori* ...low detection limits...optimum information and precision of the peak parameter estimation...completely automatic [data] processing [to obviate reliance on] operator skill" [60]. This work was reported by M. H. J. van Rijswick of the Eindhoven University of Technology.

Flow chart. "The logical framework is the crucial part of the algorithm" [61]. The flow chart shown in Fig. 14.21 [60] expresses this framework, which emphasizes the adaptability of the algorithm in the placement and nature of the inspection stage. In this inspection stage, the noise level and peak widths are continuously determined for use in setting peak thresholds and matched filter characteristics, respectively. (The peak thresholds and matched filter characteristics are described presently.) In detection, spikes having been first filtered out, normal, trace, and overlapping peaks are located. In estimation, the peaks are delineated and baseline corrections are applied so that areas can be assessed and reported.

The matched filter. The incisive tool used in the algorithm is the *matched filter*. In order of importance, there are three points to be made and understood about the matched filter in this algorithm: "The main point is the working principle of a matched filter. The second point is [the technique of modifying the matched filter] to cope with a signal that includes a baseline. The third point is [the twofold technique of optimizing] this modified [matched] filter with respect to [either] signal-to-noise ratio [or] resolution for the detection of [either] trace [or] overlapping peaks," respectively [61].

The matched filter is a mathematical technique. In it, the pattern of the signal is continuously searched over a moving time interval by comparing it with a model shape that fills a time interval of related duration. In this algorithm, the model shape is the Gaussian peak, mathematically expressed. One of the parameters of this expression is peak width, which becomes the peak width w_f of the matched filter. The matched filter width w_f is varied according to the purpose at hand within the algorithm.

To cope with a signal that includes a baseline, the matched filter is modified. The pattern of the signal is compared with the second derivative of the model Gaussian peak rather than with the model peak shape itself. This eliminates from the output of the matched filter any contribution from a linear baseline. Nevertheless, a threshold

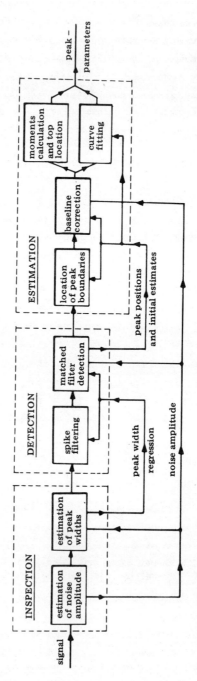

FIGURE 14.21 This is the flow chart for the high-precision van Rijswick algorithm. (Reprinted with permission from M. H. J. van Rijswick, Ref. 60, Pergamon Press, Ltd.)

Sec. 14.9 Digital Technology Applied to Gas Chromatography 377

must be imposed to screen out the contribution from random noise. Here we begin to deal with optimizing the matched filter with respect to signal-to-noise ratio for the detection of trace peaks.

Background limits. The "threshold level for sorting out spurious peaks [depends only] on the random noise amplitude and on the attenuation of the noise by the filtering" [60]. That attenuation is set by the sampling interval Δ from one data point to the next, and by the matched filter width w_f. The relationship between the random noise before (rms) and after (rms*) filtering can be shown to be as follows, where the constant equals $\sqrt{3/8\sqrt{\pi}}$:

$$\text{rms*} = \text{constant rms } \sqrt{\Delta/w_f^5} \tag{14.7}$$

"In the program the threshold level was set at 5 rms*" [60].

"For this threshold, it follows that, in order to be detectable, a peak [of width w_p] must have [the following] minimum signal-to-noise ratio before filtering" [60]:

$$(S/N)_{\min} = 5 \sqrt{\Delta/w_p} \tag{14.8}$$

After filtering, "optimum detection...is attained for the filter width w_f that maximizes the signal-to-noise ratio" at that point. This signal-to-noise ratio after filtering is defined as

$$(S/N)^* = \text{height of the filtered peak/rms*} \tag{14.9}$$

To maximize this ratio and thus to optimize peak detection, it can be shown that the modified matched filter width should be set at $w_p\sqrt{5}$.

To keep the filter width matched to the current peak width, the peak width is monitored constantly. When the peak width is found to have changed by 10%, the matched filter width is changed accordingly.

The filtered signal is recalculated with each new data point [60].

Equations (14.8) and (14.9) both suggest that the narrower a peak is for a given area, the easier it is to detect. Nevertheless, it must also have a given minimum area $A_{\min}$:

$$A_{\min} = 12 \text{ rms } \sqrt{\Delta} \, w_p \tag{14.10}$$

Given these conditions, the "start t_s of a peak is detected when the filtered signal exceeds the threshold for the first time; the end t_e when the signal returns below the threshold. The width of the detected peak is provisionally estimated as

$$w_p = \frac{(t_e - t_s)/2}{1 + K^2} \tag{14.11}$$

where $K = w_f/w_p$; the location is estimated from the maximum in the filtered signal between t_s and t_e; the area A is estimated from the amplitude h at the maximum:

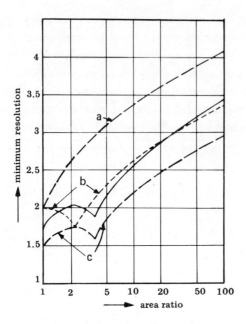

FIGURE 14.22 The "fusing limit" is the minimum resolution at which two peaks can be detected as being separate peaks. This figure shows the fusing limit as a function of the area ratio of two Gaussian peaks of equal width. Curves a, b, and c show the fusing limit loci for peak detection by peak maxima, inflection points, or second derivatives, respectively. The solid curve shows the locus for the van Rijswick algorithm. (Reprinted with permission from M. H. J. van Rijswick, Ref. 60, Pergamon Press, Ltd.)

$$A = \frac{1}{8} h (t_e - t_s)^3 \sqrt{2\pi}$$

These estimates are used as initial values in the iterative curve fitting in the case of overlapping peaks" [60].

We now turn to the optimizing of the matched filter with respect to resolution, for the detection of overlapping peaks.

Fusing limits. As usual, the resolution R_s is defined as the center-to-center separation Δt_R of two peaks divided by their average peak width w_p: $R_s = \Delta t_R / w_p$.

The *fusing limit* is defined as "the minimum resolution between two peaks at which the two can be detected separately" [60].

The fusing limit varies with the area ratio of the two peaks and also with the method of detection. As shown in Fig. 14.22 [60] for fused noise-free Gaussian peaks, detecting either inflection points (curve b) or, better, second derivative minima (curve c) yields lower fusing limits than merely distinguishing peak tops (curve a); the resolution can be smaller, yet the peaks still detected as being separate. Unfortunately, second derivatives are very sensitive to noise, which is always present in real signals. The characteristics of the modified matched filter fit nicely here.

The modified matched filter is "equivalent to the smoothing of the second derivative as the signal" [60], giving results like those shown in curve c, although at the expense of an apparent loss in resolution. However, the filter can be made to yield either optimum detection sensitivity with diminished resolution, or diminished sensitivity with optimum resolution, i.e., minimum fusing limit. Both were done. A two-step procedure was adopted.

As already mentioned, the modified matched filter has optimum sensitivity for detecting single peaks against background if used with $w_f = w_p\sqrt{5}$. When used with $w_f = 1.4 w_p$, as it was routinely to search for peaks in the first processing of each chromatogram, only 10% of this optimum sensitivity was lost. Around large peaks that had been located in the first pass, the modified matched filtering was then repeated using $w_f = 0.5 w_p$ to obtain a lower fusing limit in searching for fused peaks. The result of the second filtering is shown in the solid curve in Fig. 14.22. That curve is only 10% higher than the theoretically obtainable minimum fusing limit.

Baseline correction. The placement of the baseline correction--the interpolation of the baseline under a peak--has the greatest importance in fixing the area of that peak. The baseline correction is determined from baseline segments outside the peak boundaries.

In this algorithm, the baseline correction was generated locally "by fitting a polynomial to the bracketting background segments" adjacent "to each peak or group of overlapping peaks" [60]. (See also the technique of Caesar, which calls for a more extended interpolation of the baseline [62].)

This approach requires accurate setting of peak boundaries. Therefore, the slope of the background was assumed constant adjacent to each peak group. "The peak boundaries are then defined as the point on the leading edge and the point on the trailing edge, at the smallest distance, where the signal slopes are equal. These points are readily located by alternately calculating at each side the slope at the next point, starting from the inflection points" and moving away from the center [60].

Peak parameter estimation. The measurement of peak area was optimized by extending the peak integration limits outward on each side, one point at a time, starting from the provisionally calculated center of gravity of the peak. The extension of the integration limits is continued "until the increment of the integral is less than the change in the standard deviation" [60].

The retention time is measured from the peak top. The peak top is located by finding the maximum of a parabola that is fitted to it. The number of data samples to use in fitting the parabola is determined by finding the minimum error in the fit as the number of samples is increased. By automatically keeping the fitting region small, this approach minimizes that biasing of the parabola that tends to occur with asymmetrical peaks.

Curve fitting. Computers can sum individual peaks in relevant proportions so as to fit the overall envelope of a group of overlapping peaks. The proportions found necessary constitute area apportionment, without dropping perpendiculars or skimming.

In curve fitting, the computer is first provided with a given peak shape. The computer then repeatedly subtracts combinations of the model shape from the curve to be fitted. The process is continued until the fit is satisfactory. Unfortunately, curve fitting requires that the peak shape of all the peaks under a given envelope be known and constant. This requirement is not realistic in gas chromatography.

The peaks within a given group of unresolved peaks may not have the model peak shape [64]. Indeed, the peaks may not have either the model or any given peak shape: "...if you look at the real world you will find...a different skew factor for almost every function or group" [65]. In his comments on area apportionment algorithms, Weimann has already been quoted here to the same effect [36].

The computer could handle the technical problem easily enough. If the computer were only given a variety of peak shapes to start with, it could quickly produce numerous mixes any one of which would precisely fit the envelope. But which mix would be the correct one? Indeed, would any of them be correct? "The less known about the [peak] shape, the less significance can be attributed to the results...[With] a completely general [peak shape] model a given composite curve can be apportioned in an unlimited number of ways" [60].

The approach to curve fitting in van Rijswick's algorithm is conservative: To "look for a [peak] model that is both accurate and has as few degrees of freedom as possible" [60]. Four peak models, starting with the Gaussian, are to be tried one at a time, each model to apply to each peak within a group. The goodness of the fit and whether it shoudl be improved are assessed on statistical grounds. A given attempted fit may be improved by "the introduction of an additional peak or the extension of the Gaussian model to one of the three other models" [60].

An example of such curve fitting is shown in Fig. 14.23 [60]. Van Rijswick observed that in practice, "...curve fitting is relatively time-consuming and, therefore, only done [in]...regions in the chromatogram...where [it is] explicitly desired" [60].

We digress briefly from the van Rijswick algorithm for a final comment on curve fitting. Littlewood, Gibb, and Anderson began their curve-fitting investigations with the announced intention of "replacing...chromatography as far as possible with mathematics" [66]. But they later warned, to their credit,

> Curve-fitting techniques should be used...for quantitative analysis...only...if each and every peak to be fitted has been reliably identified before the regression analysis is used to fit parameters to them, since it is almost invariably possible to produce a virtually perfect fit to smooth data by fitting an arbitrarily large number of points to it. This important point is illustrated by [our Fig. 14.24], which shows an isolated instance in which the procedure PEAKFIND has falsely found a spurious peak appearing just before p-xylene. It is seen from the error line that the fit is excellent, and if the percentages for the spurious peak and the real p-xylene peak are added together, an excellent analysis is obtained. However, the spurious peak is not there!" [63]

Now let us return to the van Rijswick algorithm, and conclude our consideration of it.

Sec. 14.9 *Digital Technology Applied to Gas Chromatography* 381

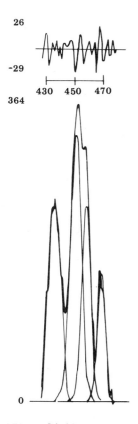

FIGURE 14.23 An example of curve fitting by the van Rijswick algorithm. In the lower figure, the thick line shows the baseline-corrected signal, and the thin lines show the fitted peaks. The upper figure shows the "residue between the signal value and the sum of the fitted peak contributions." (Reprinted with permission from M. H. J. van Rijswick, Ref. 60, Pergamon Press, Ltd.)

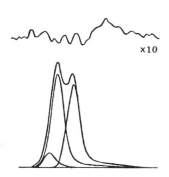

FIGURE 14.24 Curve fitting is not really trustworthy. Here, a good fit was attained, but the smallest of the three peaks was spurious. The upper line shows the error in fit, multiplied in this case by 10, but does not suggest the error. (Reprinted with permission from A. H. Anderson, T. C. Gibb, and A. B. Littlewood, Ref. 63. Copyright by the American Chemical Society.)

A few results of the van Rijswick algorithm. Figure 14.25 [60] shows a computer replot of a chromatogram from the raw digitized data *versus* a five-fold enlarged plot of the same chromatogram after baseline correction. A corresponding table customarily accompanies such a chromatogram. To offset the overconfidence often associated with computer printouts, this table includes a column giving "the estimated standard deviation or relative error" for each peak [60]. (For some of these figures, see Sec. 14.10A, Peak Measurements.)

In general, the results of the algorithm we have reviewed were found highly satisfactory. There was one basic limitation in the case of complex samples: "[In] some natural samples the chromatogram is completely crowded so that no signal segments, other than at the start and the end of the chromatogram, can be found for fitting the baseline. [In such cases,] the connection of valleys between peaks [must be used, but] this method is demonstrably lacking in accuracy in some cases" [60]. We shall see that this occasionally demonstrably inaccurate method is used routinely in most digital data processors, described next.

C. Digital Data Processors in Gas Chromatography

Memory distinguishes digital/nondigital data processors. Digital data processors place incoming digitized data into memory in digital form. From memory, these can be recalled and digitally processed in the ways we have just been reviewing.

Digital data processors differ markedly from each other in memory size, and therefore in programs. Memory size limits the changeability and extent of the programs that direct the processors.

In the least expensive processors--the computing integrators--a minimal program is embodied in permanent circuits. In the processors of middle expense range--the small or dedicated computers--the program, although still limited, is much more elaborate and also changeable. (However, such changes can usually be made only by the vendor, as opposed to the user.) In large processors, the programs and operations are limited primarily only by the cost of computer time; technically, the programs may be as elaborate as can be arranged and can be revised at will.

The Computing Integrator

It is in its use of digital technology that the computing integrator differs fundamentally from the "digital," analog-logic integrator.

For its power, presented diagrammatically in Fig. 14.26 [67], the computing integrator is astonishingly inexpensive. Its low cost reflects the commercial arrival in the 1970s of mass-produced, large-scale integrated circuits [38]. These correspond functionally to circuits with thousands (10^3 to 10^4) of transistors [68].

Early computing integrators furnished a simple printout of run parameters, retention times, and areas. Concentrations were calculated manually by means of an internally housed calculator. (The ability of a computing integrator to unload data

Sec. 14.9 Digital Technology Applied to Gas Chromatography   383

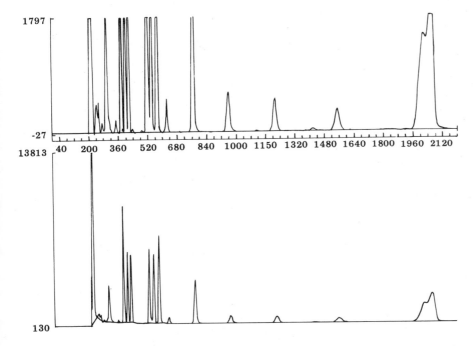

FIGURE 14.25 The plot output of a chromatogram processed by the van Rijswick algorithm. "Below: raw chromatographic signal and the fitted baseline. Above: baseline-corrected signal on a 5-fold enlarged scale." (Reprinted with permission from M. H. J. van Rijswick, Ref. 60, Pergamon Press, Ltd.)

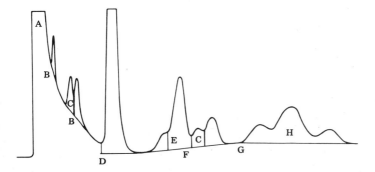

FIGURE 14.26 This figure shows the automatic interpretation of a chromatogram that is possible with a computing integrator. (A) Elimination of solvent peak; (B) Tangent skim of rider peak; (C) Fused peak split; (D) Reset to original baseline; (E) Shoulder peak split; (F) Trapezoidal baseline correction; (G) Automatic sensitivity increase for broader peaks; and (H) Integration of several peaks as a single area. (Diagram reproduced courtesy of Spectra-Physics [67].)

to a computer for postrun calculations is interesting, but problematic in utility--
the nominal availability of a computer does not mean either that it is really available or that the purchaser wishes to use it.)

The Small On-Line Computer

In the succession of devices and methods for handling data, the small on-line computer marked a fundamental break. Only with its arrival did the operator become freed from manual involvement. Only then did the cost in operator time of peak integration become independent of the complexity of the analysis (Fig. 14.27)[69].

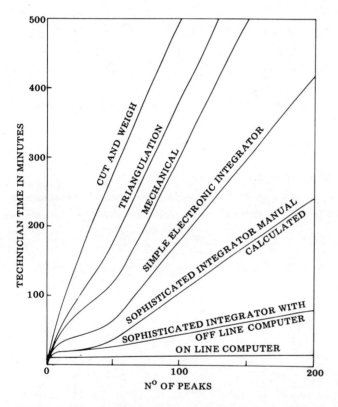

FIGURE 14.27 Chromatograms with 200 or more peaks are typical with natural mixtures. Integrating peak areas with such chromatograms tends to demand on-line computer handling of the data. (Reprinted from Lyons, Ref. 69, with permission of the author and The Institute of Petroleum, and by the courtesy of the Unilever Research Port Sunlight Laboratory.)

Before we proceed, we emphasize: This giant step did not in itself improve analytical accuracy (e.g., Ref. 42). However powerful, however consistent, a computer does only what it is told. The algorithms that direct computers do so clearly, but not always particularly well, despite the ablest of programmers. Area-apportioning algorithms, for example, simply do not do well with groups of closely fused peaks of the widely differing sizes that are to be expected. For another example, algorithms for the baseline corrections that are so crucial for accurate area measurement cannot furnish them accurately without clear adjacent baseline segments. As those who have most carefully devised them have repeatedly pointed out to us, *algorithms cannot supplant good separations*. And, as we shall see, no algorithm can help an outdated calibration [70]. The computer frees the analyst from manual involvement, *but not from responsibility*. Now let us go on.

As already mentioned, memory size determines computer power. What memory size is typical for the small on-line GC computer?

In 1969, Baumann of Varian described an on-line GC computer with a 4,096-word core memory, with 16-bit words. This memory could be expanded to 32,768 words. "An 8-k memory is required to process 10 GC channels simultaneously. However, a 4-k memory is adequate for fewer channels or for less complete processing of 10 channels..." [17].

In 1970, Collyer of BP Chemicals reported on the installation and use of a similar system using three 4,096-word units, expandable to four [45]. Deans used a 16-k core computer [71]. And in 1972, the Perkin-Elmer GC Data System used either 8,192 16-bit words or 12,288 16-bit words, depending on the number of chromatographs to be accommodated (up to eight) and the types of programs to be implemented.

Teletypewriters are the typical means of input to the computer and of output-report generation. Thus the reports can have as many columns as the typewriter can accommodate and as many lines as desired. Thus, the nature of the report also depends on the program. The user may or may not be able to change the program, depending on whether the program is proprietary, *i.e.*, whether the system was purchased intact. In any event, with such systems the programs are at least potentially variable and improvable.

The reports from these systems typically show first the parameters of the analysis and the program, *e.g.*, the instrument, the analytical method, the slope and noise thresholds if they are to be nonroutine, the reference peak designation, the type of report desired. The analysis then follows, giving a column-by-column listing of peak numbers, retention times, peak areas, area percents, and peak identification. The corresponding response factors and the quantitative analysis may also be reported in the first or immediately subsequent report. In sum, the report sets out whatever the user decides will be relevant and desirable.

Satellite and Large Computers

Satellite computers are small on-line computers of the type just described. However, they are also linked to a large, central computer. "The function of the satellites consists in process control, data acquisition and data prehandling (smoothing, data reduction etc.)" [72].

"The term 'satellite system' is used for a computer hierarchy in which the function of data evaluation is not directly coupled with the data acquisition and is set up at a separate place. The satellite computer at the place of data acquisition is as small as possible" [73].

The central computer has in its memory the program and data files that become available to the satellite computers on request. The central computer performs the data calculations and evaluation. The input-output devices such as teletypewriters are also connected to the central computer.

The arrangement is flexible rather than rigid, accommodating varying types of needs.

In any event, the programs become limited by the expense of computer time rather than by the size of the computer memory. Therefore the programs available to the user become those of a large computer, whether or not that computer is addressed through a satellite system.

These programs can then be as elaborate as desired. For instance, the high-precision van Rijswick algorithm to which we devoted such attention (Sec. 14.9B, High-Precision Off-Line Processing of Chromatograms) requires a large computer.

Also, such programs are no longer likely to be proprietary. Therefore they can be (and are) improved frequently, "modified easily corresponding to...varying analytical demands" [74]. Baudisch, for instance, described having the computer carry on a wide variety of "plausibility checks" on computerized analyses so that mistakes from whatever source can be detected and corrected, rather than hidden by overconfidence in the computer [75].

D. Automation

By "automation" we mean the complete or nearly complete handling of analyses by automatic means. These means are usually under the control of a computer.

We shall now consider automation in gas chromatography. We shall do this from the point of view of the analysis to be performed. First we consider the very-well-controlled analysis, second the general analysis.

The Very-Well-Controlled Analysis

Deans [70] and Price *et al.* [76] have described examples of "very-well-controlled" analyses. These are automated "key analyses" that, if temporarily either incorrect or unavailable "can cost, in terms of lost or spoilt production, more than the price of the analytical system" [70].

Before they are automated, the gas chromatographic separations in such analyses have been made complete, the associated automated operations--sampling, for instance-- have been carefully tested for reliability and reproducibility, and the gas chromatograph itself has been tested rather than trusted. Deans is especially instructive:

> Drift [in the calibration]...is overcome by regular calibration with mixtures o known composition. It is important that the calibration mixtures are run on th GC under exactly the same conditions, including sensitivity settings, as the samples. This calibration avoids, in our experience, the unjustified assumptio that the relative sensitivities shown on the GC panel are accurate. Calibratio

also avoids the need to use response factors from the literature, which, in our experience, can be far from the true value for any particular instrument [70].

Given these precautions, the computer is (or at least can be) further programmed to be vigilant in expecting, detecting, and even diagnosing trouble. Checking not merely the nominal facets of the operation--temperature, flow rate, etc.--the computer also monitors what the run is, as opposed to what it should be. For instance, if the baseline is not predictably close to zero at certain times, "something has gone wrong and precise quantitative analysis cannot be expected" [70]. The computer monitors the peak height of the internal standard, the separations of given pairs of peaks, the retention times of all the peaks--in sum, the computer makes constant use of all the information that constitutes "prior knowledge of the analysis" [70].

In the cases reported by Deans, the actual analysis printed out contains "only the information required and no more" [70]. On the other hand, if "the analysis is outside specifications, the specification limits and a series of asterisks are printed out on the right...side of the sheet...Each item outside specifications automatically generates a print out of appropriate diagnostic and action data...The supervisor has only to check out those with asterisks on the right hand side because these are the only ones on which action may be required" [70].

This approach overcomes the poor credibility of GC analyses when used for routine control. The approach also envisages failure of either the computer itself or of parts of the automatic system. Therefore, alternate semiautomatic or manual routes are maintained.

A principal benefit of the automation of very-well-controlled analyses is the exclusion of operator mistakes, for instance, using the wrong sample, volume of sample or internal standard, switch or valve, peak for calculation, calculational method, or result.

Another principal benefit is the immediacy of results--shortening the delay from days or weeks to the duration of the GC run.

Finally, the very-well-controlled analysis that has then been automated is likely to gain in reproducibility and precision [76]. Whether it also becomes more accurate depends on the frequency and accuracy of calibration.

The General Analysis

By a "general" analysis, we mean any analysis performed on substantially or fully automated equipment. (A general analysis merely does not merit the extreme care and attention that the very-well-controlled analysis demands.)

The general analysis is usually initiated by manual sample injection. However, it is then performed and reported automatically. The general analysis may be carried out on gas chromatographs monitored and controlled by either a remote or an internal computer.

Remote. Users of gas chromatographs monitored and controlled by a remote computer invariably express satisfaction. The comments of Guichard and Sicard are typical: "The time saved is quite remarkable...In a series of 10 analyses, the working time for a chemist is reduced from 5 hours to 20 minutes..." [58].

Schomburg and Ziegler emphasized the value of program adaptability: "Due to the high flexibility in the software...it was possible to improve the GC-data processing with this system continuously, to adapt it to new tasks, and to introduce new features like time dependent instrument control" [74].

In a very matter-of-fact paper, Lyons described the circuits and hardware necessary to automate two gas chromatographs in association with a remote computer. Problems arose in bringing the automation to fruition--for example, with one instrument, "problems were experienced by actuation of logic out of sequence by high frequency main pulses of unknown source." However, once these were solved--in the difficulty cited, "by using a filter and a high quality independent earth"--the automated, continuous supervision became "invaluable" [77]. A preparative GC, for instance, was operated and monitored continuously, 24 hr/day and 7 days/week.

Internal. Free-standing automated gas chromatographs have recently become commercially available. These devices may or may not have manual controls. For those that do not, access to them and their accessories is gained only through a keyboard. Through this keyboard, analyses are programmed, peaks labeled for automatic identification, corresponding response factors determined or entered. Programs for analyses are committed to memory--memory can be either normal internal memory or a magnetic card. In any event, these units accommodate the automatic calculation of concentrations, which are reported on an accompanying thermal printer.

What the instruments do not do, as a rule, is offer an easily extendable memory or a readily accessible, improvable algorithm. This is really no great handicap. As this chapter has already shown, and as is shown in the next section, precision and accuracy really depend not on the sophistication of the algorithms but on the care given the chromatography.

14.10 PRECISION AND ACCURACY

A. Studies of Parameters and Techniques

Gas Flow Rates for the Flame Ionization Detector

Mikkelsen showed that numerous sources of error affect precision in gas chromatography [78]. Among the errors that are general, sizeable, and easily correctable are those caused by temperature changes in flow controllers.

In 1971 Grant and Clarke showed that if a flame ionization detector is to respond reproducibly within 1%, the air and hydrogen flow rates must be not only optimized but also controlled to within 1.5%. (The carrier gas flow rate must be controlled to within 2%.) "The control of temperature in commercial equipment is normally well within the requirements [±2%]. The control of flow rates, however, leaves much to be desired. Mass flow controllers are frequently used that are affected critically by changes in ambient temperature. In addition, there is often little or no control over the supplies of hydrogen and air which have been regarded in the past as of little importance. In fact, it is important to control these more closely than the carrier gas..." [79].

Peak Measurements

Of the techniques for peak measurement, Emery [80] found that integration of actual peak areas--as opposed to the integration of synthetic signals--with an analog-logic integrator would produce an error (one standard deviation) of about 0.5% of the area measured *versus* about 1.0% with the standard Disc integrator. Precision did not vary significantly with the type of detector or column, or with the phase of the sample.

Deans [81] found that human measurement of peak heights gave better reproducibility than automatic measurement of peak areas with analog-logic integrators. Grant and Clarke [79] found the analog-logic integrator just slightly more precise than human peak height measurement, but agreed with Deans that if human judgment is called for, the human, using peak height, does better than the electronic integrator. (In agreement, see also their 1974 report [2] and the comments of Charrier *et al.* on peak height measurement by computer [7].)

To counter the overconfidence that computers engender and the feeling that algorithms can supplant separations, van Rijswick [60] caused his computer printout to carry the percent error estimated for each measured peak area. A typical printout was shown. On it, for single peaks, the relative errors ranged from 0.1% to 20% (a very small peak), with an apparently typical relative error of 0.5 to 1.0%. Curve-fitting error reflected the method of fitting the curve: Edgeworth series evaluation gave relative errors from 0.4 to 16%, and Gaussian fit, from 1.1 to 60%.

Sample Injection

"Sample injection" usually refers to the injection of liquid samples. Liquid samples are almost always injected by syringe, either manually or automatically, and either into an injection port or on-column. "This operation probably contributes a larger total error to the overall analytical error than any other single source" [82]. Herewith we report some studies relating these methods to volume reproducibility, composition constancy, and vaporization efficiency, *i.e.*, peak shape.

Price compared automatic with manual sample injection, with respect to volume reproducibility [76]. The Hewlett Packard Automatic Sampler was found to inject considerably more reproducibly than a given experienced operator. For instance, in the injection of 1 µl, normal injection produced about 50% more volume variation than automatic (4.4% *versus* 3.1%, respectively); and in the injection of 5µl, manual injection produced about twice the volume variation that automatic injection did (2.5% *versus* 1.3%, respectively).

Composition constancy during injection was tested by Mikkelsen [78]. Samples of wide boiling range are particularly susceptible to variable evaporation and fractionation during injection. As an ideal control on evaporation, Mikkelsen used indium tube melting and compared it with syringe injection. (In indium tube melting, the sample is first encapsulated in a tube of indium. Then, inserted into a hot injection port--in this case, 250°C--the indium tube not only melts almost instantaneously but also volatilizes the sample very effectively.) For the study, Mikkelsen prepared a seven-component blend of known composition and wide boiling range (from 49.7°C for 2,2-dimethyl butane to 117.5°C for 1-butanol). The chromatographic separation of these components was arranged to be complete and clean. The peak areas were measured with an analog-logic integrator.

The standard deviations of the component concentrations in 10 replicate analyses averaged 0.30% of content with injection by indium tube melting, but about 0.44% of content with manual syringe injection. With each injection method, the error was highest for the most volatile component. However, with syringe injection, the second highest error was for the least volatile component. Thus a wide boiling range in the sample may indeed be expected to decrease precision in the results.

As part of an elaborate study, Grant and Clarke compared on-column with flash vaporization injection and also found a dependence of precision on the boiling range of the sample [82]. Because on-column injection causes greater peak spreading, flash vaporization gives better precision with samples of wide boiling range. On the other hand, less peak spreading occurs with low- and medium-boiling mixtures (these test mixtures contained, respectively, methanol and toluene, and 1-propanol and cis-decalin). With these, on-column injection gives better results. Perhaps "the best compromise [for precision] would be to use on-column injection with additional heat supplied by a vaporizer" [82].

B. Comparative Analyses

Emery reported a cooperative study of quantitation [80]. On a blend circulated among the cooperating laboratories, the 16 laboratories reporting results closest to those values known to be correct showed average errors of perhaps 10% of content: For one component, 15.98% for the average found versus 14.96% known (Emery: 14.93% found); for another component, 22.44% for the average found versus 20.08% known (Emery: 20.19% found).

Overall, Emery's laboratory showed an error of about 0.5% of content. The precision of their work compares favorably with that developed in Mikkelsen's careful demonstration of attainable precision [78].

The reason that Emery's laboratory did so much better than the others is given at the end of the following section.

C. Calibration and Recalibration

Deans [81], aware of the Mikkelsen [78] and Emery [80] studies, evaluated the principal sources of error in quantitative gas chromatographic analysis. Deans showed that published response factors should not be used for quantitation within a given laboratory; that good repeatability not only does not imply good accuracy but also, instead, tends to dull critical acuity. (Repeatability means getting the same figures for a given sample from a given analyst and instrument again and again.) The best way to get accurate results is to calibrate the individual chromatograph with synthetic mixtures of known compositions that bracket the compositions being found.

Calibrations are only temporary. Therefore calibrations should accompany analyses rather than precede or follow them at any great length. In their lengthy "Study of the Effects of Instrument Variables on Accuracy and Precision in Gas Chromatography," for instance, Grant and Clarke found "numerous significant differences [in results] between sets except where the compounds concerned are chemically similar to the reference compound. These differences are clearly explained by long-term changes in the relative response of the equipment, probably in the detector. When the measured compound is similar to the reference compound..., the response changes would tend to be compensatory" [82].

Both Deans and Grant and Clarke thus agree squarely with the procedure of Emery that allowed him to do so much better than average. That procedure was this: Emery first determined the approximate compositions of the unknowns. He then accomplished the final analysis "by having three different operators run the standard sample and the known mixtures [of bracketting compositions], ...[each operator applying] his own determined response factors on each day to his sample runs" [80].

For the precision and accuracy that can be attained in gas chromatography, *care like this* remains necessary and sufficient.

REFERENCES

1. G. Charrier, M. C. Dupuis, J. C. Merlivat, J. Pons, and R. Sigelle, *Chromatographia* 5, 128 (1972).
2. D. W. Grant and A. Clarke, *J. Chromatog.* 92, 257 (1974).
3. A. Janik, *J. Chromatog. Sci.* 13, 93 (1975).
4. D. A. Podmore, *J. Clin. Pathol.* 19, 619 (1966).
5. E. Menini, *Biochem. J.* 99, 747 (1966).
6. L. P. Cawley, W. O. Musser, and H. A. Tretbar, *Amer. J. Clin. Pathol.* 48, 216 (1967).
7. G. Charrier, M. C. Dupuis, J. C. Merlivat, J. Pons, and R. Sigelle, *Chromatographia* 5, 119 (1972).
8. J. D. Cheshire and R. P. W. Scott, *J. Inst. Petrol.* 44, 74 (1958).
9. I. Hornstein, P. Crowe, and B. Ruck, *J. Chromatog.* 27, 485 (1967).
10. F. van de Craats, in *Gas Chromatography 1958* (D. H. Desty, editor), Academic, New York, 1958, pp. 248-264.
11. D. M. Rosie and R. L. Grob, *Anal. Chem.* 29, 1263 (1957).
12. A. E. Messner, D. M. Rosie, and P. A. Argabright, *Anal. Chem.* 31, 230 (1959).

13. J. V. Killhefer and E. Jungermann, *J. Amer. Oil Chem. Soc.* 37, 456 (1960).
14. R. D. Condon, P. R. Scholly, and W. Averill, in *Gas Chromatography 1960* (R. P. W. Scott, editor), Butterworths, Washington, D. C., 1960, pp. 30-45.
15. J. C. Sternberg, W. S. Gallaway, and D. T. L. Jones, in *Gas Chromatography* (N. Brenner, J. E. Callen, and M. D. Weiss, editors), Academic, New York, 1962, pp. 231-267.
16. G. Perkins, Jr., G. M. Rouayheb, L. D. Lively, and W. C. Hamilton, in *Gas Chromatography* (N. Brenner, J. E. Callen, and M. D. Weiss, editors), Academic, New York, 1962, pp. 269-285.
17. F. Baumann, D. Wallace, E. Herlicska, and J. Blesch, in *Gas Chromatography 1968* (C. L. A. Harbourn, editor), Elsevier, Amsterdam, 1969, pp. 346-382.
18. D. Ford and K. Weihman, in *Recent Advances in Gas Chromatography* (I. Domsky and J. Perry, editors), Dekker, New York, 1971, pp. 377-398.
19. L. S. Ettre and F. Kabot, *J. Gas Chromatog.* 1, 114 (1963).
20. J. Novak and J. Janak, *J. Chromatog.* 28, 392 (1967).
21. S. Dal Nogare and R. S. Juvet, *Gas-Liquid Chromatography*, Wiley, New York, 1962, p. 264.
22. *Ibid.*, p. 62.
23. W. E. Harris and H. W. Habgood, *Programmed Temperature Gas Chromatography*, Wiley, New York, 1966, Equation 4.19.
24. *Ibid.*, Equation 4.20.
25. *Ibid.*, Section 5.02.
26. M. Margosis, *Chromatographia* 7, 441 (1974).
27. H. Hachenberg, *Industrial Gas Chromatographic Trace Analysis*, Heyden, New York, 1973.
28. W. E. Harris and H. W. Habgood, *Programmed Temperature Gas Chromatography*, Wiley, New York, 1966, pp. 190-200, 271.
29. L. W. Hollingshead, H. W. Habgood, and W. E. Harris, *Can. J. Chem.* 43, 1560 (1965).
30. A. G. Altenau, R. E. Kramer, D. J. McAdoo, and C. Merritt, Jr., in *Advances in Gas Chromatography, 1965* (A. Zlatkis and L. S. Ettre, editors), Preston Technical Abstracts Co., Niles, Ill, 1966, pp. 28-35.
31. A. B. Littlewood, C. S. G. Phillips, and D. T. Price, *J. Chem. Soc. 1955*, 1480 (1955).
32. I. Hornstein and P. F. Crowe, *Anal. Chem.* 34, 1354 (1962).
33. A. Zlatkis, W. Bertsch, H. A. Lichtenstein, A. Tishbee, F. Shunbo, H. M. Liebich, A. M. Coscia, and N. Fleischer, *Anal. Chem.* 45, 763 (1973), and Figure 2.
34. *Ibid.*, Figure 10.
35. L. J. Schmauch, *Anal.Chem.* 31, 225 (1959).
36. B. Weimann, *Chromatographia* 7, 472 (1974).
37. *Previews and Reviews*, 7-67, Varian Aerograph, 2700 Mitchell Drive, Walnut Creek, Calif., 1967.
38. A. T. Leung and J. M. Gill, *Res./Devel.* 25, 36 (1974).
39. J. T. Shank and H. E. Persinger, *J. Gas Chromatog.* 5, 631 (1967).
40. R. E. Anderson, *Chromatographia* 5, 105 (1972).
41. S. N. Chesler and S. P. Cram, *Anal. Chem.* 43, 1922 (1971).
42. P. Sutre and J. P. Malenge, *Chromatographia* 5, 141 (1972).

References

43. J. M. Gill, *J. Chromatog. Sci. 10*, 1 (1972).
44. D. A. Leathard, in *New Developments in Gas Chromatography* (J. H. Purnell, editor), Wiley-Interscience, New York, 1973, pp. 29-86.
45. L. M. Collyer, L. H. C. Hawkins, and G. H. Thomson, in *Gas Chromatography 1970* (R. Stock, editor), The Institute of Petroleum, London, 1971, pp. 247-279.
46. J. Baudisch, *Chromatographia 5*, 79 (1972).
47. N. Guichard and G. Sicard, *Chromatographia 5*, 83 (1972), p. 95.
48. J. S. Fok and E. A. Abrahamson, *Chromatographia 7*, 423 (1974).
49. U. Busch, *Chromatographia 5*, 63 (1972).
50. G. A. Schlereth and K. Greiner, *Chromatographia 5*, 70 (1972).
51. J. D. Hettinger, J. R. Hubbard, J. M. Gill, and L. A. Miller, *J. Chromatog. Sci. 9*, 710 (1971).
52. A. Savitzky and M. J. E. Golay, *Anal. Chem. 36*, 1627 (1964).
53. R. N. Schmidt and W. E. Meyers, *Introduction to Computer Science and Data Processing*, 2nd ed., Holt, Rinehart, and Winston, New York, 1970, p. 392.
54. J. G. Karohl, in *Gas Chromatography 1968* (C. L. A. Harbourn, editor), The Institute of Petroleum, London, 1969, pp. 359-366.
55. R. N. Schmidt and W. E. Meyers, *Introduction to Computer Science and Data Processing*, 2nd ed., Holt, Rinehart, and Winston, New York, 1970, p. 43.
56. A. Fozard, J. J. Frances, and A. J. Wyatt, *Chromatographia 5*, 130 (1972).
57. F. Baumann, A. C. Brown, and M. B. Mitchell, *J. Chromatog. Sci. 8*, 20 (1970).
58. N. Guichard and G. Sicard, *Chromatographia 5*, 83 (1972), p. 92.
59. R. J. Maggs and A. S. Mead, in *Gas Chromatography 1970* (R. Stock, editor), The Institute of Petroleum, London, 1971, pp. 310-330.
60. M. H. J. van Rijswick, *Chromatographia 7*, 491 (1974).
61. M. H. J. van Rijswick, Personal communication, 1978.
62. F. Caesar and M. Klier, *Chromatographia 7*, 526 (1974).
63. A. H. Anderson, T. C. Gibb, and A. B. Littlewood, *Anal. Chem. 42*, 434 (1970), p. 439.
64. R. Annino, in *Advances in Gas Chromatography*, Vol. 15, (J. C. Giddings, E. Grushka, J. Cazes, and P. R. Brown, editors), Dekker, New York, 1977, pp. 33-68.
65. F. Baumann, D. Wallace, E. Herlicska, and J. Blesch, in *Gas Chromatography 1968*, (C. L. A. Harbourn, editor), Elsevier, Amsterdam, 1969, p. 381 (Levy).
66. A. B. Littlewood, T. C. Gibb, and A. H. Anderson, in *Gas Chromatography 1968*, (C. L. A. Harbourn, editor), The Institute of Petroleum, London, 1969, pp. 297-318.
67. J. Hettinger, *Autolab Technical Bulletin 111-74*, Spectra-Physics, Santa Clara, Calif.
68. A. L. Robinson, *Science 186*, 1102 (1974).
69. J. G. Lyons, in *Gas Chromatography 1970* (R. Stock, editor), The Institute of Petroleum, London, 1971, pp. 302-309.
70. D. R. Deans, *J. Chromatog. Sci. 9*, 729 (1971).
71. D. R. Deans, in *Gas Chromatography 1970* (R. Stock, editor), The Institute of Petroleum, 1971, pp. 292-301.
72. H. Ch. Broecker and H. G. W. Müller, *Chromatographia 7*, 432 (1974).

73. F. Grohe, W. Hesse, R. Kaiser, and K. H. Schneckenburger, in *Gas Chromatography 1970* (R. Stock, editor), The Institute of Petroleum, London, 1971, pp. 247-256.
74. G. Schomburg and E. Ziegler, *Chromatographia 5*, 96 (1972).
75. J. Baudisch and H.-D. Papendick, *Chromatographia 7*, 549 (1974).
76. J. G. W. Price, J. C. Scott, and L. O. Wheeler, *J. Chromatog. Sci. 9*, 722 (1971).
77. J. G. Lyons, *Chromatographia 5*, 156 (1972).
78. L. Mikkelsen, *J. Gas Chromatog. 5*, 601 (1967).
79. D. W. Grant and A. Clarke, *Anal. Chem. 43*, 1951 (1971).
80. E. M. Emery, *J. Gas Chromatog. 5*, 596 (1967).
81. D. R. Deans, *Chromatographia 1*, 187 (1968).
82. D. W. Grant and A. Clarke, *J. Chromatog. 97*, 115 (1974).

AUTHOR INDEX

Numbers in parentheses are reference numbers and indicate that an author's work is referred to although his name is not cited in the text. Numbers in italics give the page on which the complete reference is listed.

A

Abel, E. W., 137, *140*
Abrahamson, E. A., 362(48), 366(48), 367(48), *393*
Abramson, F. P., 331(131), 333(131), 334, 335(131), 336(131), *340*
Acker, L., 208(116), *222*
Ackman, R. G., 127(23), *139*
Adlard, E. R., 224(7), *250*
Aggarwal, V., 276(157), *291*
Ahuja, S., 252(6), *287*
Albro, P. W., 273(129), *291*
Alexander, G., 206(107, 108, 109), *222*
Allen, K. G., 263(82), *290*
Alley, C. C., 264(87, 88), 271(119), *290*
Alonso, J. A., 323(112), *339*
Alpert, N. L., 305(35), 307, *337*
Altenau, A. G., 347(30), 349(30), *392*
Altschuller, C. H., 254(12), 286(12), *288*
Amos, R., 44, *53*, 101, *113*
Amundson, C. H., 32(2), *36*
Anderson, A. H., 380(63, 66), 381, *393*
Anderson, D. G., 68(27), *76*
Anderson, R. E., 360(40), *392*

Andreatch, A. J., 316(82), 317(82), *338*
Anfinsen, J. R., 256(38), *288*
Angelini, P., 301(29), 302, 324(29), *337*
Änggård, E., 254(22), 256(22), 271, *288*
Annino, R., 380(64), *393*
Ansel, R. E., 68(27), *76*
Anthony, D. S., 175(59), *177*
Anthony, G. M., 283(201), *293*
Argabright, P. A., 342(12), *391*
Argauer, R. J., 175(60), *177*
Armstrong, R. J., 254(24), 257, *288, 289*
Artigas, F., 323(112), *339*
Askew, W. C., 162(28), *176*
Aston, F. W., 317, *338*
Atkinson, E. R., 256(39), *288*
Atwater, B. L., 336(143), *340*
Aue, W. A., 137(60, 61), 138, 141, 163(33), *176*, 206(106), *222*
Auer, M., 163(36), *176*
Averill, W., 192(69), 195(85), 221, 342(14), *392*
Axelson, M., 280(191), 281(191), *292*
Axen, U., 261(60), *289*
Ayres, B. O., 53(23), *54*

B

Baczynskyj, L., 261(60), *289*
Badieu, P., 280(190), *292*
Badings, H. T., 190(58), 192(58), 204(58), 205(58), 206(58), *220*
Bailey, E., 285(212), *293*
Baillart, J., 323(112), *339*
Baillie, T. A., 281(194), *293*
Baker, W. J., 120(8), 128(8), *139*
Bambrick, W. E., 316(82), 317(82), *338*
Banner, A. E., 301(30), 303, *337*
Barrall, E. M., 133(47), *140*
Barrer, R. M., 135(48, 49, 50), *140*
Barrett, J., 275, *291*
Barron, R., 324(121), 326(121), *339*
Barta, A. L., 254(10), 262(10), *288*
Bartle, K. D., 191(66, 67), 204(91), *220, 221*
Bartlet, J. C., 309(53), *337*
Bassette, R., 312(62), *338*
Baudisch, J., 362(46), 386(75), *393, 394*
Baumann, F., 342(17), 363(17), 365(17, 57), 380(65), 385(17), *392, 393*
Becker, R. S., 171(51), 172(54), *177*
Beckey, H. D., 324(121), 326(121), *339*
Bednas, M. E., 58(4), *75*
Begg, C. G., 286, *294*
Beggs, D., 326(123), *339*
Belvedere, G., 277(164), *292*
Bennett, C. E., 179(17), 189(17), *219*, 223(1), 237(1), 238(1), *250*
Bens, E. M., 117(6), 118, 119(7), *139*
Bente, H. B., 163(34), *176*
Bente, P., 216(134), 217(134), 218(134), *222*
Bentley, R., 179(18), 189(18), *219*, 277(167), 278(170), 279(167), *292*
Beran, M. J., 110(24), *113*
Bergheim-Irps, E., 276(156), 277(156), *291*
Bergstedt, L., 204(91), *221*
Beroza, M., 316(76, 77, 78), *338*
Bertsch, W., 350(33), 351(33), *392*
Bertz, A. M., 301(24), *337*
Best, F. W., 207(112), *222*
Betker, W. R., 252(3), *287*
Biemann, K., 331(129), 332(129),

Biemann, K. (continued)
334, *340*
Biggs, R. H., 175(59), *177*
Billeb, K. A., 189(51), 192(51), 193(51), 194(82), 195(82), 196, 217(135), 218(135), *220, 221, 222*
Bischoff, J., 163(35), *176*
Black, D. R., 197(87), 201, *221*
Blaisdell, B. E., 327(127), *340*
Blau, K., 252(7), 271, *287, 290*
Blesch, J., 342(17), 363(17), 365(17), 380(65), 385(17), *392, 393*
Blomberg, L., 191(66), 205(104), *220, 222*
Blos, G., 194(81), *221*
Blumer, M., 193(79), 194(79), 208(79, 115), *221, 222*
Bocek, P., 156(21), *176*, 180(19, 20), 214(19), *219*
Bodin, J. I., 256(41), *288*
Boelsma-Van Haute, E., 263(78), *289*
Bohemen, J., 99, *112*, 124(13), 125(13), 126(13), 136(13), *139*, 179(14, 15), 180(14), 181, 186(14), 194(14), *219*
Bokhoven, C., 149(10), 171(10), *176*
Bombaugh, K. J., 135, *140*
Bomstein, J., 131(36), *140*
Bonelli, E. J., 320(106), *339*
Boogaerts, T., 192(72), *221*
Borke, M. L., 126(18), *139*
Bortner, T. E., 174(56), *177*
Bossi, A., 277(164), *292*
Botlock, N., 277, 278(170), *292*
Bottcher, C. J. F., 263(78), *289*
Bouche, J., 192(71), *221*
Bournot, P., 280(190), 281(190), *292*
Bowman, P. B., 258(52), *289*
Bowman, R. L., 127(25), *140*
Bradford, B. W., 58(3), *75*
Braestrup, C., 256, 257(44), *289*
Brandt, W. W., 126(15), *139*
Braun, W. H., 139(68), *141*
Brenner, N., 15(18), *16*
Brewer, H. B., Jr., 326(124), *340*
Briegleb, G., 172(52, 53), *177*
Brittain, G. D., 254(18, 19), 278(18, 19), *288*
Brobst, K. M., 278, *292*
Brochmann-Hanssen, E., 274(146), 275, *291*
Broderick, J. J., 297(4), *336*
Broecker, H. C., 385(72), *393*
Brook, J. L., 175(59), *177*
Brooks, C. J. W., 281(194), 283

Author Index

Brooks, C. J. W. (continued)
    (201, 202), *293*
Brooks, J. B., 264(87, 88), 271,
    290
Brötell, H., 254(25, 27), 258
    (25), 268(25), *288*
Brown, A. C., III, 208(122),
    209(122), 210(122), *222*,
    365(57), *393*
Brown, L. W., 258(52), *289*
Brown, S. C., 157(24), *176*
Brownlee, K. A., 80(9), *96*
Brunner, T. R., 331(133), *340*
Burgett, C. A., 163(34), *176*
Burgher, R. D., 127(23), *139*
Burke, J. A., 165(44), 166, *177*
Burlingame, A. L., 319(96, 97),
    326(97), 327(97), *339*
Busch, U., 362(49), 363(49), *393*
Buscher, H.-P., 278(172), *292*
Bussemas, H. H., 271(113), *290*
Butler, L., 66, 67(20), *75*, 103,
    113, 137(64), *141*
Butler, W., 56(2), 57(2), 73(2),
    75(2), *75*
Buttery, R. G., 46(16), 51(22),
    53, *54*, 197(87), 201, *221*
Butts, W. C., 261, *289*

C

Caesar, F., 299(15), *336*, 379
    (62), *393*
Calam, D. H., 258(51), *289*
Calcote, H. F., 161, *176*
Callear, A. B., 178(7), *218*
Callen, J. E., 15(18), *16*
Calouche, S. I., 256(39), *288*
Cameron, A. E., 319(99), *339*
Cancalon, P., 273(140), *291*
Caprioli, R. M., 322(109), 323
    (109), *339*
Carman, P. C., 180(21), 214(21),
    219
Caro, J. H., 286(221, 222, 223),
    293
Casagrande, D. J., 273, *291*
Casals-Stenzel, J., 278(172),
    292
Casu, B., 309, *337*
Cattabeni, F., 254(23), 256(23),
    271(23), *288*
Cavagnol, J. C., 252(3), *287*
Cavalloti, L., 309, *337*
Cawley, L. P., 341, *391*
Chafetz, L., 275(152), *291*
Chait, E. M., 323, 324(115), 326
    (115), *339*
Chalkley, D. E., 58(3), *75*
Chalmers, R. A., 264(93), *290*

Chambaz, E. M., 254(15), 280(15,
    187), 281(15), 282(15), 283
    (197, 200), *288*, *292*, *293*
Chang, S. Y., 286(225), *293*
Chao, G. S., 58(6), *75*
Chapman, J. R., 285(212), *293*
Charrier, G., 341(1), 365(7), 389
    (7), *391*
Chen, E., 165(46, 47), 166(47),
    168, 171(47, 50), 172(46, 47,
    50), *177*
Cheshire, J. D., 342(8), *391*
Chesler, S. N., 361(41), 364(41),
    392
Ching, H. C., 259(53), *289*
Chiu, J., 105(14), *113*
Choudberry, G., 264(87), *290*
Chovin, P., 61, 69(12), *75*, 214,
    222
Christopherson, S. W., 263(79),
    289
Clark, S. J., 284(204), *293*
Clarke, A., 34(3), *36*, 341(2),
    342(2), 389(2, 79, 82), 390
    (82), 391(82), *391*, *394*
Clarke, D. D., 267(103), *290*
Clerc, J. T., 327(128), 335(128),
    340
Coates, V. J., 15(18), *16*
Cochrane, W. P., 285, *293*
Cohen, I. C., 286(219, 220), *293*
Coleman, A. E., 65, 67(17), 68
    (17), *75*
Coleman, H. J., 316(80, 81), *338*
Collyer, L. M., 362(45), 364(45),
    385(45), *393*
Colthup, N. B., 305(37), *337*
Condon, R. D., 342(14), *392*
Cooke, W. D., 131(31, 32), *140*
Coscia, A. M., 350(33), 351(33),
    392
Costa, E., 254(23), 256(23), 271
    (23, 115), *288*, *290*
Coupe, N. B., 313(69, 70, 71),
    314, *338*
Cowling, T. R. S., 148, *176*
Craig, B. M., 224(5), *250*
Cram, S. P., 13(5, 6), *16*, 208
    (122), 209(122), 210(122),
    222, 361(41), 364(41), *392*
Cramers, C. A., 180(19, 20), 205
    (100), 214(19), *219*, *221*
Cravello, G. D., 275(152), *291*
Craven, D. A., 280(183), *292*
Crippen, R. C., 296(3), *336*
Cronin, D. A., 206, 207(111),
    222
Crosby, D. G., 258(48), *289*
Crowe, P. F., 342(9), 350(32),
    351, *391*, *392*

Cummins, L. M., 254(37), 267(37), 268, 284(37), *288*
Curtius, H. C., 127, *140*, 311(56), *338*
Cutié, S., 139(75), *141*
Cvetanovic, R. J., 178(7), *218*

D

Daemen, J. M. H., 287, *294*
Dal Nogare, S., 79, *96*, 105(14), *113*, 149(8), *176*, 237(39), 244(50), *251*, 345(21), *392*
Daly, L. H., 305(37), *337*
Damico, J. N., 324(121), 326(121), *339*
Dandeneau, R., 216(133, 134), 217(133, 134), 218(134), *222*
Danielczik, E., 304(34), *337*
Daniels, F., 228(16), *250*
Daniewski, M., 262(66), 287(235), *289*, *294*
Dankelman, W., 287, *294*
David, D. J., 142, *175*
Davies, A. J., 172(55), 173(55), *177*
Davies, L., 300(22), *337*
Davis, A., 127(22), *139*
Davis, C. M., 165(45), 166, 171(45), *177*
Davis, P. L., 102(8), *113*
Davis, T. W. M., 309(53), *337*
Dayringer, H. E., 336(142), *340*
DeAlencar, J. W., 150(14), *176*
Deans, D. R., 385(70), 386(70), 387(70), 389, 390, *393*, *394*
Dechtiaruk, W. A., 276(160), 277(160), *292*
Decora, A. W., 178(6), *218*
Defaye, G., 283(200), *293*
DeFord, D. D., 53(23, 24), *54*
Degen, P. H., 268(106), *290*
De Goey, J., 191(62), *220*
DeGroot, M., 334(140), 335(140), 336(140), *340*
Dehennin, L., 284(206), *293*
Delaforge, M., 280(190), 281(190), *292*
De Ligny, C. L., 110(26), 111(26), *113*
Dempster, A., 317, *338*
Derrick, P. J., 319(96), *339*
DeStefano, J. J., 137(63), *141*
Desty, D. H., 15(17), *16*, 44, 45, 53, 100, 108, *112*, 156, 157, 159(17), 160(17), *176*, 178(3), 180(3), 188, 189(3), 190(3), 191(41, 59, 60, 65), 192(75), 193(41), 194(40),

Desty, D. H., (continued)
195, 197(3), 198, 203(3), 204(97, 98), 205(42), 213(3), 216(3), *218*, *220*, *221*, 224(4), 225, *250*
Devaux, P. G., 281(195), 284(208), *293*
Dewar, R. A., 156, 157(23), *176*
Dewar, M. J. S., 58(7), *75*
DeWet, W. J., 232, *250*
DeWolfe, R. H., 266(98), *290*
Deyl, Z., 273, *291*
Dhont, J. H., 298(5), *336*
Dienes, G. L., 178(2), *218*
Dijstra, A., 149(10), 171(10), *176*
Dijkstra, G., 191(62), *220*
Dimbat, M., 313(68), *338*
Dimitrov, C., 103(10), *113*, 247, 249(61, 62), *251*
Dineen, G. U., 178(6), *218*
Dinerstein, R. A., 149(9), *176*, 203, *221*
Ditshuneit, K., 194(81), *221*
Djikstra, A., 331(134), *340*
Domsky, I. I., 15, *16*
Donelli, M. G., 277(164), *292*
Donike, M., 270, 287, *290*, *294*
Dorough, H. W., 285, *293*
Drew, C. M., 117(6), 118, 119, *139*
Drozd, J., 252(5), *287*
Dubetz, D. K., 275(151), *291*
Duchamp, D. J., 261(60), *289*
Duffield, J. J., 106, *113*, 236(29), *250*
Dünges, W., 276(156, 158), 277(156), *291*, *292*
Dunlop, A. S., 43, 44, *53*
Dupuis, M. C., 341(1, 7), 365(7), 389(7), *391*
Durbin, D. E., 129, *140*
Dutton, A., 285(213), *293*
Dutton, G. S., 277, *292*

E

Eaborn, C., 285(211), *293*
Eadie, M. J., 275(151), *291*
Eagles, J., 265(96), *290*
Eaton, D. L., 131(39), 133(39), 137(62), *140*, *141*
Eberle, A. J., 259(53), *289*
Edlund, D. O., 256(38), *288*
Edwards, D. J., 271, *290*
Edwards, W. M., 286(222), *293*
Eggers, D. E., 319(99), *339*
Ehrsson, H., 254(25, 27, 33, 34), 258(27), 267(33), 268(25), 277, 279, 286, *288*, *292*, *294*

Eisler, W. J., 323(111), *339*
Elliott, R. M., 301(30), 303, *337*
Elmenhorst, H., 204(96), *221*
Emery, E. M., 389, 390(80), 391(80), *394*
Engewald, W., 207(113), 218(113), *222*
Erdtmansky, P., 272, *291*
Erley, D. S., 260(56), *289*
Ettre, L. S., 14, *16*, 101(4), *112*, 188(38), 189(49, 50, 51), 190(54), 191(63), 192(49, 51, 70, 74), 193(51), 194(51), 194(38, 49, 82), 195(82, 85, 86), 196, *220*, *221*, 249(64), *251*, 298(9, 10), 299(10), *336*, 342(19), *392*
Evans, C. D., 153(15), *176*
Evans, M. B., 127, 128(27), *140*
Evans, R. T., 254(36), 277, 285, *288*
Exley, D., 285(213), *293*
Eyem, J., 273(135), 274(135), *291*

F

Fagerson, I. S., 15(18), *16*
Fales, H. M., 261, 281, *289*, 324, 326(121, 123, 124), *339*, *340*
Fanter, D. L., 316, *338*
Farmer, J. B., 317(87), 318(87), 319(87), 320(87, 105, 108), 321(108), 322(108), *338*, *339*
Feit, C. A., 252(2), *287*
Feit, E. D., 312(59, 60), 323(60), *338*
Fenimore, D. C., 164(41), 165(41, 45), 166(41), 167(41), 168(41), 169(41), 170(41), 171(45, 48), 172(41), 173(41), 174(41), *177*
Fett, E. R., 136(55), *140*
Field, F. H., 326(116), *339*
Fies, W. F., 322(109), 323(109), *339*
Filbert, A. M., 102(7), *113*, 128, 131(37, 38), 132(38, 43), 137(62), *140*, *141*, 178(8), *218*
Finkbeiner, H., 272(125), *291*
Finkelstein, J. A., 278(177), 280(177), *292*
Firestone, D., 263(77), *289*
Fischer, I., 284(203), *293*
Fishbein, L., 273(129), *291*
Fleischer, N., 350(33), 351(33), *392*
Fok, J. S., 362(48), 366, 367(48), *393*
Ford, D., 342(18), 362(18), 363

Ford, D., (continued)
(18), 365(18), 366(18), 367(18), 368(18), 369, 374(18), *392*
Ford, L. A., 263(76), *289*
Forss, D. A., 301(29), 302, 324(29), *337*
Foster, L. L., 271, *290*
Fouda, H., 258(48), *289*
Fowler, L., 15(18), *16*
Fozard, A., 364(56), 368(56), 369(56), 370, 371, 372, *393*
Frances, J. J., 364(56), 368(56), 369(56), 370(56), 371, 372, *393*
Frederick, D. H., 131(31, 32), *140*
Fredericks, K. M., 312(61), *338*
Freeman, H. P., 286(221, 222), *293*
Friedel, R. A., 277(168), *292*
Frigerio, A., 277(164), *292*
Fryer, J. F., 245(58), *251*

G

Gallaway, W. S., 156, 160(18), 161(18), *176*, 342(15), *392*
Galli, G., 260, *289*
Gardiner, W. L., 280, *292*
Garzo, G., 206(108), *222*
Geach, G. J., 156, 159(17), 160(17), *176*
Geffner, J., 80(8), *96*
Gehrke, C. W., 163(33), *176*, 254(28), 266(28), 272(122, 123, 124), 273(122, 123, 124, 126, 137, 141), 274(123), 280(183), 286(225), *288*, *291*, *292*, *293*
Geiss, F., 334(140), 335(140), 336(140), *340*
Gelpi, E., 187, *219*, 271, *290*, 323(112), *339*
Gent, P. L., 236(33), 237, *251*
Gerasimoff, M., 73, *76*
Gerber, J. N., 323(113), *339*
Gercken, G., 264(85, 86), *290*
German, A. L., 49, 50(21), *54*, 192(78), 205(78, 105), 208(78, 105), 209, *221*, *222*
Gershengorn, M. C., 274(143), *291*
Gerstl, R., 277(163), *292*
Gibb, T. C., 380(63, 66), 381, *393*
Giddings, J. C., 14, *16*, 21, *26*, 77(3), 80(7), *96*, 99(2), 109, 110(21, 22, 23, 25), 111(27), *112*, *113*, 120(10, 11), 121

Giddings, J. C. (continued)
(10), 122(10, 11), 123, 131
(33), 132(41, 42), 136, *139*,
140, 180(23), *219*, 230(22,
25), 236(22, 25, 32), 242
(22, 42), 244(22, 32, 47),
245(22), *250*, *251*
Gill, J. M., 358(38), 360(38),
361, 362(51), 363(51), 364
(51), 365(51), 366(51), 367
(51), 369(51), 373, 382(38),
392, *393*
Gillen, D. G., 42(7), *53*
Giovanniello, T. J., 275(150),
291
Gitlow, S. E., 267(103), *290*
Giuffrida, L., 162, 163, 164,
176, 262(71), *289*, 301(23),
307(40), *337*
Glaser, E. R., 197, *221*
Glass, R. L., 263(79), *289*
Gleispach, H., 254(17), 257(17),
283(17), *288*
Glick, S. D., 264(94), *290*
Glotfelty, D. E., 286(221, 222),
293
Glueckauf, E., 42, *53*
Goedert, M., 299(16), *336*
Golay, M. J. E., 188, 189(46),
190(53), 192(53, 76), *220*,
221, 363(52), 365(52), 366
(52), 367, *393*
Goldup, A., 15(17), *16*, 44, 45,
53, 100, 108, *112*, 156, 157,
159(17), 160(17), *176*, 178
(3), 180(3), 188(3, 40, 41),
189(3), 190(3), 191(41, 59),
193(41), 195(41), 197(3),
198, 203(3), 204(3), 213(3),
216(3), *218*, *220*
Gooding, K. M., 264(91), *290*
Goodwin, E. S., 300(22), *337*
Gordy, A., 274(143), *291*
Goto, J., 252(4), *287*
Gough, T. A., 313(72), 315, 316
(72), *338*
Goulden, R., 300(22), *337*
Gove, P. B., 77(4), *96*
Grant, D. W., 34(3), *36*, 216,
222, 229(18), *250*, 341(2),
342(2), 389(2, 79, 82), 390
(82), 391(82), *391*, *394*
Grasselli, J., 307(41), *337*
Gray, P., 148, *176*
Greeley, R. H., 254(35), 262, 276,
288, *289*
Green, C. R., 207(112), *222*
Greer, M., 264(90), *290*
Gregory, N. L., 164(40), 166,
176

Greiner, K., 362(50), *393*
Griffiths, J. H., 237(37), *251*
Griggs, L. J., 278(177), 280
(177), *292*
Grimett, M. R., 286, *294*
Grob, G., 205(102), 206(102),
207, 208, *221*
Grob, K., 204(92, 93), 205(102,
103), 206(102), 207, 208,
(103, 121), 210(103), 211
(103, 121), 212, 218(136),
221, *222*
Grob, K., Jr., 208(121), 211
(121), *222*
Grob, R. L., 15, *16*, 149(12),
176, 342(11), *391*
Groenings, S., 230(23), 231, *250*
Grohe, F., 386(73), *394*
Grossman, D., 69(29), *76*
Grotch, S. L., 331(135), *340*
Gruber, V. F., 287, *294*
Grunwald, C., 254(9), 262(9), *288*
Grushka, E., 137(57), *140*
Guadagni, D. G., 197(87), 201,
221
Gualario, V. M., 276(159), *292*
Gudzinowicz, B. J., 142, 162(1),
175
Guichard, N., 362(47), 366(58),
388(58), *393*
Guiochon, G., 150(13), *176*, 191,
204(97), 205(61), 214, 216,
220, *221*, *222*, 299(16), 312,
336, *338*
Gunner, S. W., 280(184), *292*
Gunther, F. A., 142, 162(2),
174, *175*
Gyllenhaal, O., 254(27), 258(27),
288

H

Haahti, E. O. A., 47, 48(17), *54*
Habgood, H. W., 12(3), 14, *16*,
38(1), *53*, 230(21), 231(21),
232(21), 233(21), 234, 235,
236(21, 31, 34), 237(24, 38),
238(38), 240, 241(38), 242
(43), 243(31), 244(38, 53),
245(58), 246(55), *250*, *251*,
347(23, 28, 29), *392*
Hachenberg, H., 347, *392*
Hachey, D. L., 326(119), 327
(119), 328(119), 329, 330,
331(119), *339*
Hair, M. L., 102(7), *113*, 128,
131(37, 38), 132(38), *140*
Haken, J. K., 65, 68(15), 75, 298
(11), *336*

Halang, W., 194(80), *221*
Halasz, I., 131(35), 137(58), *140*, 178(4, 9), 180(22), 188(22), 189(47, 48), 192(47, 48), 214(22, 125), 215, 216(22), *218*, *219*, *220*, *222*, 236(36), *251*
Hamberg, M., 261, *289*
Hamilton, W. C., 342(16), *392*
Hamlin, W. E., 254(29), 260(29), *288*
Hammond, E. C., 120(9), 121(9), *139*
Hansen, R. P., 224(6), *250*
Harbourn, C. L. A., 15(17), *16*
Hardy, J. P., 273(130), *291*
Haresnape, J. N., 204(97), *221*
Harper, J. M., 120(9), 121(9), *139*
Harris, W. E., 12(3), 14, *16*, 38(1), *53*, 230(21), 231(21), 232(21), 233(21), 234, 235, 236(21, 31, 34), 237(24, 38), 238(38), 240, 241(38), 242(43), 243(31), 244(38, 53), 245(58), 246(55), *250*, *251*, 347(23, 28, 29) *392*
Hartkopf, A., 69(29), *76*
Hartmann, C. H., 142, 144(3), 150(3), 152(3), *175*
Hartvig, P., 269(107, 108, 109, 110), *290*
Harvey, D. J., 58(3), *75*, 254(21), 256(40), 279, 286(231), 287, *288*, *292*, *294*
Harvey, T. M., 326(120), *339*
Hastings, C. R., 137(60, 61), 138, *141*, 206(106), *222*
Hatch, F., 326(118), *339*
Haumann, J. R., 323(111), *339*
Hawke, J. C., 224(6), *250*
Hawkes, S. J., 56(2), 57(2), 64(18), 66(18), 67(20, 21), 69, 73(2), 75(2), *75*, *76*, 103, 107, 109, *113*, 131(33), 137(64), *140*, *141*, 181(28), 188(36), *219*
Hawkins, L. H. C., 362(45), 364(45), 385(45), *393*
Hayes, W. V., 133(46), *140*
Healy, J. W., 127(24), *140*
Heckman, R. A., 207(112), *222*
Hedin, P. A., 311(57, 58), *338*
Heenan, M. P., 257(46), *289*
Hefendehl, F. W., 316(83), 317(83), *338*
Heine, E., 180(22), 188(22), 214(22, 125), 215, 216(22), *219*, *222*, 236(36), *251*
Heines, V., 1(1), *15*

Heller, S. R., 332(139), *340*
Henly, R. S., 186(32), *219*
Herington, E. F. G., 228, 229, *250*
Herlicska, E., 342(17), 363(17), 365(17), 380(65), 385(17), *392*, *393*
Hermann, E. A., 139(68), *141*
Hernandez, M., 217(135), 218(135), *222*
Hersh, C. K., 135(51), *140*
Hertel, R. H., 336(141), *340*
Hertz, H. S., 331(129), 332(129), 334, *340*
Herzog, R., 319(91), *339*
Hesse, W., 386(73), *394*
Hettinger, J. D., 362(51), 363(51), 364(51), 365(51), 366(51), 367(51), 369(51), 373, 382(67), 383(67), *393*
Hildebrand, J. H., 55, 56(1), 57(1), 70(32), *75*, *76*, 228(17), *250*
Hill, R. M., 277(161), *292*
Hinds, G. P., 224(3), *250*
Hintze, U., 264(85, 86), *290*
Hishta, C., 102, *113*, 131(30, 31, 36), *140*
Hiskes, R., 216(134), 217(134), 218(134), *222*
Hites, R. A., 331(129), 332(129), 334, *340*
Hoff, J. E., 312(59, 60), 323(60), *338*
Hoffman, N. E., 254(16), 257(16), 264(91), *288*, *290*
Hoffman, R. L., 153(15), *176*
Hofmann, M., 68(25), *75*
Holden, K. G., 278(177), 280(177), *292*
Holland, J. F., 323(113), *339*
Hollander, C. S., 274(143), *291*
Hollingshead, L. W., 236(31), 243(31), *250*, 347(29), *392*
Hollis, O. L., 133(45, 46), 134, *140*
Holst, J. J., 230(23), 231, *250*
Hoodless, R. A., 163(37), *176*
Hooper, W. D., 275(151), *291*
Hopkins, R. L., 316(80), *338*
Horning, E. C., 47, 48(17, 18, 19), 49, 50(20, 21), *54*, 192(78), 205(78, 105), 208(78, 105), 209, *221*, *222*, 252, 254(1, 15, 31, 32), 263(83), 266(99), 267(1, 31, 32), 268(32), 269, 279(179), 280(15, 187), 281(15, 192, 193, 194, 195, 196), 282(15), 283(197, 198), 284

Horning, E. C. (continued)
  (208), *287*, *288*, *290*, *292*, *293*
Horning, M. G., 48(19), 50(21), *54*, 252, 254(1, 21), 259 (54), 263(83), 264(92), 267 (1), 269, 277(161), 279, 281 (192, 193, 195), 286(231), 287, *287*, *288*, *289*, *290*, *292* *293*, *294*
Hornstein, I., 46(13), 51(13), *53*, 178(1), 190(57), *218*, *220*, 302(31), 304, *337*, 342(9), 350(32), 351, *391*, *392*
Horvath, C., 131(35), *140*, 178 (4, 9), 189(47, 48), 192(47, 48), *218*, *219*, *220*
Hougen, F. W., 208, *222*
Howlett, M. D. D., 308(44), *337*
Hoyland, J. R., 332(138), *340*
Hubbard, J. R., 362(51), 363(51), 364(51), 365(51), 366(51), 367(51), 369(51), 373, *393*
Hufschmidt, M., 194(80), *221*
Hunt, D. F., 326(120, 122), *339*
Hurrell, R. A., 44, *53*, 101, *113*
Hurst, G. S., 174(56), *177*
Hurst, R. E., 278(175), *292*
Husek, P., 273(133, 134), 274 (133, 134), *291*
Husmann, H., 189(45), 205(101), 206(101, 110), 207(101), *220*, *221*, *222*, 299(19), *336*

I

Innes, W. B., 316(82), *338*
Isenhour, T. L., 63(14), 64(14), 68(14), 69(29), *75*, *76*, 331 (136), *340*
Issenberg, P. I., 46(13), 51(13), *53*, 178(1), 190(57), *218*, *220*, 302(31), 304, *337*
Iverson, J. L., 263(74, 80), *289*

J

Jacin, H., 278(176), *292*
Jaeger, H., 194(81), *221*
James, A. T., 19(6), *26*, 114, 126 (2), 127(1), *139*, 179, *219*
James, D. H., 237(37), *251*
Janak, J., 114(5), *139*, 156(21), 162(32), *176*, 309, *337*, 345 (20), *392*
Janik, A., 341(3), *391*
Jenden, D. J., 322(110), 323(110), *339*

Jennings, W. G., 190, 197(88), 202, 205, *220*, *221*
Johnson, C. F., 276(160), 277 (160), *292*
Johnson, D. B., 254(21), 286 (231), *288*, *294*
Johnson, E. G., 319(93), *339*
Johnson, J. F., 133(47), *140*
Johnson, R. L., 139(73), *141*
Jojola, R., 131(34), *140*
Jones, C. E. R., 313(69, 70, 71, 72), 314, 315, 316(72), *338*
Jones, D. T. L., 156, 160(18), 161(18), *176*, 342(15), *392*
Jones, J. K. N., 280(184), *292*
Jones, N. S., 273(132), *291*
Jones, R., 271(119), *290*
Jones, W. L., 88, *96*, 107, *113*
Jönsson, J., 273(135), 274(135), *291*
Jordan, R. L., 68(23), *75*, 248, 249(63), *251*
Joynes, P. L., 172(55), 173(55), *177*
Jungermann, E., 342(13), *392*
Justice, J. B., 63(14), 64(14), *75*
Juvet, R. S., Jr., 13(5, 6), *16*, *79*, *96*, 149(8), *176*, 237(39), *251*, 345(21), *392*

K

Kaalio, H., 261, *289*
Kaberline, S. L., 331(133), *340*
Kabot, F., 342(19), *392*
Kaiser, D. C., 284(209), *293*
Kaiser, F. E., 272(122), 273 (122), *291*
Kaiser, R. E., 191(64), 192(64), 208(114), *220*, *222*, 309(55), 310, 311, *338*, 386(73), *394*
Kaitaranta, J., 261(62), *289*
Kambara, T., 232, 233, *250*
Kammereck, R., 309, *337*
Kananen, G., 275, *291*
Kapila, S., 206(106), *222*
Karlsson, K.-A., 272, *290*
Karmen, A., 127(24, 25), *140*, 162, *176*
Karohl, J. G., 363(54), 366(54), 369(54), *393*
Karoum, F., 254(23), 256(23), 271, *288*
Kaufman, H. R., 44(10), 53, 58 (6), *75*
Kawahara, F. K., 254(26), 258, *288*

Kealy, M. P., 197(87), 201, *221*
Keiser, W. E., 305(35), 307, *337*
Kekwick, R. G. O., 246, *251*
Kellaway, P., 277(161), *292*
Kelly, R. W., 260, 261(58), 283(199), *289, 293*
Kelly, W., 301(30), 303, *337*
Keprt, L., 156(21), *176*
Kerrin, S. L., 273(130), *291*
Keulemans, A. I. M., 28(1), 36, 194, *221*
Khare, B. N., 273(139), *291*
Killhefer, J. V., 342(13), *392*
Kimble, B. J., 319(96, 97), 326(97), 327(97), *339*
King, G., 252(7), *287*
Kirkland, J. J., 137(63), *141*
Kiss, F., 120(12), *139*
Klaucke, C., 207(113), 218(113), *222*
Klebe, J. F., 272(125), *291*
Klein, P. D., 299, 300, 323(111), 326(119), 327(119), 328(119), 329, 330, 331(119), *336, 339*
Klemm, H.-P., 264(86), *290*
Klier, M., 379(62), *393*
Klingman, J. D., 273(140), *291*
Klinkenberg, A., 80(11, 12), 96, 106(16), *113*, 179(13), *219*
Klör, H.-U., 194(81), *221*
Knapman, C. E. H., 14(7), 16, 135(53), *140*
Knedel, M., 284(203), *293*
Knight, H. S., 127(21), *139*
Knight, J. B., 320(106), *339*
Knöppel, H., 334(140), 335(140), 336(140), *340*
Knowles, J. A., 267(102), *290*
Knowles, M. E., 265(96), *290*
Ko, H., 263(84), *290*
Kocher, C. W., 139(69), *141*
Kodama, H., 232, 233, *250*
Koenig, W. A., 286(226), *293*
Koffer, J. T., 150(14), *176*
Kolb, B., 163(35, 36), *176*
Kolloff, R. H., 180(24), *219*
Korolczuk, J., 262, 287(235), *289, 294*
Koshy, K. T., 284(209), *293*
Kovats, E., 60(10), 72(35), 75, 76, 298(6, 7, 8), 299(12), *336*
Kowblansky, M., 275(152), *291*
Kramer, R. E., 347(30), 349(30), *392*
Krejci, M., 162(32), *176*
Kruppa, R. F., 186(32), *219*
Kubeczka, K. H., 308(43), *337*
Kuchmy, B., 274(143), *291*
Kugler, E., 194(80), *221*

Kuksis, A., 256(43), *289*
Kuo, K. C., 254(28), 266(28), 272(122), 273(122, 141), *288, 291*
Kwantes, A., 178(5), *218*

L

Lackard, R. G., 254(9), 262(9), *288*
Laine, R. A., 279(178), *292*
Lakings, D. B., 286(225), *293*
Lam, T. F., 331(133), *340*
Lambert, M. A., 273(136), 274(136), *291*
Lamkin, W. H., 273(132), *291*
Lamparski, L. L., 139(67, 70, 72, 73, 75), *141*
Lang, R. E., 276(159), *292*
Langer, S. H., 70(30), 71(30), 76, 104(13), *113*, 124(13), 125(13), 126(13), 136(13), *139*, 179(15), *219*, 277(168), *292*
Langhorst, M. L., 139(71, 75), *141*
Langlais, R., 194(80), *221*
Langlois, W. E., 244(50), *251*
Langvardt, P. W., 139(68, 74), *141*
Larsen, F. N., 137(61), *141*
Lawson, A. E., 146, 147(6), 150(6), *175*
Lawson, A. M., 286(228), *293*
Leary, J. J., 63, 64(14), 68(14), 69(29), *75, 76*
Leathard, D. A., 295(2), 300(20), *336, 337, 361, 393*
Lebbe, J., 61, 69(12), *75*
Lee, E. H., 120(8), 128(8), *139*
LeFevre, H. F., 271(115), *290*
Leimer, K., 272(124), 273(124), *291*
Lekova, K., 73, *76*
Lertratanangkoon, K., 277(161), *292*
Leung, A. T., 358(38), 360(38), 382(38), *392*
Levi, L., 309(53, 54), *337, 338*
Levitt, M. J., 262(68), *289*
Levy, R. L., 312, *338*
Lewis, J. S., 70(33, 34), 72(33), *76*
Lhuguenot, J.-C., 271(118), *290*
Lichtenstein, H. A., 350(33), 351(33), *392*
Liddle, J. A., 264(88), *290*
Liebich, H. M., 350(33), 351(33), *392*
Light, J. F., 68(23), *75*, 248,

Light, J. F., (continued)
  249(63), *251*
Ling, L., 46(16), *53*, 197(87),
  201, *221*
Linko, R. R., 261(62), *289*
Lipsky, S. R., 164(38), *176*, 217
  (135), 218(135), *222*
Littlewood, A. B., 15(17), *16*,
  18, *26*, 224(9), 228(9), *250*,
  304(32), 307, *337*, 349(31),
  380(63, 66), 381, *392*, *393*
Lively, L. D., 342(16), *392*
Lofberg, R. T., 287, *294*
Loo, Y. H., 264(92), *290*
Lott, C. E., Jr., 278, *292*
Lovelady, H. G., 271, *290*
Lovelock, J. E., 164(41, 42, 43),
  165(41, 47), 166(41, 47), 167
  (41), 168(41), 169(41), 170
  (41), 171(47), 172(41, 47,
  55), 173(41, 55), 174(41),
  175(39), *176*, *177*
Low, M. J. D., 301(25), *337*
Lowry, S. R., 63(14), 64(14), 68
  (14), *75*
Loyd, R. J., 53(23), *54*
Luisi, M., 280, *292*
Lundin, R. E., 46(16), *53*, 197
  (87), 201, *221*
Luukkainen, T., 261, 281, *289*
Luyten, J. A., 192(73), *221*
Lyman, K. J., 286(226), *293*
Lyons, J. G., 384(69), 388(77),
  393, *394*

M

McAdoo, D. J., 347(30), 349(30),
  392
McCaffrey, I., 127(25), *140*
McCallum, N. K., 254(24), 257
  (46), *288*, *289*
McCloskey, J. A., 286(226, 228),
  293
McCue, M., 331(132), *340*
Macdonell, H. L., 131(39), 133
  (39), *140*
McDowell, C. A., 317(86), 318
  (86), 319(90), 320(86), *338*,
  339
McErlane, K. M., 280(189), *292*
McEwen, C. N., 326(120), *339*
McFadden, W. H., 14, *16*, 308,
  317(88), 319(88, 98, 101),
  320(88), 107), 331(130),
  337, *338*, *339*, *340*
MacGee, J., 263(82), 274(148),
  275(148), *290*, *291*

McKeag, R. G., 208, *222*
McKinney, R. W., 68(23), *75*, 248,
  249(63), *251*
McLafferty, F. W., 336(141, 142,
  143), *340*
McLaren, I. H., 319(100), *339*
MacLean, I., 283(201, 202), *293*
McMurray, W. J., 217(135), 218
  (135), *222*
McReynolds, W. O., 63, *75*
McWilliams, I. G., 156(19), 157
  (19, 22), 158(19, 22), 159,
  160, *176*
Madani, C., 283(200), *293*
Magee, R. J., 15(16), *16*
Maggs, R. J., 172(55), 173(55),
  177, 368, 369(59), 372, *393*
Makita, M., 179(18), 189(18),
  219, 277(167), 279(167), *292*
Malenge, J. P., 361(42), 384
  (42), *392*
Malinowski, E. R., 331(132), *340*
Mamer, O. A., 278(173), *292*
Mangold, H. R., 309, *337*
Mantz, A. W., 301(26, 27), *337*
Margosis, M., 258, *289*, 347(26),
  392
Marshall, J. L., 204(95), *221*
Martin, A. J. P., 19(6), *26*, 114,
  126(19), 127(1), *139*, 179
  (17), 189(17), *219*, 223(1),
  237(1), 238(1), *250*
Martin, R. L., 58(8), 73(8), 74,
  75, 112(28), *113*, 179(16),
  219
Martinez, F. W., Jr., 179(17),
  189(17), *219*, 223(1), 237(1),
  238(1), *250*
Martini, A., 277(164), *292*
Matin, S. B., 254(32), 267(32),
  268(32), *288*, *290*
Mattauch, J., 319(91, 92), *339*
Maume, B. F., 271(118), 280(187,
  190), 281(190), *290*, *292*
Maume, G., 280(187), *292*
Mazor, L., 249(64), *251*
Mead, A. S., 368, 369(59), 372,
  393
Meier, S., 204(96), *221*
Melcher, R. G., 139(74), *141*
Mellstrom, B., 254(34), 277(34),
  288
Menini, E., 341, *391*
Merlivat, J. C., 341(1, 7), 365
  (7), 389(7), *391*
Merritt, C., Jr., 301(29), 302,
  304(33), 324(29), *337*, 347(30),
  349(30), *392*

Messerly, J. P., 102(9), *113*, 131(30, 31), *140*
Messner, A. E., 342(12), *391*
Metaxas, J., 274(149), *291*
Metcalfe, L. D., 127(20), *139*
Metwally, M. M. E., 32(2), *36*
Meyers, W. E., 331(137), *340*, 363(53), 364(55), *393*
Mielniczuk, Z., 262(66), *289*, 287(235), *294*
Mikkelsen, L., 388, 389, 390 (78), *394*
Millen, W., 67(21), *75*
Miller, J. M., 146, 147(6), 150(6), *175*
Miller, L. A., 362(51), 363(51), 364(51), 365(51), 366(51), 367(51), 369(51), 373, *393*
Miller, M., 311(56), *338*
Milne, G. W. A., 324(121), 326 (121, 123, 124), *339, 340*
Minyard, J. P., 311(57, 58), *338*
Miranda, B. T., 131(32), *140*
Mitchell, M. B., 365(57), *393*
Mitsuma, T., 274(143), *291*
Miwa, T. K., 299, *336*
Mlejnek, O., 262(67), *289*
Moeckel, W. E., 278(177), 280 (177), *292*
Moelwyn-Hughes, E. A., 77, *96*, 224(10), 228(10), 230(20), 235(28), *250*
Moffat, A. C., 254(31, 32), 267 (31, 32), 268(32), *288*
Molander, M., 269(108), *290*
Möller, M. R., 300(21), *337*
Moodie, I. M., 273, *291*
Mooney, E. F., 64(18), 66(18), *75*
Morgan, E. D., 284(210), *293*
Morita, H. K., 254(20), 262(20), *288*, 301(26), *337*
Morozowich, W., 254(29), 260 (29), *288*
Morrison, I. M., 278(174), 279 (174), *292*
Mortimer, J. V., 236(33), 237, *251*
Moscatelli, E. A., 263(81), *290*
Moshy, R. J., 278(176), *292*
Moss, A. M., 252, 254(1), 267(1), 269, 281(193), *287, 293*
Moss, C. W., 273(136), 274(136), *291*
Mrochek, J. E., 254(13), 278 (13), 279(13), *288*
Muggli, R. Z., 126(16), *139*

Müller, H. G. W., 385(72), *393*
Muni, I. A., 254(12), 286(12), *288*
Munson, M. S. B., 325, 326(116, 117, 118), *339*
Murty, N. L., 224(5), *250*
Musser, W. O., 341, *391*
Myers, M. M., 99(2), *112*
Myher, J. J., 256(43), *289*

N

Naegeli, P. R., 327(128), 335 (128), *340*
Nagal, Y., 326(124), *340*
Nakamoto, H., 272(123), 273(123), 274(123), *291*
Nakanishi, K., 305(39), 306, *337*
Nanbara, T., 252(4), *287*
Neher, M. B., 332(138), *340*
Niecheril, J. C., 254(12), 286 (12), *288*
Nestrick, T. J., 139(67, 68, 69, 70, 71, 72, 75), *141*
Neto, C. C., 150(14), *176*
Nickless, G., 137(59), *140*
Nicosia, S., 260, *289*
Nier, A. O., 319(93, 94, 95), *339*
Nigam, I. C., 309, *337, 338*
Nihei, N. N., 274, *291*
Nikelly, J. G., 192(77), 193(77), 208(115), *221, 222*
Noebels, H. J., 15(18), *16*
Norcup, J., 286(220), *293*
Norem, S. D., 189(49), 192(49), 194(49), *220*
Novak, J., 156(21), *176*, 345 (20), *392*
Novotny, M., 191(66), 204(91, 94), *220, 221*
Nowlin, J., 277(161), *292*

O

Ober, S. S., 244, *251*
Ogilvie, J. D., 224(3), *250*
Ohline, R. W., 131(34), *140*
Okamoto, T., 316(79), *338*
Oke, T. O., 274(146), 275, *291*
Olsen, R. W., 319(97), 326(97), 327(97), *339*
Onaka, T., 316(79), *338*
Orlowski, M., 274(144, 145), *291*
Osiewicz, R., 275(154), 276(157), *291*
Osmond, C. A., 254(10), 262(10), *288*

Ottenstein, D. M., 114(3, 4), 115 (3), 116(3), 117(3), 120(3), 121(3), 124(4), 125(3, 4), 126(4, 14, 18), *139*, 180(25) *219*

P

Pacakova, V., 286(227), *293*
Pace-Asciak, C., 259(55), 260(55), *289*
Palyi, G., 206(108), *222*
Pan, T., 273(132), *291*
Pantarotto, C., 277(164), *292*
Papa, L. J., 261, *289*
Papendick, H.-D., 386(75), *394*
Parcher, J. F., 69(29), *76*, 184 (31), *219*
Parker, D. A., 204(95), *221*
Parsons, J. S., 265(97), *290*
Patterson, P. L., 174(57), *177*
Patton, W. D. M., 256(40), *288*
Paul, W., 320(103, 104), *339*
Pearson, E. B., 146(5), *175*
Pecci, J., 275(150), *291*
Peil, A., 334(140), 335(140), 336 (140), *340*
Peralta, E., 187(35), *219*, 271 (116), *290*
Perkins, G., Jr., 342(16), *392*
Perrett, R. H., 124(13), 125(13), 126(13), 136(13), *139*, 179 (15), *219*
Perry, J. A., 15, *16*, 216(130), 222, 252(2), *287*
Perry, J. H., 80(10), *96*
Perry, M. B., 254(14), 278(174), 279(14, 174), 280(14, 184, 185), *288*, *292*
Perry, S. G., 15(17), *16*, 313(69, 70), 316, *338*
Persinger, H. E., 68(26), *76*, 360 (39), 361(39), *392*
Pesyna, G. M., 336(142), *340*
Peteranetz, K. A., 254(16), 257 (16), *288*
Petsev, N., 103(10), *113*, 247, 249(61), 62), *251*
Pettit, B. C., 164(41), 165(41), 166(41), 167(41), 168(41), 169(41), 170(41), 172(41), 173(41), 174(41), *177*
Pettit, G. R., 277(162), *292*
Pfaffenburger, C. D., 50(21), *54*, 205(105), *222*, 279(179), *292*
Pfeiffer, C. D., 139(69), *141*
Phillips, C. S. G., 224(9), 228(9), (9), 237(37), *250*, *251*, 349 (31), *392*

Pierce, A. E., 254, 277(8), *288*
Pirogova, Y. I., 181(29), *219*
Pisand, J. J., 326(124), *340*
Podmore, D. A., 341, *391*
Poelman, J., 334(140), 335(140), 336(140), *340*
Polgar, A. G., 230(23), 231, *250*
Pollak, M., 120(12), *139*
Pollard, F. H., 137(59), *140*
Polvani, F., 280, *292*
Pons, J., 341(1, 7), 365(7), 389 (7), *391*
Poole, C. F., 284(210), *293*
Porter, A. H., 264(90), *290*
Porter, W. H., 256(42), 280(42), *288*
Pospichal, O., 120(12), *139*
Pospisil, P., 163(36), *176*
Post, A., 278(177), 280(177), *292*
Potts, W. J., Jr., 305(36), 307, *337*
Prausnitz, J. M., 66(19), *75*
Pretorius, V., 232, *250*
Price, D. T., 224(9), 228(9), *250*, 349(31), *392*
Price, J. G. W., 386(76), 389 (76), *394*
Prost, M., 280(190), 281(190), *292*
Prudencio, M., 323(112), *339*
Purcell, J. E., 189(49, 50, 51), 192(49, 51), 193(51), 194(49, 82), 195(82), 196, 217(135), 218(135), *220*, *221*, *222*
Purnell, J. H., 14, 15, *16*, 18(3), 21(3), 22(3), *26*, 40, *53*, 77, 79, 96, 99, 101(5), *112*, 124(13) 125(13), 126(13), 136(13), *139*, 179(14, 15), 180(14), 181, 186 (14), 194(14), 194(14), 213 (123, 124), *219*, *222*, 224(8), 225, 228(8, 15), 229(8), *250*

Q

Quinn, C. P., 101(5), *112*

R

Rabinowitz, M. P., 256(41), *288*
Radford, R. D., 126(17), *139*
Raether, M., 320(104), *339*
Rainey, W. T., 254(13), 278(13), 279(13), *288*
Rall, H. T., 316(80, 81), *338*

Ranfft, K., 277(163), *292*
Rapp, U., 204(96), *221*
Raulin, F., 273, *291*
Ray, N. H., 224(2), *250*
Redweik, U., 127(26), *140*
Rehak, V., 286(227), *293*
Reichardt, C., 60(9), *75*
Reid, R. C., 66(19), *75*
Reiffsteck, A., 284(206), *293*
Reisberg, P., 256(41), *288*
Reschke, R. F., 102(9), *113*, 131 (30, 31), *140*
Rice, C. B. F., 308, *337*
Rice, H. L., 277(162), *292*
Richards, H. M., 197(88), 202, *221*
Richardson, T., 32(2), *36*
Riess, W., 268(106), *290*
Rijks, J. A., 180(19, 20), 214 (19), *219*
Rijnders, G. W. A., 178(5), *218*
Roaldi, A., 127(22), *139*
Robb, E. W., 274(147), *291*
Roberts, T. R., 319(94), *339*
Robertson, J. H., 258(50), *289*
Robinson, A. L., 382(68), *393*
Roboz, J., 254(30), 264(30), *288*
Roehl, T. J., 272, *291*
Roeraade, J., 191(68), 192(68), 204(68), 206(68), *220*
Rogers, L. B., 106, *113*, 236(29), *250*
Rohrschneider, L., 60(11), 61(11), 62(11), *75*
Rooney, T., 216(134), 217(134), 218(134), *222*
Röper, H., 264(85), *290*
Rosie, D. M., 149(12), *176*, 342 (11, 12), *391*
Rosmus, J., 273, *291*
Ross, W. D., 175(61), *177*
Rotzche, H., 68(25), *75*
Rouayheb, G. M., 342(16), *392*
Rowland, M., 254(32), 267(32), 268(32), *288*, *290*
Royer, M. E., 263(84), *290*
Ruck, B., 342(9), *391*
Ruelius, H. W., 267(102), *290*
Ruhl, H. D., 301(24), *337*
Russell, C. P., 131(33), *140*
Russell, D. S., 58(4), *75*
Russell, G. F., 197(88), 202, *221*
Russell, P. T., 259(53), *289*
Ruthven, C. R. J., 254(23), 256 (23), 271(23), *288*
Rutten, G. A. M. F., 192(73), 206(107, 109), *221*, *222*
Ruyle, C. D., 163(33), *176*, 273 (126), *291*

Ruzicka, J. H. A., 286(220), *293*
Ryan, J. F., III, 326(122), *339*
Ryan, J. M., 178(2), *218*
Ryhage, R., 326(125), *340*

S

Sabatier, A., 299(16), *336*
Saha, N. C., 120(10, 11), 121 (10), 122(10, 11), 123, *139*
Saharabudhe, M., 309(53, 54), *337*, *338*
Said, A. S., 215, *222*, 244(52), *251*
St. John, P. A., 175(59), *177*
Sakauchi, N., 283(198), *293*
Samuelsson, B., 261, *289*
Sandler, M., 254(23), 256(23), 271(23), *288*
Sandra, P., 205(99), *221*
Sangster, I., 283(201), *293*
Sargent, M., 163(37), *176*
Sarmiento, R., 316(78), *338*
Saroff, H. A., 127(24), *140*
Saunders, D. G., 285(217), *293*
Savitzky, A., 363(52), 365(52), 366(52), 367, *393*
Scanlon, J. T., 42(7), *53*
Scarpellino, R. J., 304(34), *337*
Schauenberg, H., 334(140), 335 (140), 336(140), *340*
Schauer, R., 278(172), *292*
Scheinthal, B., 275(152), *291*
Schewe, L. R., 254(19), 278(19), *288*
Schlereth, G. A., 362(50), *393*
Schmauch, L. J., 149(9), *176*, 203, *221*, 357(35), *392*
Schmidt, R. N., 331(137), *340*, 363 (53), 364(55), *393*
Schneckenburger, K. H., 386(73), *394*
Schneider, W., 188(43, 44), *220*
Scholler, R., 284(206), *293*
Scholly, P. R., 342(14), *392*
Scholz, R. G., 126(15), *139*
Schomburg, G., 189, 205(101), 206(101, 110), 207, *220*, *221*, *222*, 299(14, 19), *336*, 386(74), 388(74), *394*
Schröder, U., 204(96), *221*
Schroeder, J. P., 58(7), *75*
Schulte, E., 208(116), *222*
Schulz, P., 262, *289*
Schumacher, G., 280(191), 281(191), *292*
Schupp, O. E., III, 70(33, 34), 72(33), *76*
Schwedt, G., 271(113), *290*

Scoggins, M. W., 266(100), *290*
Scott, J. C., 386(76), 389(76), *394*
Scott, R. L., 55, 56(1), 57(1), 70(32), *75*, *76*, 228(17), *250*
Scott, R. P. W., 15(17), *16*, 181, 187, *219*, 236(30), *250*, 342(8), *391*
Scotto, L., 264(92), *290*
Sebastian, I., 137(58), *140*
Sedvall, G., 254(22), 256(22), 271, *288*
Segura, J., 187(35), *219*, 271(116), *290*
Seiber, J. N., 258, *289*
Selby, S. M., 103(12), *113*
Shabtai, J., 58(5), *75*
Shank, J. T., 68(26), *76*, 360(39), 361(39), *392*
Shapshak, P., 273(139), *291*
Sharkey, A. G., 277(168), *292*
Shaw, S. R., 254(29), 260(29), *288*
Sheehan, R. J., 70, *76*, 104(13), *113*
Shelby, S. M., 149(11), *176*
Sheppard, A. J., 263(74, 76), *289*
Sherman, W. R., 279(181), *292*
Sherwood, T. K., 66(19), *75*
Shewbart, K. L., 273(128), *291*
Shorland, P. B., 224(6), *250*
Shtaerman, M. Y., 181(29), *219*
Shunbo, F., 350(33), 351(33), *392*
Shurlock, B. C., 295, 300(20), *336*, *337*
Sicard, G., 362(47), 366(58), 388(58), *393*
Siegfried, J., 127(26), *140*
Sievers, R. E., 175(61), *177*
Sigelle, R., 341(1, 7), 365(7), 389(7), *391*
Silverman, R. W., 322(110), 323(110), *339*
Simmonds, P. G., 164(41), 165(41), 166, 167, 168, 169, 170, 172(41), 173, 174, *177*
Simmons, M. C., 224(3), *250*
Sisenwine, S. F., 267(102), *290*
Sjenitzer, F., 80(12), *96*
Sjöquist, J., 273(135), 274(135), *291*
Sjovall, J., 280(191), 281(191), *292*
Slanski, J. M., 278(176), *292*
Sloneker, J. H., 277, *292*
Smead, D. L., 186(32), *219*
Smith, C. E., 296(3), *336*
Smith, D. H., 163(34), *176*

Smith, E. D., 126(17), *139*, 273(128), *291*
Smith, J. F., 127, 128(27), *140*
Smith, R. V., 257(45), 263, *289*
Snavely, M. K., 307(41), *337*
Snyder, L. R., 136(56), *140*
Soderquist, C. J., 258(48), *289*
Solomon, H. M., 276(160), 277(160), *292*
Solow, E. B., 274, *291*
Soltzberg, L. J., 331(133), *340*
Soukup, R. J., 304(34), *337*
Sparagana, M., 307, *337*
Spencer, S. F., 180(24), *219*
Sprinkle, T. J., 264(90), *290*
Stafford, M., 263(83), *290*
Stahl, E., 308, *337*
Stalling, D. L., 163(33), *176*, 273(126), *291*
Staszewski, R., 114(5), *139*
Stehl, R. H., 139(70, 72, 73), *141*
Steinmann, B., 127(26), *140*
Steinwedel, N., 320(103), *339*
Sternberg, J. C., 156, 160(18), 161(18), *176*, 188(37), 198, 199(37), 203, *219*, 342(15), *392*
Stevens, M. P., 312, *338*
Stillwell, R. N., 277(161), 286(226, 228), *292*, *293*
Stillwell, W. G., 277(161), *292*
Stock, R., 15(17), *16*, 308, *337*
Stocklinski, A. W., 257(45), *289*
Stockwell, P. B., 313(71), 314, *338*
Stone, W. G., 174(56), *177*
Story, M. S., 320(106), 322(109), 323(109), *339*
Street, H. V., 267, 275(155), *290*, *291*
Stringham, L. R., 274(143), *291*
Struppe, H. G., 215(29), *222*
Suffet, I. H., 197, *221*
Sugden, T. M., 160(25), 161(25), *176*
Sullivan, J. E., 254(19), 278(19), *288*
Summers, T. R., 274(149), *291*
Sunshine, I., 275(154), 276(157), *291*
Surace, M., 280, *292*
Sutre, P., 361(42), 384(42), *392*
Svojanovsky, V., 162(32), *176*
Swanton, W. T., 44, 45, *53*, 178(3), 180(3), 188(3), 189(3), 190(3), 197(3), 198, 203(3), 204(3), 213(3), 216(3), *218*
Sweeley, C. C., 179(18), 189(18), *219*, 261, 277, 279(167, 178),

Author Index

Sweeley, C. C., (continued)
  289, 292, 323(113), 339
Swift, S. M., 301(29), 302, 324
  (29), 337
Synge, R. L. M., 179(12), 219
Szafranek, J., 279(179), 292
Szczepanik, P. A., 326(119), 327
  (119), 328(119), 329, 330,
  331(119), 339
Sze, Y. L., 126(18), 139
Szymanski, H. A., 305(35), 307,
  337

T

Tacker, M. M., 286(228), 293
Tadmor, J., 107(19), 113
Takacs, J., 249(64), 251
Takeda, H., 273(137), 291
Takeuchi, T., 312, 338
Tambawala, H., 277(169), 292
Tanaka, F. S., 285, 293
Taylor, A., 286(221), 293
Taylor, G. W., 43, 44, 53
Taylor, P. L., 283(199), 293
Taylor, R., 312(61), 338
Tedder, J. M., 264(89), 290
Teranishi, R., 46(13), 51(13),
  53, 178(1), 190(57), 218,
  220, 302(31), 304, 337
Tesarik, K., 162(32), 176
Thenot, J.-P., 50(21), 54, 205
  (105), 208(105), 222, 263(83)
  266(99), 281(196), 290, 293
Thomas, B. S., 285(211, 214), 293
Thompson, A. C., 311(57, 58), 338
Thompson, C. J., 316(80, 81), 338
Thomson, G. H., 362(45), 364(45),
  385(45), 393
Thorstenson, J. H., 285, 293
Tindle, R. C., 163(33), 176
Tishbee, A., 350(33), 351(33),
  392
Tjoa, S. S., 278(173), 292
Trash, C. R., 65, 66(16), 67(16),
  75, 249(60), 251
Travis, B., 264(95), 290
Treble, R. D., 163(37), 176
Tretbar, H. A., 341, 391
Truter, E. V., 308, 309(47), 337
Tsai, S. L., 263, 289
Tsuboyama, K., 286(228), 293
Tsuge, S., 63(14), 64(14), 68(14),
  75, 312, 338
Tsuji, K., 258(50), 289
Tswett, M. S., 1, 16
Tufts, L. E., 127(22), 139
Tumlinson, J. H., 311(57, 58),
  338

Tung, R., 172(54), 177
Turner, B. C., 286(222, 223), 293
Turner, L. P., 261, 289
Tyler, J. H., 275(151), 291
Tyler, S. A., 299, 300, 336

U

Uden, P. C., 137(59), 140
Urone, P., 184(31), 219

V

Vanatta, L. E., 285(217), 293
van de Craats, F., 342(10), 391
van Deemter, J. J., 80, 96, 106
  (16), 113, 179, 194, 219, 221
Vandenheuvel, W. J. A., 47, 48
  (17, 18), 54, 287, 294
van der Pol, J. J. G., 190(58),
  192(58), 204(58), 205(58),
  206(58), 220
VanDerSlik, A. L., 284(209), 293
Vane, F., 259(54), 289
VanGent, C. M., 263(78), 289
Vanluchene, E., 205(99), 221
Van Marlen, G., 331(134), 340
van Rijswick, M. H. J., 375(60,
  61), 376, 377(60), 378(60),
  379(60), 380(60), 381, 382
  (60), 383, 389, 393
Van Swaay, M., 15(17), 16
van Tiggelen, A., 160(26), 161
  (26), 176
Vaughan, G. A., 229(18), 250
Venkataragnavan, R., 336(142,
  143), 340
Venturella, V. S., 276(159), 292
Verneer, E. A., 205(100), 221
Versino, B., 334, 335(140), 336
  (140), 340
Verstappe, M., 192(72), 221
Verzele, M., 192(71, 72), 205
  (99), 221
Vessman, J., 269(107, 108, 109,
  110), 290
Vestal, M. L., 326(123), 339
Vilceanu, R., 262, 289
Villwock, R. D., 336(141), 340
Viska, J., 120(12), 139
Vissers, H., 334(140), 335(140),
  336(140), 340
Vogh, J. W., 261, 289
Vogt, W., 284, 293
Vollmin, J. A., 204(93), 221
Von Minden, D. L., 286(226), 293
Vyakhirev, D. A., 181(29), 219

W

Waalkes, T. P., 254(28), 266(28), 286(225), *288*, *293*
Walker, J. Q., 316(73, 74), *338*
Wall, F. T., 225(13), 230(19), *250*
Wall, R. F., 15(18), *16*, 120(8), 128(8), *139*
Wallace, D., 342(17), 363(17), 365(17), 380(65), 385(17), *392*, *393*
Walle, T., 254(25, 33), 258(25), 267(33), 268(25), 279(180), *288*, *290*, *292*
Walls, F. C., 319(97), 326(97), 327(97), *339*
Walsh, J. T., 301(29), 302, 304(33), 324(29), *337*
Walton, D. R. M., 285(211), *293*
Ward, C. C., 316(80, 81), *338*
Ward, D. N., 273(132), *291*
Wassink, J. G., 190(58), 192(58), 204(58), 205(58), 206(58), *220*
Watson, A. J., 164(43), *177*
Watson, E., 254(30), 264(30, 94, 95), *288*, *290*
Watts, R. B., 246, *251*
Watts, R. W. E., 264(93), *290*
Weast, R. C., 103(12), *113*, 149(11), *176*
Weatherford, J. W., 273(132), *291*
Webb, A. C., 280(185), *292*
Weeke, F., 189(45), 205(101), 206(101), 207(101), *220*, *221*, 299(19), *336*
Wehrli, A., 72(35), *76*
Weihman, K., 342(18), 362(18), 363(18), 365(18), 366(18), 367(18), 368(18), 369, 374(18), *392*
Weimann, B., 358(36), 361(36), 362(36), 365(36), 366(36), 367(36), 368(36), 370(36), 380(36), *392*
Weisbach, J. E., 278(177), 280(177), *292*
Weiss, A. H., 277(169), *292*
Weiss, M. D., 15(18), *16*
Wells, W. W., 277(167), 279(167), *292*
Welsch, T., 207(113), 218, *222*
Welss, W. W., 179(18), 189(18), *219*
Welti, D., 308(44), *337*
Wender, I., 277(168), *292*

Wentworth, W. E., 165(46, 47), 166(47), 168, 171(47, 48, 49,
West, J. C., 262, *289*
Westbrook, J. J., 274(147), *291*
Westlake, W. E., 142, 162(2), 174, *175*
Wheals, B. B., 286(219, 220), *293*
Wheeler, L. O., 386(76), 389(76), *394*
White, D. M., 272(125), *291*
White, E. R., 278(177), 280(177), *292*
White, R. G., 305(38), 307, *337*
Whitham, B. T., 43, *53*
Whitnash, C. H., 312(62), *338*
Whyman, B. H. F., 156, 157, *176*, 188(40), *220*, 224(4), 225(4), *250*
Wiberley, S. E., 305(37), *337*
Wick, E. L., 46(13), 51(13), *53*, 178(1), 190(57), *218*, *220*, 302(31), 304, *337*
Wickramasinghe, J. A. F., 254(29), 260(29), *288*
Widmark, G., 204(91), *221*
Wien, R. G., 285, *293*
Wikström, S., 279(180), *292*
Wiley, W. C., 319(100), *339*
Wilk, S., 254(30), 264(30, 94, 95), 267(103), 274(144, 145), *288*, *290*, *291*
Wilkins, C. L., 331(133), *340*
Williams, C. M., 254(11), 264(11, 90), *288*, *290*
Willner, J., 271, *290*
Winkler, H. U., 324(121), 326(121), *339*
Wolfe, C. J., 316(73, 74), *338*
Wolfe, L. S., 259(55), 260(55), *289*
Wolfensberger, M., 127(26), *140*
Woodford, F. P., 263(78), *289*
Worthing, A. G., 80(8), *96*
Wotiz, H. H., 284(204), *293*
Wright, P. G., 148, *176*
Wyatt, A. J., 364(56), 368(56), 369(56), 370(56), 371, 372, *393*

Y

Yang, F. J., 208, 209(122), 210(122), *222*
Young, N. D., 323(113), *339*
Young, R. M., 276(157), *291*

Z

Zarembo, J. E., 278(177), 280 (177), *292*
Zerenner, E. H., 216(133), 217 (133), *222*
Ziegler, E., 386(74), 388(74), *394*
Zieserl, J. F., 261(60), *289*
Zinbo, M., 279(181), *292*
Zion, T. E., 277(161), *292*
Zlatkis, A., 44(10), *53*, 58(6), *75*, 164(41), 165(41), 166(41), 167(41), 168(41), 169(41), 170(41), 171(48), 172(41), 173(41), 174(41), *177*, 204(94), *221*, 350(33), 351, *392*
Zuiderweg, F. J., 80(11), *96*, 106(16), *113*, 179(13), *219*
Zulaica, J., 312, *338*
Zumwalt, R. W., 254(28), 266(28), 272(122, 123), 273(122, 123, 127, 141), 274(123), 286(225), *288*, *291*, *293*

SUBJECT INDEX

A

A term, see van Deemter equation: A term
Acetylacetone, 69
Acids, carboxylic, see Carboxylic acid derivatization
Acids: Phenolic, Hydroxy, and Keto: Derivatization
   detection by electron capture
     pentafluoropropionic anhydride, 264
   detection by flame ionization
     BSTFA:TMCS, 264
     oxime, then BSTFA:TMCS, 264
Adsorbability, 59
Adsorption, 124
   gas-liquid interface, 58
   vs. temperature, 230
Adsorptivity
   diatomaceous earth, 124-125
   glass, 131, 132
     Ca^{++}, 132
     sodium silicate, 132
Air
   in flame ionization detection, 158
   in hot wire detection, 152
Alcohol derivatization
   detection by electron capture
     pentafluoropropionyl preferred, 256
   detection by flame ionization
     alditol acetates, 256
     boronic acids, 256
     silylation, 256
Alditol acetate formation, 280
Algorithms
   area apportionment
     central caution, 366
     curve fitting, 371, 379, 380
     dropped perpendicular, 369, 372
     resolved peak, 369
     skimming, 369
     solution, general: none, 370
   baseline, 366
   data smoothing, 365

Algorithms (continued)
   data smoothing (continued)
     group averaging, 365
     group size vs. peak width, 366
     definition, 331, 364
     drift, 366
     vs. good separations, 366, 385
     vs. human capabilities, 364
     limitations in GC, 366
     noise level, 365
   peak
     area, 368
     definition, 367, 368
     detection, 367
     end, 368
     identification, 374
     maximum, 367
     model shapes in, 375, 380
     skewed, 373
   vs. personal responsibility, 385
   problems in GC, 364
   spikes, 365
   van Rijswick, 375-382
     background limits, 377
     baseline correction, 379
     curve fitting, 379-381
     fusing limits, 378
     logical framework, 375
     matched filter, 375, 379
     peak area, minimum, 377
     peak recognition, 377
     peak width as dynamic variable, 375, 377
     signal-to-noise ratio, 377
     spurious peak rejection, 377
Amines derivatization, 265
   detection by electron capture
     comparison, sensitizations, 267, 268
     PFB and HFP derivatives vs. amine structure, 268

Amines derivatization (continued)
  detection by flame ionization
    methylation with DMF-DMA, 266
    Hofmann degradation, 267
    stability vs. derivatization, 257
Amines, primary: derivatization
  detection by flame ionization
    DMF-DMA, 266
    silylation, 266, 267
    trifluoroacetylation, 266
Amines, tertiary: derivatization
  detection by electron capture
    ring trifluoroacylation, 268
    via quaternary intermediate, 269
Amino acids: the N-TFA, n-butyl reaction, 273
Anomerization, 278
Antibiotics derivatization
  detection by flame ionization
    silylation usual, 258
Assumptions of gas chromatography, 27, 79
Attenuator, 155
Aue deactivation vs. temperature, 218

B

B term, see van Deemter equation: B term
Band movement, in PTGC, 244
Barbiturates, 275
Barbiturates: derivatization
  acetal formation, 276
  Dünges procedure in, 276
  flash alkylation, 274
    drawbacks, 275, 276
    improvements using TMAH, 275
    Solow procedure, 274
  Greeley reaction, 276
  Hofmann degradation, 267
  sequential methylation, silylation, 277
Baseline
  as error source, 366
  vs. sample size, 366
Beckmann fission, oxime to nitrile, 281
Benzyl cyanide, 69
Boiling point, vs. carbon number in homologous series, 229
Bonded stationary phase, 136-139
  Aue-type, 137-138
  bleeding and, 137
  brushes, 137
  and hydrolysis, 137
  Nestrick-Dow-hnu, 139
Boronic acids in derivatization, 283
Brushes, 137
BSA, 254
BSA-TMCS, 254
BSTFA, 254
  in acid silylation, 262
  invented for amino acid silylation, 273

C

C term, see van Deemter equation: C term
Calcining, 116
Calibration necessity, 390, 391
Capacity factor, see Partition ratio
Carbohydrate
  anomerization rates, 278
  anomers, 278
Carbohydrates: derivatization, see Derivatization, carbohydrates
Carbon skeletal determination, 316
Carbonyl derivatization, 261, 262
Carboxylic acid derivatization
  detection by electron capture
    trichloroethylation, 263
  detection by flame ionization
    butylation, Greeley reaction, 262
    methylation, 262
    silylation, 262
  detection by thermionic flame ionization
    phosphonic acid esters, 262, 263
Carrier gas, 1
  depth, 107, 109
    and particle size, 107, 109
  flow
    in open tubular columns, 194
    pressure required, vs. particle diameter, 180
  flow control
    importance in quantitative analysis, 388
    in temperature programming, 239
  flow rate, vs. TCD sensitivity, 150
  thermal conductivities, 149
  velocity, 5, 84, 86
    optimum, 5, 98, 180
Catecholamines and indoleamines: derivatization
  detection by electron capture
    penta- and hepta-fluoro derivatives preferred, 271
  detection by flame ionization
    and mass fragmentography
      N-TFA O-TMS, 270-271
      pentafluoropropionic anhydride, 271
      TSIM, then BSA:TMCS, 269
    ideal sequenced derivatization, 269, 270
    structures, names, abbreviations, 270
Chemical ionization (CI) mass spectra, 324, 326

Subject Index

Chromatography, 1
Chromosorb P, 117, 118, 119, 120, 122, 124, 128
  efficiency vs. loading, 128
Chromosorb G, 119, 122, 123, 124
  efficiency vs. loading, 128-129
Chromosorb W, 119, 121, 122, 124
  efficiency vs. loading, 128
Cigarette smoke, middle fraction, 212
Clausius-Clapeyron equation, 224, 225
Column, 1
  characteristic β, 22
  conditioning, 247
  efficiency
    vs. diameter, 44, 178, 181
    improvement with packed columns, 101
    vs. loading, 44, 101, 105-106, 128, 180
    vs. particle size, 98-99, 180
    vs. particle size distribution, 98-99, 180
    vs. pore size, 120-122
    vs. pressure drop, 180
    vs. temperature, 232-236
  envelope, 178
  evaluation, 187
  history
    emphasis, 180
    R. P. W. Scott, 180
  length, 178
  open tubular, see Open tubular columns
  overload, 28
  packed
    history, 179, 180
    modern, 181
  performance
    comparisons, 215, 216
    evaluation, 211
    measures, see Column performance measures
  support, 178
  temperature
    a best temperature for each solute, 232
    vs. column efficiency, 232
    vs. solvent efficiency, 230
  testing parameters, 187
  use, general, 178
Column performance measures
  comparisons, Halasz and Heine, 216
  effective plates, 213
  performance parameter, 214
  permeability, 214
  Said expression, 215
  theoretical plates, 213
Computer identification of mass spectra, 331-336
Concentration at peak apex, 346
Concentration sensitivity in detectors, 143
Coulombic forces, 55
Coupling, A and C_2, 110

Correlated analyses, efficacy, 51
Cristobalite, 116
Curve fitting, reasons for unreliability of, 380, 381

D

Deadband, 354
Deactivation, see Diatomaceous earth
Debye forces, 10
Decibel, 355
Derivative, properties for GC, 252
Derivatization, 38
  acids
    carboxylic acids, see Carboxylic acid derivatization
    fatty acids, see Fatty acid derivatization; see also Methylation
    phenolic, hydroxy, and keto acids, see Phenolic, hydroxy, and keto acids derivatization
    sulfonic acids, see Sulfonic acids derivatization
  amides, imides
    barbiturates, see Barbiturates and see also Barbiturates derivatization
    pentafluoroderivatives, 277
  amines
    catecholamines and indoleamines, see Catecholamines and indoleamines derivatization
    detection by electron capture, see Amines derivatization, by electron capture
    guanido compounds, 272
    nitrosoamines, 271
    overall comments, 265
    phosphoryl amines, 272
    primary, see Amines, primary, derivatization
    simple, 265
    tertiary, 265-269
  amino acids
    criteria for suitable derivatives of, 272
    detection by electron capture, 274
    2,4-dinitrophenyl derivatives, 273
    N-trifluoroacetyl butyl ester derivative of choice, 273
    phenyl thiohydantoins, 273
    silylation not method of choice, 272
  carbamates and ureas
    methylation, 285
    trifluoroacetylation, 285

Derivatization (continued)
  carbohydrates
    alditol acetates, 270, 280
    amino sugars, 279, 280
    anomeric center elimination, 278
    anomerization, 278
    glucuronides, 279
    oxidation to acid before silylation, 279
    oximation before silylation, 279
    silylation using TSIM, 278
    sugar phosphates, 279
    Sweeley silylation, 277
    trifluoroacetylation, 278
  carbonyls
    methyl oximes, 261
    phenylhydrazones, 262
  epoxides, reaction with TMCS, 286
  glycoproteins, 280
  hydroxy compounds
    alcohols, see Alcohol derivatization
    antibiotics, see Antibiotics derivatization
    phenols, see Phenol derivatization
    prostaglandins, see Prostaglandin derivatization
  hydroxyamines, salts as initial reactants, 270
  ideal, 252
  ideal sequenced, 252, 269, 270
  imidazoles, 286
  indoles, 286
  nucleic acids, 286
  phenol alkylamines, 270
  phosphonates, 287
  phosphoric acids, 287
  purines, pyrimides, nucleosides, and nucleotides, 286
  sequenced, 252
  silylation
    reagents, comparison, 281-283
    flophemesyl reagent, 284
  steroids, see Steroids derivatization
  sulfonamides, 287
  thiols
    silylation, 287
    trifluoroacetylation, 287
  ureas, 277
Detectability D, 143
Detection by electron capture, sensitizing derivative groups, 254, 255
Detectors, 142-177
  electron capture, see Electron capture detector
  flame ionization, see Flame ionization detector
  foolproof, 146
  overload, 27
  mass spectrometer, see Mass spectrometer
  requirements for wide use, 142-146
    ease of use, 146
    linear dynamic range, 144-145

Detectors (continued)
  requirements for wide use (continued)
    response time, 145-146
    sensitivity, 143
    stability, 143, 144, 145
    temperature range, 145
    volume, effective internal, 145
  thermal conductivity, see Thermal conductivity detector
  thermionic, see Thermionic detector
  volume, effective internal, 203
DeWet-Pretorius equation, 232
Dexsil, 67
Diatomaceous earth, 114-131
  adsorptivity, 124-126
    vs. loading, 126
  area per volume, 124
  Chromosorb G, 119, 122, 123, 124, 128, 129
  Chromosorb P, 117, 118, 119, 120, 122, 124, 128
  Chromosorb W, 119, 121, 122, 124, 128
    friability, 121
  deactivation, 125-127
    acid washing, 125
    by carrier gas, 127
    by sample, 127
    stationary phase alteration, 126
    suffixes, 125-126
    surface-active agents, 126
    silylation, 125
  densities, packed, 124
  efficiency vs. loading, 128
  Gas-Chrom S, 122
  loadings, equivalent, 129-131
    equations, 130
  pink, 116
  pores, 118-119
  pore size, 120-124
    and column efficiency, 120
    vs. loading, 128
    preferred properties of, 120-121, 127-129
  silylation, 125
  structure, 114-119
    primary, 115-116
    secondary, 115-116, 117, 118
  support suffixes, 125-126
  thermal insulator, 33
  Type I, 116-122
    adsorptivity, 124-125
    manufacture, 116
  Type II, 117-122
    inertness, 124, 126
  white, 117
Diatomite, 114
Diatoms, 114, 115, 116
  and diffusion, 114

Subject Index

Diazomethane, 260
Diffusion, 77-79
　chamber, 199
　coefficient, 78
　and diatomaceous earths, 114
　Einstein's law of, 79, 84
　Fick's first law of, 78
　gas, 7
　　and resistance to mass transfer in the gas phase, 107
　liquid, 8, 103
　　of polymeric stationary phases, 66-67, 103
　vs. viscosity, 66-67
Digital integrators, 360
　deficiencies, 361
　drift corrector, 361
　linear dynamic range, 360
　memory, 360
　slope detector, 360
Digital technology in gas chromatography
　algorithms, see Algorithms
　central lesson, 366
　computers
　　large, 385
　　small, 384
　demand for, vs. mixture complexity, 384
　digital data processors: the computing integrator, 382, 383
　reports, 385
　satellite systems, 385
　separation requirements, 366, 385
　signal handling, 362
　signal sampling, 362
　　analog, 363
　　digital, 363
　　integration to pulse, 363
　　pulse formation, 363
　　pulse rate, minimum, 364
　signal transmission
　　analog, 362
　　digital, 362
　　linear dynamic range, 362, and noise in, 363
　small on-line computer, 384
Dimethoxypropane, 260
Dimethyldichlorosilane, 125
Dimethylformamide, 43-44, 56
N,N-dimethylformamide-dimethylacetal, 266
Dimethylsilicones, 70
Dimethylsulfolane, 44, 73
Dioxane, 60, 61
Dipole forces, 10, 56-57
Direction of gas chromatography, 37, 52-53
Direction-focusing mass spectrometry, 317-319
　first-order, 318
Disc integrator, exemplary chart, 359
Dispersion forces, 10, 57
Distillation, theoretical plate in, 4, 10
DNA silylation, 286

Drying agents
　BSA, 271
　dimethoxypropane, 260
Drying for silylation, 277
Drying heat-labile substances, 260
Durapak, 137

E

Eddy diffusion, 7, 83-84
Effective carbon number, 161
Effective plates, 213
Effective plate production rate, 213
Effective theoretical plates, 4, 17
Einstein's Diffusion Law, 79, 84
Electron
　affinity, 171
　impact (EI) mass spectra, 324, 325
　thermal, 165
Electron capture
　coefficient, 171
　dissociative mode, 171
　nondissociative mode, 171
Electron capture detector
　carrier gas in, 168, 174
　carrier gas selection, vs. construction, 174
　constant-current pulse modulated, 173
　construction, 165, 166
　current, standing, 165
　electron capture, 166
　　vs. carrier gas concentration, 168
　　in zero field conditions, 166
　electron concentration, vs. pulse frequency, 169
　invention, 164
　linear dynamic range, 171
　　vs. reactions after electron capture, 171
　mechanisms, 171
　mode of action, 165
　pulse amplitude in, 167
　pulse modulated, linear dynamic range, 174
　pulse sampling in, 166
　response
　　vs. molecular structure, 174, 175
　　vs. temperature, 171
　selectivity, 174, 175
　sensitivity, 166
　　vs. interval between pulses, 170
　　vs. pulse frequency, 173
　　vs. pulse period, 167

Electron capture detector (continued)
  sensitivity (continued)
    vs. source activity, 170
    temperature dependence, 171, 172
  sensitizing derivatizing groups, 254, 255
  signal in, 165
  sources
    tritium, 165
    nickel-63, 165
  temperature limits, 165
  thermostatting required, 172
Electron micrographs
  scanning
    Chromosorb G, 123
    Chromosorb P, 117, 118, 119
    Chromosorb W, 122
  transmission
    Chromosorb P, 120
    Chromosorb W, 121
Emulphor-O, 71-72
Equilibrium, 11
  assumption of instantaneous, in GC, 21, 80
Evaluation of column performance, 211

F

Fatty acid derivatization
  detection by electron capture, trichloroethylation in, 264
  methylation
    BF_3-methanol, 263
    carbonyldiimidazole, 263
    on injection, 263
Fick's first law of diffusion, 78
Film thickness of stationary phase, 8, 100, 101, 105-106
Firebrick, 6, 124-125
Flame ionization detector, 156
  air: precise control needed for quantitative analysis, 388
  air/hydrogen ratio, 158
  area percent as weight percent, 342
  construction, 156
    and water condensation, 159
  detectability, 143
  detection of non-burnables, 162
  effective internal volume, 145
  efficiency, 159
  electrodes, response factors and, 156
  hydrogen: precise flow control needed for quantitative analysis, 388
  ionization mechanisms, 160, 161
  linear dynamic range, 156
    electrode shape in, 156
    vs. positive space charge, 157
    vs. voltages used, 157
  noise, 156, 159
  overloading, 162
  proportional region, 157

Flame ionization detector (continued)
  response
    vs. hydrogen flow rate, 158
    vs. molecular structure, 161
  sensitivity, 156
    mass, 143
    optimization, 158
  voltages in, 157
  volume, effective internal, 156
Flavor, 46
  analysis, artifacts in, 51
  trace components, 51
Flow controllers
  deficiencies for quantitative analysis, 388, 389
  temperature control needed for precise quantitative analysis, 389
Flow inequalities within the packed column, causes of, 109
Fluorene picrate, 43
Flux calcining, 117
Forward search, 331
Fused quartz, origin, 217
Fused silica
  purity, 217
  manufacture, 217
  surface tensile strength, 217
Fused silica open tubular columns, 216-218
  coating of on outside, polyimide, 217
  deactivation, 218
  flexibility, 216
  inertness, 217
  vs. packed columns, 218
Fusing limit, 378

G

Gas
  diffusivity, 7, 84, 100, 101, 107-109
  holdup
    time, 17
    volume, 19, 20
    universality for solutes, 84
  makeup, effect on apparent internal volume, 145
  sampling, 29-30
  thermal conductivities, 149
  thermal conductivity, mechanism, 148-149
Gas chromatography-mass spectrometry, 301-303, 317-336
  match, 301
    interface, 302-303
Gas chromatography-thin layer chromatography, 308-311
  coupling, 309-311
  history, 309

Gas-liquid interface, 58
Gas-solid chromatography, 135-136
Glass beads, 131-133
 composition, 132
 Corning GLC-110, 133
 roughened, 132
  composition, 132
Glass open tubular columns
 adsorption on, 205
  and catalytic decomposition, 205
  tests for, 205
 apolar coatings, 205
 commercial, 205
 deactivation, 206
  particle creation, 208
  particle injection, 208
 drawing machine for, 204
 sample injection
  on-column, 208
  splitless sampling, 208
Gaussian distribution
 properties of, 80, 81
 standard deviation, 80, 81
 variance, 80
Gaussian peak shape, 79-82
 properties of, 80, 81
 variance, 80, 81
Glueckauf separation factor, 42

H

Heat of vaporization
 Clausius-Clapeyron equation, 224, 225
 and temperature, 224
 and vapor pressure, 224
HEETP, 213
Height equivalent to a theoretical plate, 4-5, 81, 97
Helium
 in open tubular columns, 194
 thermal conductivity, 149
Henry's Law, 226, 227
 constant, inverse of the partition coefficient, 226, 227
HETP, 213
Hexamethyldisilazane, 125
Hildebrand-Scatchard solubility parameter, 70
HMDS-TMCS, 254
Hofmann degradation, 267
Hot-wire thermal conductivity detector, 143 146-155
 air in, 152
 carrier gases, selection for, 148-149
 circuitry in, 151
 design, 147
 dual detector use, 153
 filament
  current in, 147
  resistances, 152
  linear dynamic range, 150

Hot-wire thermal conductivity detector (continued)
 mode of action, 148
 properties, 146
 quantitative analysis, area percent and weight percent, 149
 sensitivity S, 143, 147, 150
  concentration, 143
  vs. flow rate, 150
 stability, design for, 147
Hydrogen
 as carrier gas in open tubular columns, 194
 in flame ionization detector, 158
 thermal conductivity, 149
Hydrogen bonding, 10-11, 56-57

I

Ideal analysis, 52
Ideal sequenced derivatization, 269, 270
 Donike, 270
 Horning, 269
 Schwedt, 271
Identification, see Qualitative analysis
Induced dipole forces, 10, 57
Induction forces, 10, 57
Inertness, evidence of, 217
Infrared spectra, nonadaptability to computer identification, 305, 306
Infrared spectrometer, difficulties as identifying detector, 303, 304
Infrared spectroscopy, 305
 richness of spectra, 305
 tool of choice for identifying trapped fractions, 305
Initial temperature, 240
Input impedance, 356
Instrumentation
 column oven design, 239
 dual-column, dual-detector, 239
Integrators, 358
 analog-logic, see Digital integrators
 analog, disadvantages of, 358
 Disc, 358
  action, 359
  mode of action, 358
 recorder-driven, 358
  disadvantages of, 358
Interference, 355
 longitudinal, 355
 rejection, 356
 transverse, 355
Internal diameter inconstancy, effective in band broadening, 203

Intrinsic resolution, 244, 245
  vs. initial temperature, 245
  vs. program, 245

J

Jones-van Deemter equation, 82-92
  A term, 83, 97, 98-99
    revised, 110
  B term, 98, 100
  C term, 100
  C_G term, 88-91, 107-109
  C_{LP} term, 85-88, 100-107
  C_2 term, 91-92, 109-110
  plot, 97-98
    minimum, 98
    slope, 98
  reduced plate-height equation, 111

K

Kambara-Kodama equation, 232, 233
Keesom forces, 10-11, 56-57
Kieselguhr, 114
Kovats Retention Index, 60-61
  identification by
    favorable examples, 299
    preclusive requirements for sole use of, 298
  precision of, 299
  in temperature programming, precautions in use of, 299

L

Linear dynamic range, 144
Liquid crystals, 58
Liquid phase diffusivity, 8, 66-67, 86, 101, 103
Loading, 43, 44, 101
  low, 48, 105
  minimum, 102
London forces, 10, 57
LRGC-LRMS: an exemplary study, 327-331

M

Mass fragmentography, derivatization for alcohols, 357
Mass, in mass transfer, 85
Mass sensitivity in detectors, 143
Mass spectrometry
  computer identification of mass spectra, see also Computer search methods
    forward search, 331
    probability based matching, 336
    reverse search, 331-336
  computer search methods
    vs. computer capabilities, 333
    differences between forward and reverse search, 333-334

Mass spectrometry (continued)
  computer search methods (continued)
    forward search, 331
    mass weighting, 336
    proportionate spectrum subtraction, 336
    reverse search, 331-336
    spectrum condensation, 331-333
  ionization methods
    chemical ionization (CI), 325 326
    electron impact (EI), 324-325
    instrumentation for, 326
  mass separation techniques
    direction-focusing sectored magnetic field, 317-319
    quadrupole, 320-323; see also Quadrupole mass spectrometry
    time of flight, 319-320
  multiple ion monitoring, 323
  operation, 301
  resolution
    definition, 318
    vs. scanning speed, 319
    sectored magnetic field, best potential for, 319
  scanning speed
    vs. resolution, 319
    vs. sensitivity, 319
  sectored magnetic field, scanning methods, 319
  spectra, 301
McReynolds constants, see Rohrschneider constants
Mesh, 180
Metals, trace detection by electron capture, 175
Methyl oximes, 261
  can be further derivatized, 261
  protect against carbonyl enolization, 262
Methyl silicones, 65
Methylation
  BF_3-methanol, 263
  carbonyldiimidazole + methanol, 263
  by diazomethane, 262
  on injection, 263
Mixing chamber, 199
Mixtures, complexity of, 52
Molecular diffusion term, 7, 84-85
Molecular randomness, 299
Molecular sieves, 134-135
  mode of action, 135
  uses, 135
Multiple ion monitoring, 322, 323
Multiple path term, 7

Subject Index

N

Nitrogen
  in open tubular columns, 194
  thermal conductivity, 149
Nonequilibrium in gas chromatography, 80
Nucleosides, 286

O

Ohm's law, 353
On-column injection, 33-34
Open tubular columns
  carrier gas flow rates in, 194
  coating
    dynamic method, 191
    static method, 192
  design requirements, 198
  efficiency
    internal diameter key parameter in, 191
    parameters, 191
  equation for, 190
  flow velocities, 194
  fused silica, see Fused silica open tubular columns
  glass, see Glass open tubular columns
  history, 188, 189
  hydrogen as carrier gas, 194
  invention, by M. J. E. Golay, 188
  manufacture, 191
  performance characteristics, 197
  PLOT, 192
    overloading tolerance, 196
    preparation methods, 192-194
  sample injection, 194
Orientation forces, 10-11, 56-57
Overload
  column, 28
  detector, 27
  flame ionization detector, 162
β,β'-oxydipropionitrile, 61, 69, 73

P

Packing
  constant, 83
  ideal characteristics, 136, 139
  particle
    inertness, 6
    size, 6-7, 98-99, 107, 112
    size distribution, 6-7, 98-99, 107, 180
  preparation methods
    filtration, 182
    fluidization, 186
    simple addition, 181
    slurry, 181
    solvent stripping, 184
  particle size, and resistance to mass tranfer in the gas phase, 107
Partition coefficient, 11, 20-22, 104
  constancy, 21, 79

Partition ratio, 22-24
  measurement and calculation, 24, 103
Partition ratio function, 104-107, 109
  vs. partition coefficient, 106-107
  vs. phase ratio, 105-106
Peak
  area, error in assessing, 366
  broadening, see Spreading
  height, 341
  shape, 4, 79-82, 124
  spreading, see Spreading
  width
    vs. allowable group size, 366
    in isothermal GC, 11, 37
    measurement, 28
    relative, 82
    in temperature programming, 12, 38
Pen jitter, 355
Performance parameter, 214
Permeability, 214
Personal reponsibility in GC, 390, 391
Phase ratio β, 22, 104-106
Phenol derivatization
  detection by electron capture
    ethers vs. esters, 258
    heptafluorobutyrate stability to water, 258
    pentafluoroaryl derivative best, 257
  detection by flame ionization
    for hindered hydroxyls: TSIM, 257
    for nonhindered hydroxyls: BSTFA, 254
  detection by flame photometry
    diethylphosphate esters, 257
Phenol alkylamine derivatization, ideal sequenced, through salts, 270
Plate number, 37
  required for a given separation, 42, 179
  vs. selectivity, 53
  vs. separation factor, 42
Plug
  introduction, 3-4, 28-29
Polarity, 59-60, 63
  test mixtures, 217
Polarizability, 57
Polyethylene glycols, 57, 66-67, 68, 70
Polymer pyrolysis, 312-316
  bar chromatograms, 313-315
  carrier gas flow during, 313
  desirable characteristics, 312
  elements of good practice, 313
  instrument design, 313
  primary and secondary reactions, 312, 313

Ponca crude, 45
Porapak, 133
Porous polymer beads, 133-134
  discovery, 134
  efficiency, vs. pore size, 133
  mode of action, 133
  properties, 133
  retention, vs. loading, 133
Position servomechanism, 357
  lag, inherent, 357
Potentiometric recorder, 352
  in balance, 353
  balanced
    infinite impedance, 353
    no current drawn from source, 353
  balancing action, 354
  deadband, 354
    elimination, 355
    testing for, 354
    terrace appearance, 355
  diagram, functional, 352
  dynamic response, 356
  gain
    adjustment, 355
    control, 355
  input impedance, 356
  integrators, 359
  interference, 355
    shielding, 356
  as position servomechanism, 357
  span, 354
Pressure
  controllers, in open tubular column use, 194
  drop vs. column efficiency, 80
  gradient correction factor j, 18-19, 20
  programming vs. maximum operating temperature, 248
  required
    open tubular columns, 180
    packed columns, 180
    usual, 180
Probability based matching, 336
Program r/F, 240
  optional, 244
Programmed temperature gas chromatography, 11-12
  band movement vs. temperature, 244
  history, 237-240
  plate number calculation, 244
  program r/F, optimal, 244
  retention index, 246
  terminology
    initial temperature, 240
    intrinsic resolution, 244, 245
    program, 240
    retention temperature, 240, 241
    significant temperature, 241, 242
  trace detection, 242-244
Propylene carbonate, 69
Prostaglandin derivatization
  n-butyl boronate in, 259

Prostaglandin derivatization (continued)
  detection by electron capture: pentafluorobenzyl ester the derivative of choice, 260
  detection by flame ionization an adjunctive technique only, 259
  gas chromatography-mass spectrometry
    boronation, to raise m/e, 261
    isotopes to increase stability, 261
    methylation, boronation, silylation, 260
    methylation, then silylation, 260
  methoximes in, 259
  silylation in, 259
  stringencies of
    detectors both ultraselective and ultrasensitive, 259
    internal standards, 259
    multiple derivatization, 259
  structures, 259

Q

QF-1, 47, 64, 67
Quadrupole mass spectrometry, 320-323
  construction of quadrupole, 320, 321
  mode of operation, 321, 322
  multiple ion monitoring, 323
  resolution, 322
    dynamic variability, 322
  tuning frequently required, 322
  voltage magnitudes, a-c vs. d-c, 322
Qualitative analysis
  after collection, see Qualitative analysis: after collection
  during elution, see Qualitative analysis: during elution
  gas chromatography-mass spectrometry, 301-303, 317-336; see also Mass spectrometry
    computer identification of mass spectra, 306, 331-336
    desired characteristics, 319
    electron and chemical ionization, 324-326
    quadrupole as mass spectrometer of choice for GC-MS, 323
    resolution areas, low and high, 326-327
    HRGC-LRMS, 335-336
    LRGC-LRMS, 327-331
  identifying detectors
    infrared spectrometry, 303-304
    mass spectrometry, 300-303, 317-336

Subject Index

Qualitative analysis (continued)
  by inference from pretreatment, see
      Qualitative analysis: by inference
      from pretreatment
  by Kovats Retention Index
    precision of index, 299
    preclusive requirements for exclusive
      use of, 298
Qualitative analysis: after collection
  derivative formation, 304
  infrared spectroscopic methods
    handling condensates, 307
    trapping peaks, 305, 307
Qualitative analysis: during elution
  standards: direct use, 295
    complementary columns, 300
    mixture (or unknown) plus standard, 296
    standard alone, 296, 297
    standard mixture of known components, 297
  standards: indirect use, 298
    definition, 296, 298
    Kovats Retention Index, 298, 299
    requirements, 298
    selective detectors, 301
Qualitative analysis: by inference from
    pretreatment, 311
  polymer pyrolysis, 312-316
    bar chromatograms, 313-315
    desired characteristics, 312
    elements of good practice, 313
    reactions, primary and secondary, 312-313
  reaction gas chromatography
    carbon skeletal determinations, 316
    chemical absorption methods, 317
    pyrolysis of pure compounds, 316
    syringe reactions, 312
Quantitative analysis, 341
  acceptable, 366
  algorithms, 366-385
  area percent, 342
  area percent as weight percent, 342
  automated analyses
    general, 387
    very well controlled, 386
  automatic: problems, 361
  automation, 386-388
  computers, program adaptability in, 388
  digital integrators
    deficiencies, 361
    linear dynamic range, 360
    slope detectors, 361
  digital technology, see Digital technology in gas chromatography
  internal standards, 344
    properties, 344, 345
  linear dynamic range
    vs. sample size, 346
  peak area, 342
    vs. baseline, 366
    error in determining, 366
  peak height, 341
    disadvantages, 342

Quantitative analysis (continued)
  peak height (continued)
    precision, 341
    utility, 341
  precision
    computers and imprecision, 389
    vs. peak measurements techniques, 389
    vs. sample injection, 389, 390
  precision and accuracy, 388
    attainable, 390, 391
    probable, 390
  recorder, potentiometric, see
      Potentiometric recorder
  response factors, 343
    mole percent, 343
    weight percent, 343, 345
  signal handling
    recorder display, 362
    transmission, 362
  trace detection, 347
    isothermal elution, 347

R

Reaction gas chromatography, 316
  carbon skeletal determinations, 316
  in situ reactions
    catalytic deoxygenation, 316
    chemical absorption, 317
  pyrolysis of pure compounds, 316
Recorders, potentiometric, see
    Potentiometric recorder
Radial flow inequalities, causes
    of, 109
Reduced plate height
  equation, 111
  fundamental inequality in, 111
  particle diameter in, 111
Relative peak width, 82
Relative retention, 24, 298
Relative retention time, 13
Resistance to mass transfer, 8, 85-92
  in column efficiency: balanced
      and *low*, for high efficiency, 235
  in gas phase, 8, 88-91, 107-109
  in liquid stationary phase, 8, 85-88, 100-107
    and loading, 101
    vs. partition ratio function, 103
    and stationary phase proportion, 101
    and stationary phase thickness, 100-101
Resolution, 37-54
  derivation, 39-41
  intrinsic, 244, 245

Resolution (continued)
  vs. partition ratio, 40
  vs. plate number, 40, 41, 46
  vs. pressure programming, 236
  vs. relative retention, 40, 41
  vs. selectivity, 46
  vs. temperature programming, 38, 236
  vs. temperature program rate, 38
  time-dependent, 215
Resolution, intrinsic
  vs. initial temperature, 245
  vs. program, 245
Response factors, see Quantitative analysis, response factors
Response time, 356
  vs. peak width, 146, 203
Retention dispersion, 72
Retention Index difference ΔI, 60, 63
Retention ratio, vs. column temperature, 236
Retention temperature, 240, 241
  reproducibility, 241
Retention time, 2
  adjusted, 18, 19
  corrected, 19
  in isothermal GC, 11
  observed, 19
  in open tubular columns, 194
  relative, 13
  in temperature programming, 12
Retention volume
  adjusted, 18, 19, 20
  vs. carbon number in homologous series, 224
  corrected, 19
  net, 19, 20, 227
  observed, 19
  specific, 224
Reverse search, 331-336
RNA silylation, 286
Rohrschneider constants, 61-62
  McReynolds, 63
  probes, 62-63

S

Safrole, 69
Sample injection
  on-column, 33-34
    vs. flash vaporization, 390
Sample introduction, 3-4, 27-36
  dead spaces in, 4
  gases, 29-30
  on-column injection, 33-34
  in open tubular columns, 194, 195
  liquids, 31-34
  period, 29
  plug, 3-4, 28-29
  port, 35-36
    design, 35-36
    temperature, 28
    vaporizer tube, 36

Sample introduction (continued)
  size, maximum acceptable, 9, 27-28, 71, 73, 346
  solids, 34
  splitless sampling, 208-211
  temperature, 3, 28
  for trace enrichment, 242-244
Sample size, maximum permissible, 9, 27-28, 71, 73, 346
Selectivity, 10-11, 43, 230
  and adsorption at gas-liquid interface, 58, 230
  gas-solid vs. gas-liquid, 135
  vs. plate number, 53
  temperature dependence, 58, 70, 230
Sensitivity, 143
  adequate, 143
  concentration, 143
  mass, 143
Separability, 37, 42
  vs. column temperature, 236
  vs. rate of temperature increase, 236
Separation factor, 42
  vs. plate number, 42
Significant temperature, 241, 242
  vs. retention temperature, 242
Silanol groups, 125
Silylating reagents
  comparison, 254
  structures, 255
Silylation, 125, 254
  of aqueous samples, 277
Single focusing mass spectrometers, 318
Solids, sample introduction, 34
Solid support, see Diatomaceous earth
  pore size distribution, 121
Solubility, 10-11, 55-58
  chemical bonding, 10
  coulombic forces, 55-56
  Debye forces, 10
  dispersion forces, 10, 57
  hydrogen bonding, 10-11, 56-57
  induced dipole, 10, 57
  Keesom forces, 10, 56-57
  London forces, 10, 57
  orientation forces, 10, 56-57
  specific interaction, 10-11, 58
Solute, 1
  motion in each phase, 103
Solvent
  efficiency, 230
    increase with temperature decrease, 231
  selectivity, 230
Solvent selection, 70-75
  like dissolves like, 71, 73
  retention dispersion, 72
Span, 354

Specific retention volume, 224
Splitless sampling, 208
 explanation, 211
 temperature requirement relaxation, 210
 trace detection in, 210
Splitters, 195, 197
Spreading
 causes
  extra-column, 198
  in-column, see van Deemter equation
 vs. discontinuous change in inner diameters, 203
 in laminar flow, 203
 requirements for Desty's work, 204
 under stopped flow, 103
Squalane, 61, 64
Stability in detectors, 143
Standard deviation, 80
Stationary phase, 55-76
 Apiezon L, 65, 70
 bonded, see Bonded stationary phase
 Carbowax 20-M, 64, 66, 68
 characterization, 59-63
  Kovats Retention Index, 60-61
  McReynolds constants, 63
  nearest-neighbor, 63-64
  probes, 62-63
  Rohrschneider constants, 61-62
  standards, 61
 DC-710, 64, 66, 67
 DEGA, 64, 68, 70
 DEGS, 64, 68, 70, 73
 Dexsil, 67
 dimethylformamide, 43-44, 56
 dimethylsulfolane, 44, 73
 Emulphor-O, 71-72
 fluorene picrate, 43
 OV-1, 66, 70
 OV-3, 64, 66, 67
 OV-7, 64-66, 67
 OV-11, 67
 OV-17, 70
 OV-22, 64, 66
 OV-101, 66
 OV-210, 67, 70
 OV-225, 68, 71
 OV-275, 69
 β,β'-oxydipropionitrile, 61, 69, 73
 polarity, 59-60, 63
 polyesters, 68, 70
 polyethylene glycols, 57, 66-67, 68, 70
 preferred, 63-70
  saturated hydrocarbons, 65
   silicones, 65-68, 70
   six, 69-70
   twelve, 63-69
 proliferation, 59, 63
 QF-1, 47, 64, 67
 SE-30, 47, 64, 65, 66, 70
 selection, 70-75
 SF96-200, 66, 67
 Silar-5CP, 68

Stationary phase (continued)
 Silar 10C, 69, 70
 vs. solubility, 66
 SP-2100, 70
 SP-2250, 70
 SP-2340, 70
 SP-2401, 67-70
 squalane, 61, 64, 65
 TCEP, 64, 69
 temperature, maximum operating, 243
 vapor pressure, vs. temperature, 247
 volatility, vs. temperature programming, 239
 XE-60, 64, 68
Steroids, 47-50
 metabolic profile, 48-50
 separation history, 47-50
 structures, numbering of, 282
Steroids: derivatization
 detection by electron capture and mass spectrometry
  flophemesyl ethers, 284
  halomethyldimethylsilyl ethers, 285
  HFB variability, 284
  PFB oxime: 5 pg, 284
 detection by flame ionization and mass spectrometry, 281
  alkyl oximes, 281
  boronation, 283
  MO-TMS derivative, 280, 281
  oxime-to-nitrile fission, avoidance of, 281
 detection by thermionic flame ionization
  dimethyl phosphine ester, 284
Stopped-flow
 in on-column injection, 34
 spreading in, 103
Stripper, rotary, 185
Subtraction methods, 317
Suffixes, for supports, 125
Sulfonic acids: derivatization
 chlorination with $PCl_5 + PCl_3$, 265
 silylation by TSIM, 265
Supports, 6-7, 114-141, 178; see also Diatomaceous earth
 adsorption, 6-7, 124-125
 diatomaceous earth, 114-131
 molecular sieves, 134-135
 packed densities, 124
 porous polymer beads, 133-134
 glass beads, 131-133
  roughened, 132
 particle diameters, 98-99, 107, 112, 180-181
 preferred, properties of, 127
 Teflon, 131

Syringe
  gas, 29-30
  plunger-in-barrel, 31-32
  plunger-in-needle, 31-33
Syringe reactions, 312

T

Teflon as support, 131
Temperature
  injection, 28
  optimum column, 38
Temperature programming, see Programmed temperature gas chromatography
Temperature selection vs. solvent efficiency, 230
Tenax GC, 350
Theoretical plate, 4, 10
  calculation of, 5
    derivation of, 81-82
  in distillation, 4, 10
  effective, 18, 41
  first, as receiver of injected sample, 27, 79
  height of, 4
Thermal conductivity
  gases
    mechanism, 148-149
    selected, 149
Thermal conductivity detector, see Hot-wire thermal conductivity detector
  area percent as weight percent, 342
Thermal stability of stationary phase, 247
Thermal electrons, 165
Thermionic detector
  discovery, 162
  ionization efficiency, 163
  selectivity, 162, 164
  specificity, 163, 164
Thin layer chromatography, 308, 309
Time of flight mass spectrometer, 319, 320
Time in gas phase, all molecules, 84
Tortuosity constant, 84
Trace enrichment, 242-244
  factor, 348
  vs. initial temperature, 243
  injection onto cold column, 350
  precolumn, cool, 350
Trifluoropropyl methyl silicones, 70
Trouton's rule, 228
TSIM, 254, 255, 260, 265, 278, 282, 283
Tungsten temperature coefficient of resistance, 152

V

Valve, gas sampling, 29-30
van Deemter equation, 6-9, 82-95
  A term, 6-7, 83-84, 97, 98-99
    and particle diameter, 98-99
    and particle size distribution, 98-99

van Deemter equation (continued)
  B term, 7-8, 84-85, 98, 100
  C term, 8, 85-92, 98
    derivation, 82-95
  eddy diffusion term, 7, 83-84
  molecular diffusion term, 7, 84-85
  multiple path term, 7
  velocity distribution term, 91-92, 109-110
Vapor pressure
  vs. carbon number in homologous series, 228, 229
  Clausius-Clapeyron equation, 224
  vs. temperature, 224
  for any stationary phase, 247
Variance, 80
  additivity of, 81
Velocity distribution term, 91-92, 109-110
Viscosity, 66-67
Volatile stripping, 350
Voltage divider, 155

W

Wall coated open tubular column, 100
Water, the chromatographing of, 133
Water and silylation, 277
Wheatstone bridge, 151, 154